THE CAUCHY-

This lively, problem-oriented text is designed to coach readers toward mastery of the most fundamental mathematical inequalities. With the Cauchy–Schwarz inequality as the initial guide, the reader is led through a sequence of fascinating problems whose solutions are presented as they might have been discovered — either by one of history's famous mathematicians or by the reader. The problems emphasize beauty and surprise, but along the way readers will find systematic coverage of the geometry of squares, convexity, the ladder of power means, majorization, Schur convexity, exponential sums, and the inequalities of Hölder, Hilbert, and Hardy.

The text is accessible to anyone who knows calculus and who cares about solving problems. It is well suited to self-study, directed study, or as a supplement to courses in analysis, probability, and combinatorics.

J. Michael Steele is C. F. Koo Professor of Statistics at the Wharton School, University of Pennsylvania. He is the author of more than 100 mathematical publications, including the books *Probability Theory and Combinatorial Optimization* and *Stochastic Calculus and Financial Applications*. He is also the founding editor of the *Annals of Applied Probability*.

MAA PROBLEM BOOKS SERIES

Problem Books is a series of the Mathematical Association of America consisting of collections of problems and solutions from annual mathematical competitions; compilations of problems (including unsolved problems) specific to particular branches of mathematics; books on the art and practice of problem solving, etc.

Committee on Publications
Gerald Alexanderson, *Chair*

Roger Nelsen *Editor*

Irl Bivens	Clayton Dodge
Richard Gibbs	George Gilbert
Gerald Heuer	Elgin Johnston
Kiran Kedlaya	Loren Larson
Margaret Robinson	Mark Saul

A Friendly Mathematics Competition: 35 Years of Teamwork in Indiana, edited by Rick Gillman

The Inquisitive Problem Solver, Paul Vaderlind, Richard K. Guy, and Loren C. Larson

Mathematical Olympiads 1998–1999: Problems and Solutions from Around the World, edited by Titu Andreescu and Zuming Feng

Mathematical Olympiads 1999–2000: Problems and Solutions from Around the World, edited by Titu Andreescu and Zuming Feng

Mathematical Olympiads 2000–2001: Problems and Solutions from Around the World, edited by Titu Andreescu, Zuming Feng, and George Lee, Jr.

The William Lowell Putnam Mathematical Competition Problems and Solutions: 1938–1964, A. M. Gleason, R. E. Greenwood, and L. M. Kelly

The William Lowell Putnam Mathematical Competition Problems and Solutions: 1965–1984, Gerald L. Alexanderson, Leonard F. Klosinski, and Loren C. Larson

The William Lowell Putnam Mathematical Competition 1985–2000: Problems, Solutions, and Commentary, Kiran S. Kedlaya, Bjorn Poonen, and Ravi Vakil

USA and International Mathematical Olympiads 2000, edited by Titu Andreescu and Zuming Feng

USA and International Mathematical Olympiads 2001, edited by Titu Andreescu and Zuming Feng

USA and International Mathematical Olympiads 2002, edited by Titu Andreescu and Zuming Feng

THE CAUCHY–SCHWARZ MASTER CLASS

An Introduction to the Art of Mathematical Inequalities

J. MICHAEL STEELE

University of Pennsylvania

THE MATHEMATICAL ASSOCIATION OF AMERICA

CAMBRIDGE UNIVERSITY PRESS
Cambridge, New York, Melbourne, Madrid, Cape Town, Singapore,
São Paulo, Delhi, Dubai, Tokyo, Meixco City

Cambridge University Press
32 Avenue of the Americas, New York, NY 10013-2473, USA

www.cambridge.org
Information on this title: www.cambridge.org/9780521546775

© J. Michael Steele 2004

This publication is in copyright. Subject to statutory exception
and to the provisions of relevant collective licensing agreements,
no reproduction of any part may take place without the written
permission of Cambridge University Press.

First published 2008
7th printing 2010

A catalog record for this publication is available from the British Library.

Library of Congress Cataloging in Publication Data

Steele, J. Michael.
The Cauchy-Schwarz master class : an introduction to the art of mathematical inequalities / J. Michael Steele.
p. cm. – (MAA problem books)
Includes bibliographical references and index.
ISBN 0-521-83775-8 (hardback) – ISBN 0-521-54677-X (pbk.)
1. Inequalities (Mathematics) I. Title. II. MAA problem book series.
QA295.S78 2004
512.9'7 – DC22   2004040402

ISBN 978-0-521-83775-0 Hardback
ISBN 978-0-521-54677-5 Paperback

Cambridge University Press has no responsibility for the persistence or accuracy of URLS for external or third-party Internet Web sites referred to in this publication and does not guarantee that any content on such Web sites is, or will remain, accurate or appropriate.

Contents

		page
Preface		ix
1	Starting with Cauchy	1
2	The AM-GM Inequality	19
3	Lagrange's Identity and Minkowski's Conjecture	37
4	On Geometry and Sums of Squares	51
5	Consequences of Order	73
6	Convexity — The Third Pillar	87
7	Integral Intermezzo	105
8	The Ladder of Power Means	120
9	Hölder's Inequality	135
10	Hilbert's Inequality and Compensating Difficulties	155
11	Hardy's Inequality and the Flop	166
12	Symmetric Sums	178
13	Majorization and Schur Convexity	191
14	Cancellation and Aggregation	208
Solutions to the Exercises		226
Chapter Notes		284
References		291
Index		301

Preface

In the fine arts, a master class is a small class where students and coaches work together to support a high level of technical and creative excellence. This book tries to capture the spirit of a master class while providing coaching for readers who want to refine their skills as solvers of problems, especially those problems dealing with mathematical inequalities.

The most important prerequisite for benefiting from this book is the desire to master the craft of discovery and proof. The formal requirements are quite modest. Anyone with a solid course in calculus is well prepared for almost everything to be found here, and perhaps half of the material does not even require calculus. Nevertheless, the book develops many results which are rarely seen, and even experienced readers are likely to find material that is challenging and informative.

With the Cauchy–Schwarz inequality as the initial guide, the reader is led through a sequence of interrelated problems whose solutions are presented as they might have been discovered — either by one of history's famous mathematicians or by the reader. The problems emphasize beauty and surprise, but along the way one finds systematic coverage of the geometry of squares, convexity, the ladder of power means, majorization, Schur convexity, exponential sums, and all of the so-called classical inequalities, including those of Hölder, Hilbert, and Hardy.

To solve a problem is a very human undertaking, and more than a little mystery remains about how we best guide ourselves to the discovery of original solutions. Still, as George Pólya and others have taught us, there are principles of problem solving. With practice and good coaching we can all improve our skills. Just like singers, actors, or pianists, we have a path toward a deeper mastery of our craft.

Acknowledgments

The initial enthusiasm of Herb Wilf and Theodore Körner propelled this project into being, and they deserve my special thanks. Many others have also contributed in essential ways over a period of years. In particular, Cynthia Cronin-Kardon provided irreplaceable library assistance, and Steve Skillen carefully translated almost all of the figures into PSTricks. Don Albers, Lauren Cowles, and Patrick Kelly all provided wise editorial advice which was unfailingly accepted. Patricia Steele ceded vast stretches of our home to ungainly stacks of paper and helped in many other ways.

For their responses to my enquiries and their comments on special parts of the text, I am pleased to thank Tony Cai, Persi Diaconis, Dick Dudley, J.-P. Kahane, Kirin Kedlaya, Hojoo Lee, Lech Maliganda, Zhihua Qian, Bruce Reznick, Paul Shaman, Igor Sharplinski, Larry Shepp, Huili Tang, and Rick Vitale. Many others kindly provided preprints, reprints, or pointers to their work or the work of others.

For their extensive comments covering the whole text (and in some cases in more than one version), I owe great debts to Cengiz Belentepe, Claude Dellacherie, Jirka Matoušek, Xioli Meng, and Nicholas Ward.

1
Starting with Cauchy

Cauchy's inequality for real numbers tells us that

$$a_1b_1 + a_2b_2 + \cdots + a_nb_n \leq \sqrt{a_1^2 + a_2^2 + \cdots + a_n^2}\sqrt{b_1^2 + b_2^2 + \cdots + b_n^2},$$

and there is no doubt that this is one of the most widely used and most important inequalities in all of mathematics. A central aim of this course — or master class — is to suggest a path to mastery of this inequality, its many extensions, and its many applications — from the most basic to the most sublime.

THE TYPICAL PLAN

The typical chapter in this course is built around the solution of a small set of *challenge problems*. Sometimes a challenge problem is drawn from one of the world's famous mathematical competitions, but more often a problem is chosen because it illustrates a mathematical technique of wide applicability.

Ironically, our first challenge problem is an exception. To be sure, the problem hopes to offer honest coaching in techniques of importance, but it is unusual in that it asks you to solve a problem that you are likely to have seen before. Nevertheless, the challenge is sincere; almost everyone finds some difficulty directing fresh thoughts toward a familiar problem.

Problem 1.1 *Prove Cauchy's inequality. Moreover, if you already know a proof of Cauchy's inequality, find another one!*

COACHING FOR A PLACE TO START

How does one solve a problem in a fresh way? Obviously there cannot be any universal method, but there are some hints that almost always help. One of the best of these is to try to solve the problem by means of a *specific principle* or *specific technique*.

Here, for example, one might insist on proving Cauchy's inequality

1

just by algebra — or just by geometry, by trigonometry, or by calculus. Miraculously enough, Cauchy's inequality is wonderfully provable, and each of these approaches can be brought to a successful conclusion.

A PRINCIPLED BEGINNING

If one takes a dispassionate look at Cauchy's inequality, there is another principle that suggests itself. Any time one faces a valid proposition that depends on an integer n, there is a reasonable chance that mathematical induction will lead to a proof. Since none of the standard texts in algebra or analysis gives such a proof of Cauchy's inequality, this principle also has the benefit of offering us a path to an "original" proof — provided, of course, that we find any proof at all.

When we look at Cauchy's inequality for $n = 1$, we see that the inequality is trivially true. This is all we need to start our induction, but it does not offer us any insight. If we hope to find a serious idea, we need to consider $n = 2$ and, in this second case, Cauchy's inequality just says

$$(a_1 b_1 + a_2 b_2)^2 \leq (a_1^2 + a_2^2)(b_1^2 + b_2^2). \tag{1.1}$$

This is a simple assertion, and you may see at a glance why it is true. Still, for the sake of argument, let us suppose that this inequality is not so obvious. How then might one search systematically for a proof?

Plainly, there is nothing more systematic than simply expanding both sides to find the equivalent inequality

$$a_1^2 b_1^2 + 2a_1 b_1 a_2 b_2 + a_2^2 b_2^2 \leq a_1^2 b_1^2 + a_1^2 b_2^2 + a_2^2 b_1^2 + a_2^2 b_2^2,$$

then, after we make the natural cancellations and collect terms to one side, we see that inequality (1.1) is also equivalent to the assertion that

$$0 \leq (a_1 b_2)^2 - 2(a_1 b_2)(a_2 b_1) + (a_2 b_1)^2. \tag{1.2}$$

This equivalent inequality actually puts the solution of our problem within reach. From the well-known factorization $x^2 - 2xy + y^2 = (x-y)^2$ one finds

$$(a_1 b_2)^2 - 2(a_1 b_2)(a_2 b_1) + (a_2 b_1)^2 = (a_1 b_2 - a_2 b_1)^2, \tag{1.3}$$

and the nonnegativity of this term confirms the truth of inequality (1.2). By our chain of equivalences, we find that inequality (1.1) is also true, and thus we have proved Cauchy's inequality for $n = 2$.

THE INDUCTION STEP

Now that we have proved a nontrivial case of Cauchy's inequality, we are ready to look at the induction step. If we let $H(n)$ stand for the hypothesis that Cauchy's inequality is valid for n, we need to show that $H(2)$ and $H(n)$ imply $H(n+1)$. With this plan in mind, we do not need long to think of first applying the hypothesis $H(n)$ and then using $H(2)$ to stitch together the two remaining pieces. Specifically, we have

$$
\begin{aligned}
& a_1 b_1 + a_2 b_2 + \cdots + a_n b_n + a_{n+1} b_{n+1} \\
&= (a_1 b_1 + a_2 b_2 + \cdots + a_n b_n) + a_{n+1} b_{n+1} \\
&\leq (a_1^2 + a_2^2 + \cdots + a_n^2)^{\frac{1}{2}} (b_1^2 + b_2^2 + \cdots + b_n^2)^{\frac{1}{2}} + a_{n+1} b_{n+1} \\
&\leq (a_1^2 + a_2^2 + \cdots + a_n^2 + a_{n+1}^2)^{\frac{1}{2}} (b_1^2 + b_2^2 + \cdots + b_n^2 + b_{n+1}^2)^{\frac{1}{2}},
\end{aligned}
$$

where in the first inequality we used the induction hypothesis $H(n)$, and in the second inequality we used $H(2)$ in the form

$$\alpha\beta + a_{n+1} b_{n+1} \leq (\alpha^2 + a_{n+1}^2)^{\frac{1}{2}} (\beta^2 + b_{n+1}^2)^{\frac{1}{2}}$$

with the new variables

$$\alpha = (a_1^2 + a_2^2 + \cdots + a_n^2)^{\frac{1}{2}} \quad \text{and} \quad \beta = (b_1^2 + b_2^2 + \cdots + b_n^2)^{\frac{1}{2}}.$$

The only difficulty one might have finding this proof comes in the last step where we needed to see how to use $H(2)$. In this case the difficulty was quite modest, yet it anticipates the nature of the challenge one finds in more sophisticated problems. The actual application of Cauchy's inequality is never difficult; the challenge always comes from seeing *where* Cauchy's inequality should be applied and *what* one gains from the application.

THE PRINCIPLE OF QUALITATIVE INFERENCES

Mathematical progress depends on the existence of a continuous stream of new problems, yet the processes that generate such problems may seem mysterious. To be sure, there is genuine mystery in any deeply original problem, but most new problems evolve quite simply from well-established principles. One of the most productive of these principles calls on us to expand our understanding of a *quantitative* result by first focusing on its *qualitative* inferences.

Almost any significant quantitative result will have some immediate qualitative corollaries and, in many cases, these corollaries can be derived independently, without recourse to the result that first brought them to light. The alternative derivations we obtain this way often help us to see the fundamental nature of our problem more clearly. Also, much more often than one might guess, the qualitative approach even yields new quantitative results. The next challenge problem illustrates how these vague principles can work in practice.

Problem 1.2 *One of the most immediate qualitative inferences from Cauchy's inequality is the simple fact that*

$$\sum_{k=1}^{\infty} a_k^2 < \infty \text{ and } \sum_{k=1}^{\infty} b_k^2 < \infty \text{ imply that } \sum_{k=1}^{\infty} |a_k b_k| < \infty. \qquad (1.4)$$

Give a proof of this assertion that does not call on Cauchy's inequality.

When we consider this challenge, we are quickly drawn to the realization that we need to show that the product $a_k b_k$ is small when a_k^2 and b_k^2 are small. We could be sure of this inference if we could prove the existence of a constant C such that

$$xy \leq C(x^2 + y^2) \qquad \text{for all real } x, y.$$

Fortunately, as soon as one writes down this inequality, there is a good chance of recognizing why it is true. In particular, one might draw the link to the familiar factorization

$$0 \leq (x-y)^2 = x^2 - 2xy + y^2,$$

and this observation is all one needs to obtain the bound

$$xy \leq \frac{1}{2}x^2 + \frac{1}{2}y^2 \qquad \text{for all real } x, y. \qquad (1.5)$$

Now, when we apply this inequality to $x = |a_k|$ and $y = |b_k|$ and then sum over all k, we find the interesting *additive* inequality

$$\sum_{k=1}^{\infty} |a_k b_k| \leq \frac{1}{2} \sum_{k=1}^{\infty} a_k^2 + \frac{1}{2} \sum_{k=1}^{\infty} b_k^2. \qquad (1.6)$$

This bound gives us another way to see the truth of the qualitative assertion (1.4) and, thus, it passes one important test. Still, there are other tests to come.

A Test of Strength

Any time one meets a new inequality, one is almost duty bound to test the strength of that inequality. Here that obligation boils down to asking how close the new additive inequality comes to matching the quantitative estimates that one finds from Cauchy's inequality.

The additive bound (1.6) has two terms on the right-hand side, and Cauchy's inequality has just one. Thus, as a first step, we might look for a way to combine the two terms of the additive bound (1.6), and a natural way to implement this idea is to normalize the sequences $\{a_k\}$ and $\{b_k\}$ so that each of the right-hand sums is equal to one.

Thus, if neither of the sequences is made up of all zeros, we can introduce new variables

$$\hat{a}_k = a_k / \left(\sum_j a_j^2\right)^{\frac{1}{2}} \quad \text{and} \quad \hat{b}_k = b_k / \left(\sum_j b_j^2\right)^{\frac{1}{2}},$$

which are *normalized* in the sense that

$$\sum_{k=1}^{\infty} \hat{a}_k^2 = \sum_{k=1}^{\infty} \left\{ a_k^2 / \left(\sum_j a_j^2\right) \right\} = 1$$

and

$$\sum_{k=1}^{\infty} \hat{b}_k^2 = \sum_{k=1}^{\infty} \left\{ b_k^2 / \left(\sum_j b_j^2\right) \right\} = 1.$$

Now, when we apply inequality (1.6) to the sequences $\{\hat{a}_k\}$ and $\{\hat{b}_k\}$, we obtain the simple-looking bound

$$\sum_{k=1}^{\infty} \hat{a}_k \hat{b}_k \leq \frac{1}{2} \sum_{k=1}^{\infty} \hat{a}_k^2 + \frac{1}{2} \sum_{k=1}^{\infty} \hat{b}_k^2 = 1$$

and, in terms of the original sequences $\{a_k\}$ and $\{b_k\}$, we have

$$\sum_{k=1}^{\infty} \left\{ a_k / \left(\sum_j a_j^2\right)^{\frac{1}{2}} \right\} \left\{ b_k / \left(\sum_j b_j^2\right)^{\frac{1}{2}} \right\} \leq 1.$$

Finally, when we clear the denominators, we find our old friend Cauchy's inequality — though this time it also covers the case of possibly infinite sequences:

$$\sum_{k=1}^{\infty} a_k b_k \leq \left(\sum_{j=1}^{\infty} a_j^2\right)^{\frac{1}{2}} \left(\sum_{j=1}^{\infty} b_j^2\right)^{\frac{1}{2}}. \tag{1.7}$$

The additive bound (1.6) led us to a proof of Cauchy's inequality which is quick, easy, and modestly entertaining, but it also connects to a larger theme. Normalization gives us a systematic way to pass from an additive inequality to a multiplicative inequality, and this is a trip we will often need to make in the pages that follow.

ITEM IN THE DOCK: THE CASE OF EQUALITY

One of the enduring principles that emerges from an examination of the ways that inequalities are developed and applied is that many benefits flow from understanding when an inequality is sharp, or nearly sharp. In most cases, this understanding pivots on the discovery of the circumstances where equality can hold.

For Cauchy's inequality this principle suggests that we should ask ourselves about the relationship that must exist between the sequences $\{a_k\}$ and $\{b_k\}$ in order for us to have

$$\sum_{k=1}^{\infty} a_k b_k = \left(\sum_{k=1}^{\infty} a_k^2\right)^{\frac{1}{2}} \left(\sum_{k=1}^{\infty} b_k^2\right)^{\frac{1}{2}}. \qquad (1.8)$$

If we focus our attention on the nontrivial case where neither of the sequences is identically zero and where both of the sums on the right-hand side of the identity (1.8) are finite, then we see that each of the steps we used in the derivation of the bound (1.7) can be reversed. Thus one finds that the identity (1.8) implies the identity

$$\sum_{k=1}^{\infty} \hat{a}_k \hat{b}_k = \frac{1}{2} \sum_{k=1}^{\infty} \hat{a}_k^2 + \frac{1}{2} \sum_{k=1}^{\infty} \hat{b}_k^2 = 1. \qquad (1.9)$$

By the two-term bound $xy \leq (x^2 + y^2)/2$, we also know that

$$\hat{a}_k \hat{b}_k \leq \frac{1}{2} \hat{a}_k^2 + \frac{1}{2} \hat{b}_k^2 \qquad \text{for all } k = 1, 2, \ldots, \qquad (1.10)$$

and from these we see that if strict inequality were to hold for even one value of k then we could not have the equality (1.9). This observation tells us in turn that the case of equality (1.8) can hold for nonzero series only when we have $\hat{a}_k = \hat{b}_k$ for all $k = 1, 2, \ldots$. By the definition of these normalized values, we then see that

$$a_k = \lambda b_k \qquad \text{for all } k = 1, 2, \ldots, \qquad (1.11)$$

where the constant λ is given by the ratio

$$\lambda = \left(\sum_{j=1}^{\infty} a_j^2\right)^{\frac{1}{2}} \Big/ \left(\sum_{j=1}^{\infty} b_j^2\right)^{\frac{1}{2}}.$$

Here one should note that our argument was brutally straightforward, and thus, our problem was not much of a challenge. Nevertheless, the result still expresses a minor miracle; the *one* identity (1.8) has the strength to imply an *infinite* number of identities, one for each value of $k = 1, 2, \ldots$ in equation (1.11).

BENEFITS OF GOOD NOTATION

Sums such as those appearing in Cauchy's inequality are just barely manageable typographically and, as one starts to add further features, they can become unwieldy. Thus, we often benefit from the introduction of shorthand notation such as

$$\langle \mathbf{a}, \mathbf{b} \rangle = \sum_{j=1}^{n} a_j b_j \tag{1.12}$$

where $\mathbf{a} = (a_1, a_2, \ldots, a_n)$ and $\mathbf{b} = (b_1, b_2, \ldots, b_n)$. This shorthand now permits us to write Cauchy's inequality quite succinctly as

$$\langle \mathbf{a}, \mathbf{b} \rangle \leq \langle \mathbf{a}, \mathbf{a} \rangle^{\frac{1}{2}} \langle \mathbf{b}, \mathbf{b} \rangle^{\frac{1}{2}}. \tag{1.13}$$

Parsimony is fine, but there are even deeper benefits to this notation if one provides it with a more abstract interpretation. Specifically, if V is a real vector space (such as $\mathbb{R}^d$), then we say that a function on $V \times V$ defined by the mapping $(\mathbf{a}, \mathbf{b}) \mapsto \langle \mathbf{a}, \mathbf{b} \rangle$ is an *inner product* and we say that $(V, \langle \cdot, \cdot \rangle)$ is a *real inner product space* provided that the pair $(V, \langle \cdot, \cdot \rangle)$ has the following five properties:

(i) $\langle \mathbf{v}, \mathbf{v} \rangle \geq 0$ for all $\mathbf{v} \in V$,
(ii) $\langle \mathbf{v}, \mathbf{v} \rangle = 0$ if and only if $\mathbf{v} = 0$,
(iii) $\langle \alpha \mathbf{v}, \mathbf{w} \rangle = \alpha \langle \mathbf{v}, \mathbf{w} \rangle$ for all $\alpha \in \mathbb{R}$ and all $\mathbf{v}, \mathbf{w} \in V$,
(iv) $\langle \mathbf{u}, \mathbf{v} + \mathbf{w} \rangle = \langle \mathbf{u}, \mathbf{v} \rangle + \langle \mathbf{u}, \mathbf{w} \rangle$ for all $\mathbf{u}, \mathbf{v}, \mathbf{w} \in V$, and finally,
(v) $\langle \mathbf{v}, \mathbf{w} \rangle = \langle \mathbf{w}, \mathbf{v} \rangle$ for all $\mathbf{v}, \mathbf{w} \in V$.

One can easily check that the shorthand introduced by the sum (1.12) has each of these properties, but there are many further examples of useful inner products. For example, if we fix a set of positive real numbers $\{w_j : j = 1, 2, \ldots, n\}$ then we can just as easily define an inner product on $\mathbb{R}^n$ with the weighted sums

$$\langle \mathbf{a}, \mathbf{b} \rangle = \sum_{j=1}^{n} a_j b_j w_j \tag{1.14}$$

and, with this definition, one can check just as before that $\langle \mathbf{a}, \mathbf{b} \rangle$ satisfies all of the properties that one requires of an inner product. Moreover, this example only reveals the tip of an iceberg; there are many useful inner products, and they occur in a great variety of mathematical contexts.

An especially useful example of an inner product can be given by considering the set $V = C[a, b]$ of real-valued continuous functions on the bounded interval $[a, b]$ and by defining $\langle \cdot, \cdot \rangle$ on V by setting

$$\langle f, g \rangle = \int_a^b f(x) g(x)\, dx, \tag{1.15}$$

or more generally, if $w : [a, b] \to \mathbb{R}$ is a continuous function such that $w(x) > 0$ for all $x \in [a, b]$, then one can define an inner product on $C[a, b]$ by setting

$$\langle f, g \rangle = \int_a^b f(x) g(x) w(x)\, dx.$$

We will return to these examples shortly, but first there is an opportunity that must be seized.

An Opportunistic Challenge

We now face one of those pleasing moments when good notation suggests a good theorem. We introduced the idea of an inner product in order to state the basic form (1.7) of Cauchy's inequality in a simple way, and now we find that our notation pulls us toward an interesting conjecture: Can it be true that in every inner product space one has the inequality $\langle \mathbf{v}, \mathbf{w} \rangle \leq \langle \mathbf{v}, \mathbf{v} \rangle^{\frac{1}{2}} \langle \mathbf{w}, \mathbf{w} \rangle^{\frac{1}{2}}$? This conjecture is indeed true, and when framed more precisely, it provides our next challenge problem.

Problem 1.3 *For any real inner product space $(V, \langle \cdot, \cdot \rangle)$, one has for all $\mathbf{v}$ and $\mathbf{w}$ in V that*

$$\langle \mathbf{v}, \mathbf{w} \rangle \leq \langle \mathbf{v}, \mathbf{v} \rangle^{\frac{1}{2}} \langle \mathbf{w}, \mathbf{w} \rangle^{\frac{1}{2}}; \tag{1.16}$$

moreover, for nonzero vectors $\mathbf{v}$ and $\mathbf{w}$, one has

$$\langle \mathbf{v}, \mathbf{w} \rangle = \langle \mathbf{v}, \mathbf{v} \rangle^{\frac{1}{2}} \langle \mathbf{w}, \mathbf{w} \rangle^{\frac{1}{2}} \quad \text{if and only if } \mathbf{v} = \lambda \mathbf{w}$$

for a nonzero constant λ.

As before, one may be tempted to respond to this challenge by just rattling off a previously mastered textbook proof, but that temptation should still be resisted. The challenge offered by Problem 1.3 is important, and it deserves a fresh response — or, at least, a relatively fresh response.

For example, it seems appropriate to ask if one might be able to use some variation on the additive method which helped us prove the plain vanilla version of Cauchy's inequality. The argument began with the observation that $(x-y)^2 \geq 0$ implies $xy \leq x^2/2 + y^2/2$, and one might guess that an analogous idea could work again in the abstract case.

Here, of course, we need to use the defining properties of the inner product, and, as we go down the list looking for an analog to $(x-y)^2 \geq 0$, we are quite likely to hit on the idea of using property (i) in the form

$$\langle \mathbf{v} - \mathbf{w}, \mathbf{v} - \mathbf{w} \rangle \geq 0.$$

Now, when we expand this inequality with the help of the other properties of the inner product $\langle \cdot, \cdot \rangle$, we find that

$$\langle \mathbf{v}, \mathbf{w} \rangle \leq \frac{1}{2} \langle \mathbf{v}, \mathbf{v} \rangle + \frac{1}{2} \langle \mathbf{w}, \mathbf{w} \rangle. \tag{1.17}$$

This is a perfect analog of the additive inequality that gave us our second proof of the basic Cauchy inequality, and we face a classic situation where all that remains is a "matter of technique."

A Retraced Passage — Conversion of an Additive Bound

Here we are oddly lucky since we have developed only one technique that is even remotely relevant — the normalization method for converting an additive inequality into one that is multiplicative. Normalization means different things in different places, but, if we take our earlier analysis as our guide, what we want here is to replace $\mathbf{v}$ and $\mathbf{w}$ with related terms that reduce the right side of the bound (1.17) to 1.

Since the inequality (1.16) holds trivially if either $\mathbf{v}$ or $\mathbf{w}$ is equal to zero, we may assume without loss of generality that $\langle \mathbf{v}, \mathbf{v} \rangle$ and $\langle \mathbf{w}, \mathbf{w} \rangle$ are both nonzero, so the normalized variables

$$\hat{\mathbf{v}} = \mathbf{v}/\langle \mathbf{v}, \mathbf{v} \rangle^{\frac{1}{2}} \quad \text{and} \quad \hat{\mathbf{w}} = \mathbf{w}/\langle \mathbf{w}, \mathbf{w} \rangle^{\frac{1}{2}} \tag{1.18}$$

are well defined. When we substitute these values for $\mathbf{v}$ and $\mathbf{w}$ in the bound (1.17), we then find $\langle \hat{\mathbf{v}}, \hat{\mathbf{w}} \rangle \leq 1$. In terms of the original variables $\mathbf{v}$ and $\mathbf{w}$, this tells us $\langle \mathbf{v}, \mathbf{w} \rangle \leq \langle \mathbf{v}, \mathbf{v} \rangle^{\frac{1}{2}} \langle \mathbf{w}, \mathbf{w} \rangle^{\frac{1}{2}}$, just as we wanted to show.

Finally, to resolve the condition for equality, we only need to examine our reasoning in reverse. If equality holds in the abstract Cauchy inequality (1.16) for nonzero vectors $\mathbf{v}$ and $\mathbf{w}$, then the normalized variables $\hat{\mathbf{v}}$ and $\hat{\mathbf{w}}$ are well defined. In terms of the normalized variables, the equality of $\langle \mathbf{v}, \mathbf{w} \rangle$ and $\langle \mathbf{v}, \mathbf{v} \rangle^{\frac{1}{2}} \langle \mathbf{w}, \mathbf{w} \rangle^{\frac{1}{2}}$ tells us that $\langle \hat{\mathbf{v}}, \hat{\mathbf{w}} \rangle = 1$, and this tells us in turn that $\langle \hat{\mathbf{v}} - \hat{\mathbf{w}}, \hat{\mathbf{v}} - \hat{\mathbf{w}} \rangle = 0$ simply by expansion of the inner product. From this we deduce that $\hat{\mathbf{v}} - \hat{\mathbf{w}} = 0$; or, in other words, $\mathbf{v} = \lambda \mathbf{w}$ where we set $\lambda = \langle \mathbf{v}, \mathbf{v} \rangle^{\frac{1}{2}}/\langle \mathbf{w}, \mathbf{w} \rangle^{\frac{1}{2}}$.

The Pace of Science — The Development of Extensions

Augustin-Louis Cauchy (1789–1857) published his famous inequality in 1821 in the second of two notes on the theory of inequalities that formed the final part of his book *Cours d'Analyse Algébrique*, a volume which was perhaps the world's first rigorous calculus text. Oddly enough, Cauchy did not use his inequality in his text, except in some illustrative exercises. The first time Cauchy's inequality was applied in earnest by anyone was in 1829, when Cauchy used his inequality in an investigation of Newton's method for the calculation of the roots of algebraic and transcendental equations. This eight-year gap provides an interesting gauge of the pace of science; now, each month, there are hundreds — perhaps thousands — of new scientific publications where Cauchy's inequality is applied in one way or another.

A great many of those applications depend on a natural analog of Cauchy's inequality where sums are replaced by integrals,

$$\int_a^b f(x)g(x)\,dx \leq \left(\int_a^b f^2(x)\,dx\right)^{\frac{1}{2}} \left(\int_a^b g^2(x)\,dx\right)^{\frac{1}{2}}. \quad (1.19)$$

This bound first appeared in print in a *Mémoire* by Victor Yacovlevich Bunyakovsky which was published by the Imperial Academy of Sciences of St. Petersburg in 1859. Bunyakovsky (1804–1889) had studied in Paris with Cauchy, and he was quite familiar with Cauchy's work on inequalities; so much so that by the time he came to write his *Mémoire*, Bunyakovsky was content to refer to the classical form of Cauchy's inequality for finite sums simply as *well-known*. Moreover, Bunyakovsky did not dawdle over the limiting process; he took only a single line to pass from Cauchy's inequality for finite sums to his continuous analog (1.19). By ironic coincidence, one finds that this analog is labelled as inequality (**C**) in Bunyakovsky's *Mémoire*, almost as though Bunyakovsky had Cauchy in mind.

Bunyakovsky's *Mémoire* was written in French, but it does not seem to have circulated widely in Western Europe. In particular, it does not seem to have been known in Göttingen in 1885 when Hermann Amandus Schwarz (1843–1921) was engaged in his fundamental work on the theory of minimal surfaces.

In the course of this work, Schwarz had the need for a two-dimensional integral analog of Cauchy's inequality. In particular, he needed to show that if $S \subset \mathbb{R}^2$ and $f: S \to \mathbb{R}$ and $g: S \to \mathbb{R}$, then the double integrals

$$A = \iint_S f^2 \, dxdy, \quad B = \iint_S fg \, dxdy, \quad C = \iint_S g^2 \, dxdy$$

must satisfy the inequality

$$|B| \leq \sqrt{A} \cdot \sqrt{C}, \tag{1.20}$$

and Schwarz also needed to know that the inequality is strict unless the functions f and g are proportional.

An approach to this result via Cauchy's inequality would have been problematical for several reasons, including the fact that the strictness of a discrete inequality can be lost in the limiting passage to integrals. Thus, Schwarz had to look for an alternative path, and, faced with necessity, he discovered a proof whose charm has stood the test of time.

Schwarz based his proof on one striking observation. Specifically, he noted that the real polynomial

$$p(t) = \iint_S \Big(tf(x,y) + g(x,y)\Big)^2 \, dxdy = At^2 + 2Bt + C$$

is always nonnegative, and, moreover, $p(t)$ is strictly positive unless f and g are proportional. The binomial formula then tells us that the coefficients must satisfy $B^2 \leq AC$, and unless f and g are proportional, one actually has the strict inequality $B^2 < AC$. Thus, from a single algebraic insight, Schwarz found everything that he needed to know.

Schwarz's proof requires the wisdom to consider the polynomial $p(t)$, but, granted that step, the proof is lightning quick. Moreover, as one finds from Exercise 1.11, Schwarz's argument can be used almost without change to prove the inner product form (1.16) of Cauchy's inequality, and even there Schwarz's argument provides one with a quick understanding of the case of equality. Thus, there is little reason to wonder why Schwarz's argument has become a textbook favorite, even though it does require one to pull a rabbit — or at least a polynomial — out of a hat.

THE NAMING OF THINGS — ESPECIALLY INEQUALITIES

In light of the clear historical precedence of Bunyakovsky's work over that of Schwarz, the common practice of referring to the bound (1.19) as Schwarz's inequality may seem unjust. Nevertheless, by modern standards, both Bunyakovsky and Schwarz might count themselves lucky to have their names so closely associated with such a fundamental tool of mathematical analysis. Except in unusual circumstances, one garners little credit nowadays for crafting a continuous analog to a discrete inequality, or vice versa. In fact, many modern problem solvers favor a method of investigation where one rocks back and forth between discrete and continuous analogs in search of the easiest approach to the phenomena of interest.

Ultimately, one sees that inequalities get their names in a great variety of ways. Sometimes the name is purely descriptive, such as one finds with the triangle inequality which we will meet shortly. Perhaps more often, an inequality is associated with the name of a mathematician, but even then there is no hard-and-fast rule to govern that association. Sometimes the inequality is named after the first finder, but other principles may apply — such as the framer of the final form, or the provider of the best known application.

If one were to insist on the consistent use of the rule of first finder, then Hölder's inequality would become Rogers's inequality, Jensen's inequality would become Hölder's inequality, and only riotous confusion would result. The most practical rule — and the one used here — is simply to use the traditional names. Nevertheless, from time to time, it may be scientifically informative to examine the roots of those traditions.

Exercises

Exercise 1.1 (The 1-Trick and the Splitting Trick)
Show that for each real sequence $a_1, a_2, \ldots, a_n$ one has

$$a_1 + a_2 + \cdots + a_n \leq \sqrt{n}(a_1^2 + a_2^2 + \cdots + a_n^2)^{\frac{1}{2}} \qquad \text{(a)}$$

and show that one also has

$$\sum_{k=1}^n a_k \leq \left(\sum_{k=1}^n |a_k|^{2/3}\right)^{\frac{1}{2}} \left(\sum_{k=1}^n |a_k|^{4/3}\right)^{\frac{1}{2}}. \qquad \text{(b)}$$

The two tricks illustrated by this simple exercise will be our constant companions throughout the course. We will meet them in almost countless variations, and sometimes their implications are remarkably subtle.

Exercise 1.2 (Products of Averages and Averages of Products)
Suppose that $p_j \geq 0$ for all $j = 1, 2, \ldots, n$ and $p_1 + p_2 + \cdots + p_n = 1$. Show that if a_j and b_j are nonnegative real numbers that satisfy the termwise bound $1 \leq a_j b_j$ for all $j = 1, 2, \ldots, n$, then one also has the

aggregate bound for the averages,

$$1 \leq \left\{\sum_{j=1}^{n} p_j a_j\right\}\left\{\sum_{j=1}^{n} p_j b_j\right\}. \tag{1.21}$$

This graceful bound is often applied with $b_j = 1/a_j$. It also has a subtle complement which is developed much later in Exercise 5.8.

Exercise 1.3 (Why Not Three or More?)
Cauchy's inequality provides an upper bound for a sum of pairwise products, and a natural sense of confidence is all one needs to guess that there are also upper bounds for the sums of products of three or more terms. In this exercise you are invited to justify two prototypical extensions. The first of these is definitely easy, and the second is not much harder, provided that you do not give it more respect than it deserves:

$$\left(\sum_{k=1}^{n} a_k b_k c_k\right)^4 \leq \left(\sum_{k=1}^{n} a_k^2\right)^2 \sum_{k=1}^{n} b_k^4 \sum_{k=1}^{n} c_k^4, \tag{a}$$

$$\left(\sum_{k=1}^{n} a_k b_k c_k\right)^2 \leq \sum_{k=1}^{n} a_k^2 \sum_{k=1}^{n} b_k^2 \sum_{k=1}^{n} c_k^2. \tag{b}$$

Exercise 1.4 (Some Help From Symmetry)
There are many situations where Cauchy's inequality conspires with symmetry to provide results that are visually stunning. Here are two examples from a multitude of graceful possibilities.
(a) Show that for all positive x, y, z one has

$$S = \left(\frac{x+y}{x+y+z}\right)^{1/2} + \left(\frac{x+z}{x+y+z}\right)^{1/2} + \left(\frac{y+z}{x+y+z}\right)^{1/2} \leq 6^{1/2}.$$

(b) Show that for all positive x, y, z one has

$$x + y + z \leq 2\left\{\frac{x^2}{y+z} + \frac{y^2}{x+z} + \frac{z^2}{x+y}\right\}.$$

Exercise 1.5 (A Crystallographic Inequality with a Message)
Recall that $f(x) = \cos(\beta x)$ satisfies the identity $f^2(x) = \frac{1}{2}(1 + f(2x))$, and show that if $p_k \geq 0$ for $1 \leq k \leq n$ and $p_1 + p_2 + \cdots + p_n = 1$ then

$$g(x) = \sum_{k=1}^{n} p_k \cos(\beta_k x) \quad \text{satisfies} \quad g^2(x) \leq \frac{1}{2}\{1 + g(2x)\}.$$

Starting with Cauchy

This is known as the Harker–Kasper inequality, and it has far-reaching consequences in crystallography. For the theory of inequalities, there is an additional message of importance; given any functional *identity* one should at least consider the possibility of an analogous *inequality* for a more extensive class of related functions, such as the class of mixtures used here.

Exercise 1.6 (A Sum of Inversion Preserving Summands)
Suppose that $p_k > 0$ for $1 \leq k \leq n$ and $p_1 + p_2 + \cdots + p_n = 1$. Show that one has the bound

$$\sum_{k=1}^{n} \left(p_k + \frac{1}{p_k} \right)^2 \geq n^3 + 2n + 1/n,$$

and determine necessary and sufficient conditions for equality to hold here. We will see later (Exercise 13.6, p. 206), that there are analogous results for powers other than 2.

Exercise 1.7 (Flexibility of Form)
Prove that for all real x, y, α and β one has

$$(5\alpha x + \alpha y + \beta x + 3\beta y)^2$$
$$\leq (5\alpha^2 + 2\alpha\beta + 3\beta^2)(5x^2 + 2xy + 3y^2). \tag{1.22}$$

More precisely, show that the bound (1.22) is an immediate corollary of the Cauchy–Schwarz inequality (1.16) provided that one designs a special inner product $\langle \cdot, \cdot \rangle$ for the job.

Exercise 1.8 (Doing the Sums)
The effective use of Cauchy's inequality often depends on knowing a convenient estimate for one of the bounding sums. Verify the four following classic bounds for real sequences:

$$\sum_{k=0}^{\infty} a_k x^k \leq \frac{1}{\sqrt{1-x^2}} \left(\sum_{k=0}^{\infty} a_k^2 \right)^{\frac{1}{2}} \quad \text{for } 0 \leq x < 1, \tag{a}$$

$$\sum_{k=1}^{n} \frac{a_k}{k} < \sqrt{2} \left(\sum_{k=1}^{n} a_k^2 \right)^{\frac{1}{2}}, \tag{b}$$

$$\sum_{k=1}^{n} \frac{a_k}{\sqrt{n+k}} < (\log 2)^{\frac{1}{2}} \left(\sum_{k=1}^{n} a_k^2 \right)^{\frac{1}{2}}, \quad \text{and} \tag{c}$$

$$\sum_{k=0}^{n} \binom{n}{k} a_k \leq \binom{2n}{n}^{\frac{1}{2}} \left(\sum_{k=0}^{n} a_k^2 \right)^{\frac{1}{2}}. \qquad (d)$$

Exercise 1.9 (Beating the Obvious Bounds)

Many problems of mathematical analysis depend on the discovery of bounds which are stronger than those one finds with the direct application of Cauchy's inequality. To illustrate the kind of opportunity one might miss, show that for any real numbers a_j, $j = 1, 2 \ldots, n$, one has the bound

$$\left| \sum_{j=1}^{n} a_j \right|^2 + \left| \sum_{j=1}^{n} (-1)^j a_j \right|^2 \leq (n+2) \sum_{j=1}^{n} a_j^2.$$

Here the direct application of Cauchy's inequality gives a bound with $2n$ instead of the value $n+2$, so for large n one does better by a factor of nearly two.

Exercise 1.10 (Schur's Lemma — The R and C Bound)

Show that for each rectangular array $\{c_{jk} : 1 \leq j \leq m, 1 \leq k \leq n\}$, and each pair of sequences $\{x_j : 1 \leq j \leq m\}$ and $\{y_k : 1 \leq k \leq n\}$, we have the bound

$$\left| \sum_{j=1}^{m} \sum_{k=1}^{n} c_{jk} x_j y_k \right| \leq \sqrt{RC} \left(\sum_{j=1}^{m} |x_j|^2 \right)^{1/2} \left(\sum_{k=1}^{n} |y_k|^2 \right)^{1/2} \qquad (1.23)$$

where R and C are the *row sum* and *column sum* maxima defined by

$$R = \max_{j} \sum_{k=1}^{n} |c_{jk}| \quad \text{and} \quad C = \max_{k} \sum_{j=1}^{m} |c_{jk}|.$$

This bound is known as *Schur's Lemma*, but, ironically, it may be the second most famous result with that name. Nevertheless, this inequality is surely the single most commonly used tool for bounding a quadratic form. One should note in the extreme case when $n = m$, $c_{jk} = 0$ $j \neq k$, and $c_{jj} = 1$ for $1 \leq j \leq n$, Schur's Lemma simply recovers Cauchy's inequality.

Exercise 1.11 (Schwarz's Argument in an Inner Product Space)

Let v and w be elements of the inner product space $(V, \langle \cdot, \cdot \rangle)$ and consider the quadratic polynomial defined for $t \in \mathbb{R}$ by

$$p(t) = \langle \mathbf{v} + t\mathbf{w}, \mathbf{v} + t\mathbf{w} \rangle.$$

Observe that this polynomial is nonnegative and use what you know about the solution of the quadratic equation to prove the inner product version (1.16) of Cauchy's inequality. Also, examine the steps of your proof to establish the conditions under which the case of equality can apply. Thus, confirm that Schwarz's argument (page 11) applies almost without change to prove Cauchy's inequality for a general inner product.

Exercise 1.12 (Example of a Self-generalization)

Let $\langle \cdot, \cdot \rangle$ denote an inner product on the vector space V and suppose that $\mathbf{x}_1, \mathbf{x}_2, \ldots, \mathbf{x}_n$ and $\mathbf{y}_1, \mathbf{y}_2, \ldots, \mathbf{y}_n$ are sequences of elements of V. Prove that one has the following vector analog of Cauchy's inequality:

$$\sum_{j=1}^{n} \langle \mathbf{x}_j, \mathbf{y}_j \rangle \le \left(\sum_{j=1}^{n} \langle \mathbf{x}_j, \mathbf{x}_j \rangle \right)^{\frac{1}{2}} \left(\sum_{j=1}^{n} \langle \mathbf{y}_j, \mathbf{y}_j \rangle \right)^{\frac{1}{2}}. \quad (1.24)$$

Note that if one takes $n = 1$, then this bound simply recaptures the Cauchy–Schwarz inequality for an inner product space, while, if one keeps n general but specializes the vector space V to be $\mathbb{R}$ with the trivial inner product $\langle \mathbf{x}, \mathbf{y} \rangle = xy$, then the bound (1.24) simply recaptures the plain vanilla Cauchy inequality.

Exercise 1.13 (Application of Cauchy's Inequality to an Array)

Show that if $\{a_{jk} : 1 \le j \le m, \ 1 \le k \le n\}$ is an array of real numbers then one has

$$m \sum_{j=1}^{m} \left(\sum_{k=1}^{n} a_{jk} \right)^2 + n \sum_{k=1}^{n} \left(\sum_{j=1}^{m} a_{jk} \right)^2 \le \left(\sum_{j=1}^{m} \sum_{k=1}^{n} a_{jk} \right)^2 + mn \sum_{j=1}^{m} \sum_{k=1}^{n} a_{jk}^2.$$

Moreover, show that equality holds here if and only if there exist α_j and β_k such that $a_{jk} = \alpha_j + \beta_k$ for all $1 \le j \le m$ and $1 \le k \le n$.

Exercise 1.14 (A Cauchy Triple and Loomis–Whitney)

Here is a generalization of Cauchy's inequality that has as a corollary a discrete version of the Loomis–Whitney inequality, a result which in the continuous case provides a bound on the volume of a set in terms of the volumes of the projections of that set onto lower dimensional subspaces. The discrete Loomis–Whitney inequality (1.26) was only recently developed, and it has applications in information theory and the theory of algorithms.

(a) Show that for any nonnegative a_{ij}, b_{jk}, c_{ki} with $1 \le i, j, k \le n$ one

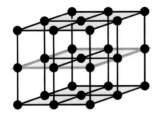

Here we have a set A with cardinality $|A| = 27$ with projections that satisfy $|A_x| = |A_y| = |A_z| = 9$.

Fig. 1.1. The discrete Loomis–Whitney inequality says that for any collection A of points in $\mathbb{R}^3$ one has $|A| \leq |A_x|^{\frac{1}{2}}|A_y|^{\frac{1}{2}}|A_z|^{\frac{1}{2}}$. The cubic arrangement indicated here suggests the canonical situation where one finds the case of equality in the bound.

has the triple product inequality

$$\sum_{i,j,k=1}^{n} a_{ij}^{\frac{1}{2}} b_{jk}^{\frac{1}{2}} c_{ki}^{\frac{1}{2}} \leq \left\{\sum_{i,j=1}^{n} a_{ij}\right\}^{\frac{1}{2}} \left\{\sum_{j,k=1}^{n} b_{jk}\right\}^{\frac{1}{2}} \left\{\sum_{k,i=1}^{n} c_{ki}\right\}^{\frac{1}{2}}. \quad (1.25)$$

(b) Let A denote a finite set of points in $\mathbb{Z}^3$ and let A_x, A_y, A_z denote the projections of A onto the corresponding coordinate planes that are orthogonal to the x, y, or z-axes. Let $|B|$ denote the cardinality of a set $B \subset \mathbb{Z}^3$ and show that the projections provide an upper bound on the cardinality of A:

$$|A| \leq |A_x|^{\frac{1}{2}}|A_y|^{\frac{1}{2}}|A_z|^{\frac{1}{2}}. \quad (1.26)$$

Exercise 1.15 (An Application to Statistical Theory)

If $p(k; \theta) \geq 0$ for all $k \in D$ and $\theta \in \Theta$ and if

$$\sum_{k \in D} p(k; \theta) = 1 \quad \text{for all } \theta \in \Theta, \quad (1.27)$$

then for each $\theta \in \Theta$ one can think of $\mathcal{M}_\theta = \{p(k; \theta) : k \in D\}$ as specifying a *probability model* where $p(k; \theta)$ represents the probability that we "observe k" when the parameter θ is the true "state of nature." If the function $g : D \to \mathbb{R}$ satisfies

$$\sum_{k \in D} g(k) p(k; \theta) = \theta \quad \text{for all } \theta \in \Theta, \quad (1.28)$$

then g is called an *unbiased estimator* of the parameter θ. Assuming that D is finite and $p(k; \theta)$ is a differentiable function of θ, show that one has the lower bound

$$\sum_{k \in D} (g(k) - \theta)^2 p(k; \theta) \geq 1/I(\theta) \tag{1.29}$$

where $I : \Theta \to \mathbb{R}$ is defined by the sum

$$I(\theta) = \sum_{k \in D} \left\{ p_\theta(k; \theta) / p(k; \theta) \right\}^2 p(k; \theta), \tag{1.30}$$

where $p_\theta(k; \theta) = \partial p(k; \theta)/\partial \theta$. The quantity defined by the left side of the bound (1.29) is called the *variance* of the unbiased estimator g, and the quantity $I(\theta)$ is known as the *Fisher information* at θ of the model $\mathcal{M}_\theta$. The inequality (1.29) is known as the Cramér–Rao lower bound, and it has extensive applications in mathematical statistics.

2
Cauchy's Second Inequality: The AM-GM Bound

Our initial discussion of Cauchy's inequality pivoted on the application of the elementary real variable inequality

$$xy \leq \frac{x^2}{2} + \frac{y^2}{2} \quad \text{for all } x, y \in \mathbb{R}, \tag{2.1}$$

and one may rightly wonder how so much value can be drawn from a bound which comes from the trivial observation that $(x-y)^2 \geq 0$. Is it possible that the humble bound (2.1) has a deeper physical or geometric interpretation that might reveal the reason for its effectiveness?

For nonnegative x and y, the direct term-by-term interpretation of the inequality (2.1) simply says that the area of the rectangle with sides x and y is never greater than the average of the areas of the two squares with sides x and y, and although this interpretation is modestly interesting, one can do much better with just a small change. If we first replace x and y by their square roots, then the bound (2.1) gives us

$$4\sqrt{xy} < 2x + 2y \quad \text{for all nonnegative } x \neq y, \tag{2.2}$$

and this inequality has a much richer interpretation.

Specifically, suppose we consider the set of all rectangles with area A and side lengths x and y. Since $A = xy$, the inequality (2.2) tells us that a square with sides of length $s = \sqrt{xy}$ must have the smallest perimeter among all rectangles with area A. Equivalently, the inequality tells us that among all rectangles with perimeter p, the square with side $s = p/4$ alone attains the maximal area.

Thus, the inequality (2.2) is nothing less than a rectangular version of the famous *isoperimetric property* of the circle, which says that among all planar regions with perimeter p, the circle of circumference p has the largest area. We now see more clearly why $xy \leq x^2/2 + y^2/2$ might be powerful; it is part of that great stream of results that links symmetry and optimality.

FROM SQUARES TO n-CUBES

One advantage that comes from the isoperimetric interpretation of the bound $\sqrt{xy} \leq (x+y)/2$ is the boost that it provides to our intuition. Human beings are almost hardwired with a feeling for geometrical truths, and one can easily conjecture many plausible analogs of the bound $\sqrt{xy} \leq (x+y)/2$ in two, three, or more dimensions.

Perhaps the most natural of these analogs is the assertion that the cube in $\mathbb{R}^3$ has the largest volume among all boxes (i.e., rectangular parallelepipeds) that have a given surface area. This intuitive result is developed in Exercise 2.9, but our immediate goal is a somewhat different generalization — one with a multitude of applications.

A box in $\mathbb{R}^n$ has 2^n corners, and each of those corners is incident to n edges of the box. If we let the lengths of those edges be $a_1, a_2, \ldots, a_n$, then the same isoperimetric intuition that we have used for squares and cubes suggests that the n-cube with edge length S/n will have the largest volume among all boxes for which $a_1 + a_2 + \cdots + a_n = S$. The next challenge problem offers an invitation to find an honest proof of this intuitive claim. It also recasts this geometric conjecture in the more common analytic language of arithmetic and geometric means.

Problem 2.1 (Arithmetic Mean-Geometric Mean Inequality)

Show that for every sequence of nonnegative real numbers $a_1, a_2, \ldots, a_n$ one has

$$\left(a_1 a_2 \cdots a_n\right)^{1/n} \leq \frac{a_1 + a_2 + \cdots + a_n}{n}. \qquad (2.3)$$

FROM CONJECTURE TO CONFIRMATION

For $n = 2$, the inequality (2.3) follows directly from the elementary bound $\sqrt{xy} \leq (x+y)/2$ that we have just discussed. One then needs just a small amount of luck to notice (as Cauchy did long ago) that the same bound can be applied twice to obtain

$$(a_1 a_2 a_3 a_4)^{\frac{1}{4}} \leq \frac{(a_1 a_2)^{\frac{1}{2}} + (a_3 a_4)^{\frac{1}{2}}}{2} \leq \frac{a_1 + a_2 + a_3 + a_4}{4}. \qquad (2.4)$$

This inequality confirms the conjecture (2.3) when $n = 4$, and the new bound (2.4) can be used again with $\sqrt{xy} \leq (x+y)/2$ to find that

$$(a_1 a_2 \cdots a_8)^{\frac{1}{8}} \leq \frac{(a_1 a_2 a_3 a_4)^{\frac{1}{4}} + (a_5 a_6 a_7 a_8)^{\frac{1}{4}}}{2} \leq \frac{a_1 + a_2 + \cdots + a_8}{8},$$

which confirms the conjecture (2.3) for $n = 8$.

Clearly, we are on a roll. Without missing a beat, one can repeat this argument k times (or use induction) to deduce that

$$(a_1 a_2 \cdots a_{2^k})^{1/2^k} \leq (a_1 + a_2 + \cdots + a_{2^k})/2^k \qquad \text{for all } k \geq 1. \quad (2.5)$$

The bottom line is that we have proved the target inequality for all $n = 2^k$, and all one needs now is just some way to fill the gaps between the powers of two.

The natural plan is to take an $n < 2^k$ and to look for some way to use the n numbers $a_1, a_2, \ldots, a_n$ to define a longer sequence $\alpha_1, \alpha_2, \ldots, \alpha_{2^k}$ to which we can apply the inequality (2.5). The discovery of an effective choice for the values of the sequence $\{\alpha_i\}$ may call for some exploration, but one is not likely to need too long to hit on the idea of setting $\alpha_i = a_i$ for $1 \leq i \leq n$ and setting

$$\alpha_i = \frac{a_1 + a_2 + \cdots + a_n}{n} \equiv A \qquad \text{for } n < i \leq 2^k;$$

in other words, we simply pad the original sequence $\{a_i : 1 \leq i \leq n\}$ with enough copies of the average A to give us a sequence $\{\alpha_i : 1 \leq i \leq 2^k\}$ that has length equal to 2^k.

The average A is listed $2^k - n$ times in the padded sequence $\{\alpha_i\}$, so, when we apply inequality (2.5) to $\{\alpha_i\}$, we find

$$\left\{ a_1 a_2 \cdots a_n \cdot A^{2^k - n} \right\}^{1/2^k} \leq \frac{a_1 + a_2 + \cdots + a_n + (2^k - n)A}{2^k} = \frac{2^k A}{2^k} = A.$$

Now, if we clear the powers of A to the right-hand side, then we find

$$(a_1 a_2 \cdots a_n)^{1/2^k} \leq A^{n/2^k},$$

and, if we then raise both sides to the power $2^k/n$, we come precisely to our target inequality,

$$(a_1 a_2 \cdots a_n)^{1/n} \leq \frac{a_1 + a_2 + \cdots + a_n}{n}. \quad (2.6)$$

A Self-Generalizing Statement

The AM-GM inequality (2.6) has an instructive *self-generalizing* quality. Almost without help, it pulls itself up by the bootstraps to a new result which covers cases that were left untouched by the original. Under normal circumstances, this generalization might seem to be too easy to qualify as a challenge problem, but the final result is so important the problem easily clears the hurdle.

Problem 2.2 (The AM-GM Inequality with Rational Weights)

Suppose that $p_1, p_2, \ldots, p_n$ are nonnegative rational numbers that sum to one, and show that for any nonnegative real numbers $a_1, a_2, \ldots, a_n$ one has

$$a_1^{p_1} a_2^{p_2} \cdots a_n^{p_n} \leq p_1 a_1 + p_2 a_2 + \cdots + p_n a_n. \tag{2.7}$$

Once one asks what role the rationality of the p_j might play, the solution presents itself soon enough. If we take an integer M so that for each j we can write $p_j = k_j/M$ for an integer k_j, then one finds that the ostensibly more general version (2.7) of the AM-GM follows from the original version (2.3) of the AM-GM applied to a sequence of length M with lots of repetition. One just takes the sequence with k_j copies of a_j for each $1 \leq j \leq n$ and then applies the plain vanilla AM-GM inequality (2.3); there is nothing more to it, or, at least there is nothing more if we attend strictly to the stated problem.

Nevertheless, there is a further observation one can make. Once the result (2.7) is established for rational values, the same inequality follows for general values of p_j "just by taking limits." In detail, we first choose a sequence of numbers $p_j(t)$, $j = 1, 2, \ldots, n$ and $t = 1, 2, \ldots$ for which we have

$$p_j(t) \geq 0, \quad \sum_{j=1}^n p_j(t) = 1, \quad \text{and} \quad \lim_{t \to \infty} p_j(t) = p_j.$$

One then applies the bound (2.7) to the n-tuples $(p_1(t), p_2(t), \ldots, p_n(t))$, and, finally, one lets n go to infinity to get the general result.

The technique of proving an inequality first for rationals and then extending to reals is often useful, but it does have some drawbacks. For example, the strictness of an inequality may be lost as one passes to a limit so the technique may leave us without a clear understanding of the case of equality. Sometimes this loss is unimportant, but for a tool as fundamental as the general AM-GM inequality, the conditions for equality are important. One would prefer a proof that handles all the features of the inequality in a unified way, and there are several pleasing alternatives to the method of rational approximation.

Pólya's Dream and a Path of Rediscovery

The AM-GM inequality turns out to have a remarkable number of proofs, and even though Cauchy's proof via the imaginative leap-forward fall-back induction is a priceless part of the world's mathematical inheritance, some of the alternative proofs are just as well loved. One

The AM-GM Inequality

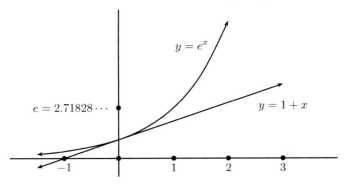

Fig. 2.1. The line $y = 1 + x$ is tangent to the curve $y = e^x$ at the point $x = 0$, and the line is below the curve for all $x \in \mathbb{R}$. Thus, we have $1 + x \leq e^x$ for all $x \in \mathbb{R}$, and, moreover, the inequality is strict except when $x = 0$. Here one should note that the y-axis has been scaled so that e is the unit; thus, the divergence of the two functions is more rapid than the figure may suggest.

particularly charming proof is due to George Pólya who reported that the proof came to him in a dream. In fact, when asked about his proof years later Pólya replied that it was the *best mathematics he had ever dreamt.*

Like Cauchy, Pólya begins his proof with a simple observation about a nonnegative function, except Pólya calls on the function $x \mapsto e^x$ rather than the function $x \mapsto x^2$. The graph of $y = e^x$ in Figure 2.1 illustrates the property of $y = e^x$ that is the key to Pólya's proof; specifically, it shows that the tangent line $y = 1 + x$ runs below the curve $y = e^x$, so one has the bound

$$1 + x \leq e^x \qquad \text{for all } x \in \mathbb{R}. \tag{2.8}$$

Naturally, there are analytic proofs of this inequality; for example, Exercise 2.2 suggests a proof by induction, but the evidence of Figure 2.1 is all one needs to move to the next challenge.

Problem 2.3 (The General AM-GM Inequality)

Take the hint of exploiting the exponential bound, and discover Pólya's proof for yourself; that is, show that the inequality (2.8) *implies that*

$$a_1^{p_1} a_2^{p_2} \cdots a_n^{p_n} \leq p_1 a_1 + p_2 a_2 + \cdots + p_n a_n \tag{2.9}$$

for nonnegative real numbers $a_1, a_2, \ldots, a_n$ and each sequence $p_1, p_2, \ldots, p_n$ of positive real numbers which sums to one.

In the AM-GM inequality (2.9) the left-hand side contains a product of terms, and the analytic inequality $1 + x \leq e^x$ stands ready to bound such a product by the exponential of a sum. Moreover, there are two ways to exploit this possibility; we could write the multiplicands a_k in the form $1 + x_k$ and then apply the analytic inequality (2.8), or we could modify the inequality (2.8) so that its applies directly to the a_k. In practice, one would surely explore both ideas, but for the moment, we will focus on the second plan.

If one makes the change of variables $x \mapsto x - 1$, then the exponential bound (2.8) becomes

$$x \leq e^{x-1} \qquad \text{for all } x \in \mathbb{R}, \tag{2.10}$$

and if we apply this bound to the multiplicands a_k, $k = 1, 2, \ldots$, we find

$$a_k \leq e^{a_k - 1} \quad \text{and} \quad a_k^{p_k} \leq e^{p_k a_k - p_k}.$$

When we take the product we find that the geometric mean $a_1^{p_1} a_2^{p_2} \cdots a_n^{p_n}$ is bounded above by

$$R(a_1, a_2, \ldots, a_n) = \exp\left(\left\{\sum_{k=1}^n p_k a_k\right\} - 1\right). \tag{2.11}$$

We may be pleased to know that the geometric mean $G = a_1^{p_1} a_2^{p_2} \cdots a_n^{p_n}$ is bounded by R, but we really cannot be too thrilled until we understand how R compares with the arithmetic mean

$$A = p_1 a_1 + p_2 a_2 + \cdots + p_n a_n,$$

and this is where the problem gets interesting.

A MODEST PARADOX

When we ask ourselves about a possible relation between A and R, one answer comes quickly. From the bound $A \leq e^{A-1}$ one sees that R is also an *upper bound* on the arithmetic mean A, so, all in one package, we have the double bound

$$\max\{a_1^{p_1} a_2^{p_2} \cdots a_n^{p_n},\ p_1 a_1 + p_2 a_2 + \cdots + p_n a_n\}$$
$$\leq \exp\left(\left\{\sum_{k=1}^n p_k a_k\right\} - 1\right). \tag{2.12}$$

This inequality inequality now presents us with a task which is at least a bit paradoxical. Can it really be possible to establish an inequality *between* two quantities when all one has is an upper bound on their maximum?

The AM-GM Inequality

MEETING THE CHALLENGE

While we might be discouraged for a moment, we should not give up too quickly. We should at least think long enough to notice that the bound (2.12) does provide a relationship between A and G in the special case when one of the two maximands on the left-hand side is equal to the term on the right-hand side. Perhaps we can exploit this observation.

Once this is said, the familiar notion of normalization is likely to come to mind. Thus, if we consider the new variables α_k, $k = 1, 2, \ldots, n$, defined by the ratios

$$\alpha_k = \frac{a_k}{A} \quad \text{where } A = p_1 a_1 + p_2 a_2 + \cdots + p_n a_n,$$

and if we apply the bound (2.11) to these new variables, then we find

$$\left(\frac{a_1}{A}\right)^{p_1} \left(\frac{a_2}{A}\right)^{p_2} \cdots \left(\frac{a_n}{A}\right)^{p_n} \leq \exp\left(\left\{\sum_{k=1}^{n} p_k \frac{a_k}{A}\right\} - 1\right) = 1.$$

After we clear the multiples of A to the right side and recall that one has $p_1 + p_2 + \cdots + p_n = 1$, we see that the proof of the general AM-GM inequality (2.9) is complete.

A FIRST LOOK BACK

When we look back on this proof of the AM-GM inequality (2.9), one of the virtues that we find is that it offers us a convenient way to identify the circumstances under which we can have equality; namely, if we examine the first step we see that we have

$$\frac{a_k}{A} < e^{(a_k/A)-1} \quad \text{unless} \quad \frac{a_k}{A} = 1, \tag{2.13}$$

and we *always* have

$$\frac{a_k}{A} \leq e^{(a_k/A)-1},$$

so we see that one also has

$$\left(\frac{a_1}{A}\right)^{p_1} \left(\frac{a_2}{A}\right)^{p_2} \cdots \left(\frac{a_n}{A}\right)^{p_n} < \exp\left(\left\{\sum_{k=1}^{n} p_k \frac{a_k}{A}\right\} - 1\right) = 1, \tag{2.14}$$

unless $a_k = A$ for all $k = 1, 2, \ldots, n$. In other words, we find that one has equality in the AM-GM inequality (2.9) if and only if

$$a_1 = a_2 = \cdots = a_n.$$

Looking back, we also see that the two lines (2.13) and (2.14) actually contain a full proof of the general AM-GM inequality. One could even argue with good reason that the single line (2.13) is all the proof that one really needs.

A LONGER LOOK BACK

This identification of the case of equality in the AM-GM bound may appear to be only an act of convenient tidiness, but there is much more to it. There is real power to be gained from understanding when an inequality is most effective, and we have already seen two examples of the energy that may be released by exploiting the case of equality.

When one compares the way that the AM-GM inequality was extracted from the bound $1+x \leq e^x$ with the way that Cauchy's inequality was extracted from the bound $xy \leq x^2/2 + y^2/2$, one may be struck by the effective role played by normalization — even though the normalizations were of quite different kinds. Is there some larger principle afoot here, or is this just a minor coincidence?

There is more than one answer to this question, but an observation that seems pertinent is that normalization often helps us focus the application of an inequality on the point (or the region) where the inequality is most effective. For example, in the derivation of the AM-GM inequality from the bound $1 + x \leq e^x$, the normalizations let us focus in the final step on the point $x = 0$, and this is precisely where $1 + x \leq e^x$ is sharp. Similarly, in the last step of the proof of Cauchy's inequality for inner products, normalization essentially brought us to the case of $x = y = 1$ in the two variable bound $xy \leq x^2/2 + y^2/2$, and again this is precisely where the bound is sharp.

These are not isolated examples. In fact, they are pointers to one of the most prevalent themes in the theory of inequalities. Whenever we hope to apply some underlying inequality to a new problem, the success or failure of the application will often depend on our ability to recast the problem so that the inequality is applied in one of those pleasing circumstances where the inequality is sharp, or nearly sharp.

In the cases we have seen so far, normalization helped us reframe our problems so that an underlying inequality could be applied more efficiently, but sometimes one must go to greater lengths. The next challenge problem recalls what may be one of the finest illustrations of this fight in all of the mathematical literature; it has inspired generations of mathematicians.

Pólya's Coaching and Carleman's Inequality

In 1923, as the first step in a larger project, Torsten Carleman proved a remarkable inequality which over time has come to serve as a benchmark for many new ideas and methods. In 1926 George Pólya gave an elegant proof of Carleman's inequality that depended on little more than the AM-GM inequality.

The secret behind Pólya's proof was his reliance on the general principle that one should try to use an inequality where it is most effective. The next challenge problem invites you to explore Carleman's inequality and to see if with a few hints you might also discover Pólya's proof.

Problem 2.4 (Carleman's Inequality)
Show that for each sequence of positive real numbers $a_1, a_2, \ldots$ one has the inequality

$$\sum_{k=1}^{\infty} (a_1 a_2 \cdots a_k)^{1/k} \leq e \sum_{k=1}^{\infty} a_k, \qquad (2.15)$$

where e denotes the natural base $2.71828\ldots$.

Our experience with the series version of Cauchy's inequality suggests that a useful way to approach a *quantitative* result such as the bound (2.15) is to first consider a simpler *qualitative* problem such as showing

$$\sum_{k=1}^{\infty} a_k < \infty \;\Rightarrow\; \sum_{k=1}^{\infty} (a_1 a_2 \cdots a_k)^{1/k} < \infty. \qquad (2.16)$$

Here, in the natural course of events, one would apply the AM-GM inequality to the summands on the right, do honest calculations, and hope for good luck. This plan leads one to the bound

$$\sum_{k=1}^{n} (a_1 a_2 \cdots a_k)^{1/k} \leq \sum_{k=1}^{n} \frac{1}{k} \sum_{j=1}^{k} a_j = \sum_{j=1}^{n} a_j \sum_{k=j}^{n} \frac{1}{k},$$

and — with no great surprise — we find that the plan does not work. As $n \to \infty$ our upper bound diverges, and we find that the naive application of the AM-GM inequality has left us empty-handed.

Naturally, this failure was to be expected since this challenge problem is intended to illustrate the *principle of maximal effectiveness* whereby we conspire to use our tools under precisely those circumstances when they are at their best. Thus, to meet the real issue, we need to ask ourselves why the AM-GM bound failed us and what we might do to overcome that failure.

Pursuit of a Principle

By the hypothesis on the left-hand side of the implication (2.16), the sum $a_1 + a_2 + \cdots$ converges, and this modest fact may suggest the likely source of our difficulties. Convergence implies that in any long block $a_1, a_2, \ldots, a_n$ there must be terms that are "highly unequal," and we know that in such a case the AM-GM inequality can be highly inefficient. Can we find some way to make our application of the AM-GM bound more forceful? More precisely, can we direct our application of the AM-GM bound toward some sequence with terms that are more nearly equal?

Since we know very little about the individual terms, we do not know precisely what to do, but one may well not need long to think of multiplying each a_k by some fudge factor c_k which we can try to specify more completely once we have a clear understanding of what is really needed. Naturally, the vague aim here is to find values of c_k so that the sequence of products $c_1 a_1, c_2 a_2, \ldots$ will have terms that are more nearly equal than the terms of our original sequence. Nevertheless, heuristic considerations carry us only so far. Ultimately, honest calculation is our only reliable guide.

Here we have the pleasant possibility of simply repeating our earlier calculation while keeping our fingers crossed that the new fudge factors will provide us with useful flexibility. Thus, if we just follow our nose and calculate as before, we find

$$\sum_{k=1}^{\infty} (a_1 a_2 \cdots a_k)^{1/k} = \sum_{k=1}^{\infty} \frac{(a_1 c_1 a_2 c_2 \cdots a_k c_k)^{1/k}}{(c_1 c_2 \cdots c_k)^{1/k}}$$
$$\leq \sum_{k=1}^{\infty} \frac{a_1 c_1 + a_2 c_2 + \cdots + a_k c_k}{k (c_1 c_2 \cdots c_k)^{1/k}}$$
$$= \sum_{k=1}^{\infty} a_k c_k \sum_{j=k}^{\infty} \frac{1}{j (c_1 c_2 \cdots c_j)^{1/j}}, \qquad (2.17)$$

and here we should take a breath. From this formula we see that the proof of the qualitative conjecture (2.16) will be complete if we can find some choice of the factors c_k, $k = 1, 2, \ldots$ such that the sums

$$s_k = c_k \sum_{j=k}^{\infty} \frac{1}{j (c_1 c_2 \cdots c_j)^{1/j}} \qquad k = 1, 2, \ldots. \qquad (2.18)$$

form a bounded sequence.

NECESSITY, POSSIBILITY, AND COMFORT

The hunt for a suitable choice of the c_k can take various directions, but, wherever our compass points, we eventually need to estimate the sum s_k. We should probably try to make this task as easy as possible, and here we are perhaps lucky that there are only a few series with tail sums that we can calculate. In fact, almost all of these come from the telescoping identity

$$\sum_{j=k}^{\infty} \left\{ \frac{1}{b_j} - \frac{1}{b_{j+1}} \right\} = \frac{1}{b_k}$$

that holds for all real monotone sequences $\{b_j : 1, 2, \ldots\}$ with $b_j \to \infty$.

Among the possibilities offered by this identity, the simplest choice is surely given by

$$\sum_{j=k}^{\infty} \frac{1}{j(j+1)} = \sum_{j=k}^{\infty} \left\{ \frac{1}{j} - \frac{1}{j+1} \right\} = \frac{1}{k} \qquad (2.19)$$

and, when we compare the sums (2.18) and (2.19), we see that s_k may be put into the simplest form when we define the fudge factors by the implicit recursion

$$(c_1 c_2 \cdots c_j)^{1/j} = j + 1 \qquad \text{for } j = 1, 2, \ldots. \qquad (2.20)$$

This choice gives us a short formula for s_k,

$$s_k = c_k \sum_{j=k}^{\infty} \frac{1}{j(c_1 c_2 \cdots c_j)^{1/j}} = c_k \sum_{j=k}^{\infty} \frac{1}{j(j+1)} = \frac{c_k}{k}, \qquad (2.21)$$

and all we need now is to estimate the size of c_k.

THE END OF THE TRAIL

Fortunately, this estimation is not difficult. From the implicit recursion (2.20) for c_j applied twice we find that

$$c_1 c_2 \cdots c_{j-1} = j^{j-1} \quad \text{and} \quad c_1 c_2 \cdots c_j = (j+1)^j,$$

so division gives us the explicit formula

$$c_j = \frac{(j+1)^j}{j^{j-1}} = j \left(1 + \frac{1}{j} \right)^j.$$

From this formula and our original bound (2.17) we find

$$\sum_{k=1}^{\infty} (a_1 a_2 \cdots a_k)^{1/k} \leq \sum_{k=1}^{\infty} \left(1 + \frac{1}{k} \right)^k a_k, \qquad (2.22)$$

and this bound puts Carleman's inequality (2.15) in our grasp. In fact, the bound (2.22) is even a bit stronger than Carleman's inequality since setting $x = 1/k$ in the familiar analytic bound $1 + x \leq e^x$ implies that

$$\left(1 + 1/k\right)^k < e \qquad \text{for all } k = 1, 2, \ldots.$$

EFFICIENCY AND THE CASE OF EQUALITY

There is more than a common dose of accidental elegance in Pólya's proof of Carleman's inequality, and some care must be taken not to lose track of the central idea. The insight to be savored is that there are circumstances where one may greatly improve the effectiveness of an inequality simply by restructuring the problem so that the inequality is applied in a situation that is close to the critical case of equality. Pólya's proof of Carleman's inequality illustrates this idea with exceptional charm, but there are many straightforward situations where its effect is just as great.

WHO WAS GEORGE PÓLYA?

George Pólya (1887–1985) was one of the most influential mathematicians of the 20th century, but his most enduring legacy may be the insight he passed on to us about teaching and learning. Pólya saw the process of problem solving as a fundamental human activity — one filled with excitement, creativity, and the love of life. He also thought hard about how one might become a more effective solver of mathematical problems and how one might coach others to do so.

Pólya summarized his thoughts in several books, the most famous of which is *How to Solve It*. The central premise of Pólya's text is that one can often make progress on a mathematical problem by asking certain *general common sense questions*. Many of Pólya's questions may seem obvious to a natural problem solver — or to anyone else — but, nevertheless, the test of time suggests that they possess considerable wisdom.

Some of the richest of Pólya's suggestions may be repackaged as the modestly paradoxical question: "What is the simplest problem that you *cannot* solve?" Here, of course, the question presupposes that one already has some particular problem in mind, so this suggestion is perhaps best understood as shorthand for a longer list of questions which would include at least the following:

- "Can you solve your problem in a special case?"

The AM-GM Inequality

- "Can you relate your problem to a similar one where the answer is already known?" and
- "Can you compute anything at all that is related to what you would really like to compute?"

Every reader is encouraged to experiment with Pólya's questions while addressing the exercises. Perhaps no other discipline can contribute more to one's effectiveness as a solver of mathematical problems.

Exercises

Exercise 2.1 (More from Leap-forward Fall-back Induction)

Cauchy's leap-forward, fall-back induction can be used to prove more than just the AM-GM inequality; in particular, it can be used to show that Cauchy's inequality for $n = 2$ implies the general result. For example, by Cauchy's inequality for $n = 2$ applied twice, one has

$$\begin{aligned}
&a_1b_1 + a_2b_2 + a_3b_3 + a_4b_4 \\
&= \{a_1b_1 + a_2b_2\} + \{a_3b_3 + a_4b_4\} \\
&\leq (a_1^2 + a_2^2)^{\frac{1}{2}}(b_1^2 + b_2^2)^{\frac{1}{2}} + (a_3^2 + a_4^2)^{\frac{1}{2}}(b_3^2 + b_4^2)^{\frac{1}{2}} \\
&\leq (a_1^2 + a_2^2 + a_3^2 + a_4^2)^{\frac{1}{2}}(b_1^2 + b_2^2 + b_3^2 + b_4^2)^{\frac{1}{2}},
\end{aligned}$$

which is Cauchy's inequality for $n = 4$. Extend this argument to obtain Cauchy's inequality for all $n = 2^k$ and consequently for all n. This may be the method by which Cauchy discovered his famous inequality, even though in his textbook he chose to present a different proof.

Exercise 2.2 (Bernoulli and the Exponential Bound)

Pólya's proof of the AM-GM inequality used the analytic bound

$$1 + x \leq e^x \quad \text{for all } x \in \mathbb{R}, \tag{2.23}$$

which is closely related to an inequality of Jacob Bernoulli (1654–1705),

$$1 + nx \leq (1+x)^n \quad \text{for all } x \in [-1, \infty) \text{ and all } n = 1, 2, \ldots. \tag{2.24}$$

Prove Bernoulli's inequality (2.24) by induction and show how it may be used to prove that $1 + x \leq e^x$ for all $x \in \mathbb{R}$. Finally, by calculus or by other means, prove one of the more general versions of Bernoulli's inequality suggested by Figure 2.2; for example, prove that

$$1 + px \leq (1+x)^p \quad \text{for all } x \geq -1 \text{ and all } p \geq 1. \tag{2.25}$$

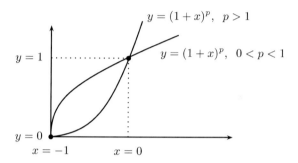

Fig. 2.2. The graph of $y = (1+x)^p$ suggests a variety of relationships, each of which depends on the range of x and the size of p. Perhaps the most useful of these is Bernoulli's inequality (2.25) where one has $p \geq 1$ and $x \in [-1, \infty)$.

Exercise 2.3 (Bounds by Pure Powers)

In the day-to-day work of mathematical analysis, one often uses the AM-GM inequality to bound a product or a sum of products by a simpler sum of *pure powers*. Show that for positive x, y, α, and β one has

$$x^\alpha y^\beta \leq \frac{\alpha}{\alpha+\beta} x^{\alpha+\beta} + \frac{\beta}{\alpha+\beta} y^{\alpha+\beta}, \qquad (2.26)$$

and, for a typical corollary, show that one also has the more timely bound $x^{2004} y + xy^{2004} \leq x^{2005} + y^{2005}$.

Exercise 2.4 (A Canadian Challenge)

Participants in the 2002 Canadian Math Olympiad were asked to prove the bound

$$a + b + c \leq \frac{a^3}{bc} + \frac{b^3}{ac} + \frac{c^3}{ab}$$

and to determine when equality can hold. Can you meet the challenge?

Exercise 2.5 (A Bound Between Differences)

Show that for nonnegative x and y and integer n one has

$$n(x-y)(xy)^{(n-1)/2} \leq x^n - y^n. \qquad (2.27)$$

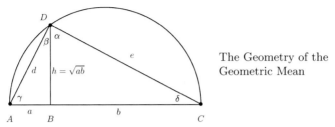

The Geometry of the
Geometric Mean

Fig. 2.3. The AM-GM inequality as Euclid could have imagined it. The circle has radius $(a+b)/2$ and the triangle's height h cannot be larger. Therefore if one proves that $h = \sqrt{ab}$ one has a geometric proof of the AM-GM for $n = 2$.

Exercise 2.6 (Geometry of the Geometric Mean)

There is indeed some geometry behind the definition of the geometric mean. The key relations were known to Euclid, although there is no evidence that Euclid specifically considered any inequalities. By appealing to the geometry of Figure 2.3 prove that $h = \sqrt{ab}$ and thereby automatically deduce that $\sqrt{ab} \leq (a+b)/2$.

Exercise 2.7 (One Bounded Product Implies Another)

Show that for nonnegative x, y, and z one has the implication

$$1 \leq xyz \implies 8 \leq (1+x)(1+y)(1+z). \tag{2.28}$$

Can you also propose a generalization?

Exercise 2.8 (Optimality Principles for Products and Sums)

Given positive $\{a_k : 1 \leq k \leq n\}$ and positive c and d, we consider the maximization problem P_1,

$$\max\{x_1 x_2 \cdots x_n : a_1 x_1 + a_2 x_2 + \cdots + a_n x_n = c\},$$

and the minimization problem P_2,

$$\min\{a_1 x_1 + a_2 x_2 + \cdots + a_n x_n : x_1 x_2 \cdots x_n = d\}.$$

Show that for *both* of these problems the condition for optimality is given by the relation

$$a_1 x_1 = a_2 x_2 = \cdots = a_n x_n. \tag{2.29}$$

These optimization principles are extremely productive, and they can provide useful guidance even when they do not exactly apply.

Exercise 2.9 (An Isoperimetric Inequality for the 3-Cube)

Show that among all boxes with a given surface area, the cube has the largest volume. Since a box with edge lengths $a, b,$ and c has surface area $A = 2ab + 2ac + 2bc$ and since a cube with surface area A has edge length $(A/6)^{1/2}$, the analytical task is to show

$$abc \leq (A/6)^{3/2}$$

and to confirm that equality holds if and only if $a = b = c$.

Exercise 2.10 (Åkerberg's Refinement)

Show that for any nonnegative real numbers $a_1, a_2, \ldots, a_n$ and $n \geq 2$ one has the bound

$$a_n \left\{ \frac{a_1 + a_2 + \cdots + a_{n-1}}{n-1} \right\}^{n-1} \leq \left\{ \frac{a_1 + a_2 + \cdots + a_n}{n} \right\}^n. \quad (2.30)$$

In a way, this relation is a *refinement* of the AM-GM inequality since the AM-GM inequality follow immediately by iteration of the bound (2.30). To prove the recurrence (2.30), one might first show that

$$y(n - y^{n-1}) = ny - y^n \leq n - 1 \qquad \text{for all } y \geq 0.$$

The key is then to make a wise choice of y.

Exercise 2.11 (Superadditivity of the Geometric Mean)

Show that for nonnegative a_k and b_k, $1 \leq k \leq n$, one has

$$\left(\prod_{k=1}^n a_k \right)^{1/n} + \left(\prod_{k=1}^n b_k \right)^{1/n} \leq \left(\prod_{k=1}^n (a_k + b_k) \right)^{1/n}. \quad (2.31)$$

This inequality of H. Minkowski asserts that the geometric mean is a *superadditive* function of its vector of arguments. Show that this inequality follows from the AM-GM inequality and determine the circumstances under which one can have equality.

For a generic hint, consider the possibility of dividing both sides by the quantity on the right side. Surprisingly often one finds that an inequality may become more evident if it is placed in a "standard form" which asserts that a given algebraic quantity is bounded by one.

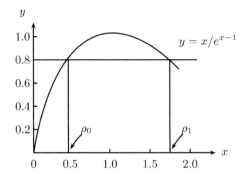

Fig. 2.4. The curve $y = x/e^{x-1}$ helps us measure the extent to which the individual terms of the averages must be squeezed together when the two sides of the AM-GM bound have a ratio that is close to one. For example, if we have $y \geq 0.99$, then we must have $0.694 \leq x \leq 1.149$.

Exercise 2.12 (On Approximate Equality in the AM-GM Bound)

If the nonnegative real numbers $a_1, a_2, \ldots, a_n$ are all approximately equal to a constant λ, then it is easy to check that both the arithmetic mean A and the geometric mean G are approximately equal. There are several ways to frame a converse to this observation, and this exercise considers an elegant method first proposed by George Pólya.

Show that if one has the inequality

$$0 < \frac{A - G}{A} = \epsilon < 1, \tag{2.32}$$

then one has the bound

$$\rho_0 \leq \frac{a_k}{A} \leq \rho_1 \quad \text{for all } k = 1, 2, \ldots, n, \tag{2.33}$$

where $\rho_0 \in (0, 1]$ and $\rho_1 \in [1, \infty)$ are two the roots of the equation

$$\frac{x}{e^{x-1}} = (1 - \epsilon)^n. \tag{2.34}$$

As Figure 2.4 suggests, one key to this result is the observation that the map $x \mapsto x/e^{x-1}$ is monotone increasing on $[0, 1]$ and monotone decreasing on $[1, \infty)$.

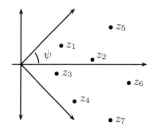

Any n points inside the complex right half plane are contained in a symmetric cone with central angle 2ψ with $0 \leq \psi < \pi$.

Fig. 2.5. The complex analog of the AM-GM inequality provides a nontrivial bound on the product $|z_1 z_2 \cdots z_n|^{1/n}$ provided that z_j, $j = 1, 2, \ldots, n$ are in the interior of the right half-plane. The quality of the bound depends on the central angle of the cone that contains the points.

Exercise 2.13 (An AM-GM Inequality for Complex Numbers)

Consider a set S of n complex numbers $z_1, z_2, \ldots, z_n$ for which the polar forms $z_j = \rho_j e^{i\theta_j}$ satisfy the constraints

$$0 \leq \rho_j < \infty \quad \text{and} \quad 0 \leq |\theta_j| < \psi < \pi/2, \qquad 1 \leq j \leq n.$$

As one sees in Figure 2.5, the spread in the arguments of the $z_j \in S$ is bounded by 2ψ. Show that for such numbers one has the bound

$$\bigl(\cos \psi\bigr) |z_1 z_2 \cdots z_n|^{1/n} \leq \frac{1}{n} |z_1 + z_2 + \cdots + z_n|. \tag{2.35}$$

Here one should note that if the z_j, $j = 1, 2, \ldots, n$ are all real numbers, then one can take $\psi = 0$, in which case the bound (2.35) recaptures the usual AM-GM inequality.

Exercise 2.14 (A Leap-Forward Fall-Back *Tour de Force*)

One can use Cauchy's leap-forward fall-back method of induction to prove that for all nonnegative $x_1, x_2, \ldots, x_m$ and for all integer powers $n = 1, 2, \ldots$ one has the bound

$$\left\{ \frac{x_1 + x_2 + \cdots + x_m}{m} \right\}^n \leq \frac{x_1^n + x_2^n + \cdots + x_m^n}{m}. \tag{2.36}$$

This is a special case of the power mean inequality which we develop at length in Chapter 8, but here the focus is on *mastery of technique*. This exercise leads to one of the more sustained applications of Cauchy's method that one is likely to meet.

3
Lagrange's Identity and Minkowski's Conjecture

The inductive proof of Cauchy's inequality used the polynomial identity

$$(a_1^2 + a_2^2)(b_1^2 + b_2^2) = (a_1b_1 + a_2b_2)^2 + (a_1b_2 - a_2b_1)^2, \qquad (3.1)$$

but that proof made no attempt to exploit this formula to the fullest. In particular, we completely ignored the term $(a_1b_2 - a_2b_1)^2$ except for noting that it must be nonnegative. To be sure, any inequality must strike a compromise between precision and simplicity, but no one wants to be wasteful. Thus, we face a natural question: Can one extract any useful information from the castaway term?

One can hardly doubt that the term $(a_1b_2 - a_2b_1)^2$ captures some information. At a minimum, it provides an explicit measure of the difference between the squares of the two sides of Cauchy's inequality, so perhaps it can provide a useful way to measure the *defect* that one incurs with each application of Cauchy's inequality.

The basic factorization (3.1) also tells us that for $n = 2$ one has equality in Cauchy's inequality exactly when $(a_1b_2 - a_2b_1)^2 = 0$; so, assuming that $(b_1, b_2) \neq (0, 0)$, we see that we have equality if and only if (a_1, a_2) and (b_1, b_2) are proportional in the sense that

$$a_1 = \lambda b_1 \text{ and } a_2 = \lambda b_2 \qquad \text{for some real } \lambda.$$

This observation has far-reaching consequences, and the first challenge problem invites one to prove an analogous characterization of the case of equality for the n-dimensional Cauchy inequality.

Problem 3.1 (On Equality in Cauchy's Bound)
 Show that if $(b_1, b_2, \ldots, b_n) \neq 0$ then equality holds in Cauchy's inequality if and only if there is a constant λ such that $a_i = \lambda b_i$ for all $i = 1, 2, \ldots, n$. Also, as before, if you already know a proof of this fact, you are invited to find a new one.

Passage to a More General Identity

Since the identity (3.1) provides a quick solution to Problem 3.1 when $n = 2$, one way to try to solve the problem in general is to look for a suitable extension of the identity (3.1) to n dimensions. Thus, if we introduce the quadratic polynomial $Q_n = Q_n(a_1, a_2, \ldots, a_n; b_1, b_2, \ldots, b_n)$ that is given by the difference of the squares of the two sides of Cauchy's inequality, then Q_n equals

$$(a_1^2 + a_2^2 + \cdots + a_n^2)(b_1^2 + b_2^2 + \cdots + b_n^2) - (a_1 b_1 + a_2 b_2 + \cdots + a_n b_n)^2,$$

and Q_n measures the "defect" in Cauchy's inequality in n dimensions, just like $Q_2 = (a_1 b_2 - b_1 a_2)^2$ measures the defect in two dimensions. We have already seen that Q_2 can be written as the square of a polynomial, and now the challenge is to see if there is an analogous representation of Q_n as a square, or possibly as a sum of squares.

If we simply expand Q_n, then we find that it can be written as

$$Q_n = \sum_{i=1}^{n} \sum_{j=1}^{n} a_i^2 b_j^2 - \sum_{i=1}^{n} \sum_{j=1}^{n} a_i b_i a_j b_j. \tag{3.2}$$

As it sits, this formula may not immediately suggest any way to make further progress. We could use a nice hint, and even though there is no hint that always helps, there is a general principle that often provides useful guidance: *pursue symmetry*.

Symmetry as a Hint

In practical terms, the suggestion to pursue symmetry just means that we should try to write our identity in a way that makes any symmetry as clear as possible. Here, the symmetry between i and j in the second double sum is forceful and clear, yet the symmetrical role of i and j in first double sum is not quite as evident. To be sure, symmetry is there, and we can make it stand out better if we rewrite Q_n in the form

$$Q_n = \frac{1}{2} \sum_{i=1}^{n} \sum_{j=1}^{n} (a_i^2 b_j^2 + a_j^2 b_i^2) - \sum_{i=1}^{n} \sum_{j=1}^{n} a_i b_i a_j b_j. \tag{3.3}$$

Now both double sums display transparent symmetry in i and j, and the new representation *does* suggest how to make progress; it almost screams for us to bring the two double sums together, and once this is done, one quickly finds the factorization

$$Q_n = \frac{1}{2} \sum_{i=1}^{n} \sum_{j=1}^{n} \left\{ a_i^2 b_j^2 - 2 a_i b_j a_j b_i + a_j^2 b_i^2 \right\} = \frac{1}{2} \sum_{i=1}^{n} \sum_{j=1}^{n} (a_i b_j - a_j b_i)^2.$$

The whole story now fits into a single, informative, self-verifying line known as *Lagrange's Identity*:

$$\left(\sum_{i=1}^{n} a_i b_i\right)^2 = \sum_{i=1}^{n} a_i^2 \sum_{i=1}^{n} b_i^2 - \frac{1}{2} \sum_{i=1}^{n} \sum_{j=1}^{n} (a_i b_j - a_j b_i)^2. \qquad (3.4)$$

Our path to this identity was motivated by our desire to understand the nonnegative polynomial Q_n, but, once the identity (3.4) is written down, it is easily verified just by multiplication. Thus, we meet one of the paradoxes of polynomial identities.

One should note that Cauchy's inequality is an immediate corollary of Lagrange's identity, and, indeed, the proof that Cauchy chose to include in his 1821 textbook was based on just this observation. Here, we went in search of what became Lagrange's identity (3.4) because we hoped it might lead to a clear understanding of the case of equality in Cauchy's inequality. Along the way, we happened to find an independent proof of Cauchy's inequality, but we still need to close the loop on our challenge problem.

EQUALITY AND A GAUGE OF PROPORTIONALITY

If $(b_1, b_2, \ldots, b_n) \neq 0$, then there exist some $b_k \neq 0$, and if equality holds in Cauchy's inequality, then all of the terms on the right-hand side of Lagrange's identity (3.4) must be identically zero. If we consider just the terms that contain b_k, then we find

$$a_i b_k = a_k b_i \qquad \text{for all } 1 \leq i \leq n,$$

and, if we take $\lambda = a_k / b_k$, then we also have

$$a_i = \lambda b_i \qquad \text{for all } 1 \leq i \leq n.$$

That is, Lagrange's identity tells us that for nonzero sequences one can have equality in Cauchy's inequality if and only if the two sequences are proportional. Thus we have a complete and precise answer to our first challenge problem.

This analysis of the case of equality underscores that the symmetric form

$$Q_n = \frac{1}{2} \sum_{i=1}^{n} \sum_{j=1}^{n} (a_i b_j - a_j b_i)^2$$

has two useful interpretations. We introduced it originally as a measure of the difference between the two sides of Cauchy's inequality, but we see now that it is also a measure of the extent to which the two vectors

$(a_1, a_2, \ldots, a_n)$ and $(b_1, b_2, \ldots, b_n)$ are proportional. Moreover, Q_n is such a natural measure of proportionality that one can well imagine a feasible course of history where the measure Q_n appears on the scene before Cauchy's inequality is conceived. This modest inversion of history has several benefits; in particular, it lead one to a notable inequality of E.A. Milne which is described in Exercise 3.8.

ROOTS AND BRANCHES OF LAGRANGE'S IDENTITY

Joseph Louis de Lagrange (1736–1813) developed the case $n = 3$ of the identity (3.4) in 1773 in the midst of an investigation of the geometry of pyramids. The study focused on questions in three-dimensional space, and Lagrange did not mention that the corresponding results for $n = 2$ were well known, even to the mathematicians of antiquity. In particular, the two-dimensional version of the identity (3.4) was known to the Alexandrian Greek mathematician Diophantus, or, at least one can draw that inference from a problem that Diophantus included in his textbook *Arithmetica*, a volume whose provenance can only be traced to sometime between 50 A.D. and 300 A.D.

Lagrange and his respected predecessor Pierre de Fermat (1601–1665) were quite familiar with the writings of Diophantus. In fact, much of what we know today of Fermat's discoveries comes to us from the marginal comments that Fermat made in his copy of the Bachet translation of Diophantus's *Arithmetica*. In just such a note, Fermat asserted that for $n \geq 3$ the equation $x^n + y^n = z^n$ has no solution in positive integers, and he also wrote "I have discovered a truly remarkable proof which this margin is too small to contain."

As all the world knows now, this assertion eventually came to be known as Fermat's Last Theorem, or, more aptly, Fermat's conjecture; and for more than three centuries, the conjecture eluded the best efforts of history's finest mathematicians. The world was shocked — and at least partly incredulous — when in 1993 Andrew Wiles announced that he had proved Fermat's conjecture. Nevertheless, within a year or so the proof outlined by Wiles had been checked by the leading experts, and it was acknowledged that Wiles had done the deed that many considered to be beyond human possibility.

PERSPECTIVE ON A GENERAL METHOD

Our derivation of Lagrange's identity began with a polynomial that we knew to be nonnegative, and we then relied on elementary algebra and good fortune to show that the polynomial could be written as a sum

Lagrange's Identity and Minkowski's Conjecture

of squares. The resulting identity did not need long to reveal its power. In particular, it quickly provided an independent proof of Cauchy's inequality and a transparent explanation for the necessary and sufficient conditions for equality.

This experience even suggests an interesting way to search for new, useful, polynomial identities. We just take any polynomial that we know to be nonnegative, and we then look for a representation of that polynomial as a sum of squares. If our experience with Lagrange's identity provides a reliable guide, the resulting polynomial identity should have a fair chance of being interesting and informative.

There is only one problem with this plan — we do not know any systematic way to write a nonnegative polynomial as a sum of squares. In fact, we do not even know if such a representation is always possible, and this observation brings us to our second challenge problem.

Problem 3.2 *Can one always write a nonnegative polynomial as a sum of squares? That is, if the real polynomial $P(x_1, x_2, \ldots, x_n)$ satisfies*

$$P(x_1, x_2, \ldots, x_n) \geq 0 \quad \text{for all } (x_1, x_2, \ldots, x_n) \in \mathbb{R}^n,$$

can one find a set of s real polynomials $Q_k(x_1, x_2, \ldots, x_n)$, $1 \leq k \leq s$, such that

$$P(x_1, x_2, \ldots, x_n) = Q_1^2 + Q_2^2 + \cdots + Q_s^2?$$

This problem turns out to be wonderfully rich. It leads to work that is deeper and more wide ranging than our earlier problems, and, even now, it continues to inspire new research.

A DEFINITIVE ANSWER — IN A SPECIAL CASE

As usual, one does well to look for motivation by examining some simple cases. Here the first case that is not completely trivial occurs when $n = 1$ and the polynomial $P(x)$ is simply a quadratic $ax^2 + bx + c$ with $a \neq 0$. Now, if we recall the method of completing the square that one uses to derive the binomial formula, we then see that $P(x)$ can be written as

$$P(x) = ax^2 + bx + c = a\left(x + \frac{b}{2a}\right)^2 + \frac{4ac - b^2}{4a}, \tag{3.5}$$

and this representation very nearly answers our question. We only need to check that the last two summands may be written as the squares of real polynomials.

If we consider large values of x, we see that $P(x) \geq 0$ implies that

$a > 0$, and if we take $x_0 = -b/2a$, then from the sum (3.5) we see that $P(x_0) \geq 0$ implies $4ac - b^2 \geq 0$. The bottom line is that both terms on the right-hand side of the identity (3.5) are nonnegative, so $P(x)$ can be written as $Q_1^2 + Q_2^2$ where Q_1 and Q_2 are real polynomials which we can write explicitly as

$$Q_1(x) = a^{\frac{1}{2}}\left(x + \frac{b}{2a}\right) \quad \text{and} \quad Q_2(x) = \frac{\sqrt{b^2 - 4ac}}{2\sqrt{a}}.$$

This solves our problem for quadratic polynomials of one variable, and even though the solution is simple, it is not trivial. In particular, the identity (3.5) has some nice corollaries. For example, it shows that $P(x)$ is minimized when $x = -b/2a$ and that the minimum value of $P(x)$ is equal to $(4ac - b^2)/4a$ — two useful facts that are more commonly obtained by calculus.

Exploiting What We Know

The simplest nontrivial case of Lagrange's identity is

$$(a_1^2 + a_2^2)(b_1^2 + b_2^2) = (a_1 b_1 + a_2 b_2)^2 + (a_1 b_2 - a_2 b_1)^2,$$

and, since polynomials may be substituted for the reals in this formula, we find that it provides us with a powerful fact: the set of polynomials that can be written as the sum of squares of two polynomials is *closed* under multiplication. That is, if $P(x) = Q(x)R(x)$ where $Q(x)$ and $R(x)$ have the representations

$$Q(x) = Q_1^2(x) + Q_2^2(x) \quad \text{and} \quad R(x) = R_1^2(x) + R_2^2(x),$$

then $P(x)$ also has a representation as a sum of two squares. More precisely, if we have

$$P(x) = Q(x)R(x) = \left(Q_1^2(x) + Q_2^2(x)\right)\left(R_1^2(x) + R_2^2(x)\right),$$

then $P(x)$ can also be written as

$$\{Q_1(x)R_1(x) + Q_2(x)R_2(x)\}^2 + \{Q_1(x)R_2(x) - Q_2(x)R_1(x)\}^2. \quad (3.6)$$

This identity suggests that induction may be of help. We have already seen that a nonnegative polynomial of degree two can be written as a sum of squares, so an inductive proof has no trouble getting started. We should then be able to use the representation (3.6) to complete the induction, once we understand how nonnegative polynomials can be factored.

Factorization of Nonnegative Polynomials

Two cases now present themselves; either $P(x)$ has a real root, or it does not. When $P(x)$ has a real root r with multiplicity m, we can write

$$P(x) = (x-r)^m R(x) \qquad \text{where } R(r) \neq 0,$$

so, if we set $x = r + \epsilon$, then we have $P(r+\epsilon) = \epsilon^m R(r+\epsilon)$. Also, by the continuity of R, there is a δ such that $R(r+\epsilon)$ has the same sign for all ϵ with $|\epsilon| \leq \delta$. Since $P(x)$ is always nonnegative, we then see that ϵ^m has the same sign for all $|\epsilon| \leq \delta$, so m must be even. If we set $m = 2k$, we see that

$$P(x) = Q^2(x) R(x) \qquad \text{where } Q(x) = (x-r)^k,$$

and, from this representation, we see that $R(x)$ is also a nonnegative polynomial. Thus, we have found a useful factorization for the case when $P(x)$ has a real root.

Now, suppose that $P(x)$ has no real roots. By the fundamental theorem of algebra, there is a complex root r, and since

$$0 = P(r) \qquad \text{implies } 0 = \overline{P(r)} = P(\bar{r}),$$

we see that the complex conjugate $\bar{r}$ is also a root of P. Thus, P has the factorization

$$P(x) = (x-r)(x-\bar{r}) R(x) = Q(x) R(x).$$

The real polynomial $Q(x) = (x-r)(x-\bar{r})$ is positive for large x, and it has no real zeros, so it must be positive for all real x. By assumption, $P(x)$ is nonnegative, so we see that $R(x)$ is also nonnegative. Thus, again we find that any nonnegative polynomial $P(x)$ with degree greater than two can be written as the product of two nonconstant, nonnegative polynomials. By induction, we therefore find that any nonnegative polynomial in one variable can be written as the sum of the squares of two real polynomials.

One Variable Down — Only N Variables to Go

Our success with polynomials of one variable naturally encourages us to consider nonnegative polynomials in two or more variables. Unfortunately, the gap between the a one variable problem and a two variable problem sometimes turns out to be wider than the Grand Canyon.

For polynomials in two variables, the zero sets $\{(x,y) : P(x,y) = 0\}$ are no longer simple discrete sets of points. Now they can take on a bewildering variety of geometrical shapes that almost defy classification.

Lagrange's Identity and Minkowski's Conjecture

After some exploration, we may even come to believe that there might exist nonnegative polynomials of two variables that *cannot* be written as the sum of squares of real polynomials. This is precisely what the great mathematician Hermann Minkowski first suggested, and, if we are to give full measure to the challenge problem, we will need to prove Minkowski's conjecture.

THE STRANGE POWER OF LIMITED POSSIBILITIES

There is an element of hubris to taking up a problem that defeated Minkowski, but there are times when hubris pays off. Ironically, there are even times when we can draw strength from the fact that we have very few ideas to try. Here, for example, we know so few ways to construct nonnegative polynomials that we have little to lose from seeing where those ways might lead. Most of the time, such explorations just help us understand a problem more deeply, but once in a while, a fresh, elementary approach to a difficult problem can lead to a striking success.

WHAT ARE OUR OPTIONS?

How can we construct a nonnegative polynomial? Polynomials that are given to us as sums of squares of real polynomials are always nonnegative, but such polynomials cannot help us with Minkowski's conjecture. We might also consider the nonnegative polynomials that one finds by squaring both sides of Cauchy's inequality and taking the difference, but Lagrange's identity tells us that this construction is also doomed. Finally, we might consider those polynomials that the AM-GM inequality tells us must be nonnegative. For the moment this is our only feasible idea, so it obviously deserves a serious try.

THE AM-GM PLAN

We found earlier that nonnegative real numbers $a_1, a_2, \ldots, a_n$ must satisfy the AM-GM inequality

$$(a_1 a_2 \cdots a_n)^{1/n} \leq \frac{a_1 + a_2 + \cdots + a_n}{n}, \tag{3.7}$$

and we can use this inequality to construct a vast collection of nonnegative polynomials. Nevertheless, if we do not want to get lost in complicated examples, we need to limit our search to the very simplest cases. Here, the simplest choice for nonnegative a_1 and a_2 are $a_1 = x^2$ and $a_2 = y^2$; so, if we want to make the product $a_1 a_2 a_3$ as simple as possible, we can take $a_3 = 1/x^2 y^2$ so that $a_1 a_2 a_3$ just equals one. The

Lagrange's Identity and Minkowski's Conjecture

AM-GM inequality then tells us that

$$1 \leq \frac{1}{3}(x^2 + y^2 + 1/x^2y^2)$$

and, after the natural simplifications, we see that the polynomial

$$P(x,y) = x^4y^2 + x^2y^4 - 3x^2y^2 + 1$$

is nonnegative for all choices of real x and y; thus, we find our first serious candidate for such a polynomial that cannot be written in the form

$$P(x,y) = Q_1^2(x,y) + Q_2^2(x,y) + \cdots + Q_s^2(x,y) \qquad (3.8)$$

for some integer s. Now we only need to find some way to argue that the representation (3.8) is indeed impossible. We only have elementary tools at our disposal, but these may well suffice. Even a modest exploration shows that the representation (3.8) is quite confining.

For example, we first note that our candidate polynomial $P(x,y)$ has degree six, so none of the polynomials Q_k can have degree greater than three. Moreover, when we specialize by taking $y = 0$, we find

$$1 = P(x,0) = Q_1^2(x,0) + Q_2^2(x,0) + \cdots + Q_s^2(x,0),$$

while by taking $x = 0$, we find

$$1 = P(0,y) = Q_1^2(0,y) + Q_2^2(0,y) + \cdots + Q_s^2(0,y),$$

so both of the univariate polynomials $Q_k^2(x,0)$ and $Q_k^2(0,y)$ must be bounded. From this observation and the fact that each polynomial $Q_k(x,y)$ has degree not greater than three, we see that they must be of the form

$$Q_k(x,y) = a_k + b_k xy + c_k x^2 y + d_k xy^2 \qquad (3.9)$$

for some constants a_k, b_k, c_k, and d_k.

Minkowski's conjecture is now on the ropes; we just need to land a knock-out punch. When we look back at our candidate $P(x,y)$, we see the striking feature that all of its coefficients are nonnegative except for the coefficient of x^2y^2 which is equal to -3. This observation suggests that we should see what one can say about the possible values of the coefficient of x^2y^2 in the sum $Q_1^2(x,y) + Q_2^2(x,y) + \cdots + Q_s^2(x,y)$.

Here we have some genuine luck. By the explicit form (3.9) of the terms $Q_k(x,y)$, $1 \leq k \leq s$, we can easily check that the coefficient of x^2y^2 in the polynomial $Q_1^2(x,y) + Q_2^2(x,y) + \cdots + Q_s^2(x,y)$ is just $b_1^2 + b_2^2 + \cdots + b_s^2$. Since this sum is nonnegative, it cannot equal -3, and, consequently, the nonnegative polynomial $P(x,y)$ cannot be written as a sum of squares of real polynomials. Remarkably enough, the AM-GM inequality has guided us successfully to a proof of Minkowski's conjecture.

SOME PERSPECTIVE ON MINKOWSKI'S CONJECTURE

We motivated Minkowski's conjecture by our exploration of Lagrange's identity, and we proved Minkowski's conjecture by making good use of the AM-GM inequality. This is a logical and instructive path. Nevertheless, it strays a long way from the historical record, and it may leave the wrong impression.

While it is not precisely clear what led Minkowski to his conjecture, he was most likely concerned at first with number theoretic results such as the classic theorem of Lagrange which asserts that every natural number may be written as the sum of four or fewer perfect squares. In any event, Minkowski brought his conjecture to David Hilbert, and in 1888, Hilbert published a proof of the existence of nonnegative polynomials that cannot be written as a sum of the squares of real polynomials. Hilbert's proof was long, subtle, and indirect.

The first explicit example of a nonnegative polynomial that cannot be written as the sum of the squares of real polynomials was given in 1967, almost eighty years after Hilbert proved the existence of such polynomials. The explicit example was discovered by T.S. Motzkin, and he used precisely the same AM-GM technique described here.

HILBERT'S 17TH PROBLEM

In 1900, David Hilbert gave an address in Paris to the second International Congress of Mathematicians which many regard as the most important mathematical address of all time. In his lecture, Hilbert described 23 problems which he believed to be worth the attention of the world's mathematicians at the dawn of the 20th century. The problems were wisely chosen, and they have had a profound influence on the development of mathematics over the past one hundred years.

The 17th problem on Hilbert's great list is a direct descendant of Minkowski's conjecture, and in this problem Hilbert asked if every nonnegative polynomial in n variables must have a representation as a sum of squares of *ratios* of polynomials. This modification of Minkowski's problem makes all the difference, and Hilbert's question was answered affirmatively in 1927 by Emil Artin. Artin's solution of Hilbert's 17th

Lagrange's Identity and Minkowski's Conjecture

problem is now widely considered to be one of the crown jewels of modern algebra.

EXERCISES

Exercise 3.1 (A Trigonometric Path to Discovery)

One only needs multiplication to verify the identity of Diophantus,

$$(a_1b_1 + a_2b_2)^2 = (a_1^2 + a_2^2)(b_1^2 + b_2^2) - (a_1b_2 - a_2b_1)^2, \qquad (3.10)$$

yet multiplication does not suggest how such an identity might have been discovered. Take the more inventive path suggested by Figure 3.1 and show that the identity of Diophantus is a consequence of the most the famous theorem of all, the one universally attributed to Pythagoras (circa 497 B.C.).

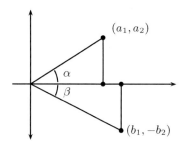

The classic identity
$$1 = \cos^2(\alpha + \beta) + \sin^2(\alpha + \beta)$$
permits one to deduce that
$(a_1^2 + a_2^2)(b_1^2 + b_2^2)$ equals
$(a_1b_1 + a_2b_2)^2 + (a_1b_2 - a_2b_1)^2$.

Fig. 3.1. In the right light, the identity (3.10) of Diophantus and the theorem of Pythagoras can be seen to be fraternal twins, though one is algebraic and the other geometric.

Exercise 3.2 (Brahmagupta's Identity)

Brahmagupta (circa 600 A.D.) established an identity which shows that for any integer D the product of two numbers which can be written in the form $a^2 - Db^2$ with $a, b \in \mathbb{Z}$ must be an integer of the same form. More precisely, Brahmagupta's identity says

$$(a^2 - Db^2)(\alpha^2 - D\beta^2) = (a\alpha + D\beta)^2 - D(a\beta + \alpha b)^2.$$

(a) Prove Brahmagupta's identity by evaluating the product
$$(a + b\sqrt{D})(a - b\sqrt{D})(\alpha + \beta\sqrt{D})(\alpha - \beta\sqrt{D})$$

in two different ways. Incidentally, the computation is probably more interesting than you might guess.

(b) Can you modify the pattern used to prove Brahmagupta's identity to give another proof of the identity (3.10) of Diophantus?

Exercise 3.3 (A Continuous Analogue of Lagrange's Identity)

Formulate and prove a continuous analogue of Lagrange's identity. Next, show that your identity implies Schwarz's inequality and finally use your identity to derive a necessary and sufficient condition for equality to hold.

Exercise 3.4 (A Cauchy Interpolation)

Show for $0 \leq x \leq 1$ and for any pair of real vectors $(a_1, a_2, \ldots, a_n)$ and $(b_1, b_2, \ldots, b_n)$ that the quantity

$$\left\{ \sum_{j=1}^{n} a_j b_j + 2x \sum_{1 \leq j < k \leq n} a_j b_k \right\}^2$$

is bounded above by the product

$$\left\{ \sum_{j=1}^{n} a_j^2 + 2x \sum_{1 \leq j < k \leq n} a_j a_k \right\} \left\{ \sum_{j=1}^{n} b_j^2 + 2x \sum_{1 \leq j < k \leq n} b_j b_k \right\}. \quad (3.11)$$

The charm of this bound is that for $x = 0$ it reduces to Cauchy's inequality and for $x = 1$ it reduces to the algebraic identity

$$\left\{ (a_1 + a_2 + \cdots + a_n)(b_1 + b_2 + \cdots + b_n) \right\}^2$$
$$= (a_1 + a_2 + \cdots + a_n)^2 (b_1 + b_2 + \cdots + b_n)^2.$$

Thus, we have an inequality that *interpolates* between two known results.

Exercise 3.5 (Monotonicity and a Ratio Bound)

Show that if $f : [0, 1] \to (0, \infty)$ is nonincreasing, then one has

$$\frac{\int_0^1 x f^2(x)\, dx}{\int_0^1 x f(x)\, dx} \leq \frac{\int_0^1 f^2(x)\, dx}{\int_0^1 f(x)\, dx}. \quad (3.12)$$

As a hint, one might consider the possibility of proving a Lagrange type identity by beginning with a double integral on $[0, 1] \times [0, 1]$ whose integrand is guaranteed to be positive by our monotonicity hypothesis.

Exercise 3.6 (Monotonicity of the Product Defect)

Show that for a pair of monotone sequences $0 \leq a_1 \leq a_2 \leq \cdots$ and $0 \leq b_1 \leq b_2 \leq \cdots$ the quantities defined by

$$D_n = n \sum_{j=1}^{n} a_j b_j - \sum_{j=1}^{n} a_j \sum_{j=1}^{n} b_j \qquad \text{for } n = 1, 2, \ldots \qquad (3.13)$$

are also monotone nondecreasing. Specifically, show that for each integer $n = 0, 1, \ldots$ one has $D_n \leq D_{n+1}$.

Exercise 3.7 (The Four-Letter Identity via Polarization)

For any real numbers a_j, b_j, s_j and t_j, $1 \leq j \leq n$, there is an identity due independently to Binet and Cauchy which states that

$$\sum_{j=1}^{n} a_j s_j \sum_{j=1}^{n} b_j t_j - \sum_{j=1}^{n} a_j b_j \sum_{j=1}^{n} s_j t_j = \sum_{1 \leq j < k \leq n} (a_j b_k - b_j a_k)(s_j t_k - s_k t_j).$$

This generalizes Lagrange's identity, as one can check by setting $s_j = b_j$ and $t_j = a_j$, but it is much more informative to know that the Cauchy–Binet identity may be obtained as a corollary of the much simpler result of Lagrange.

In fact, the passage is quite straightforward, provided one knows how to exploit the *polarization transformation*

$$f(u) \mapsto \frac{1}{4}\{f(u+v) - f(u-v)\}.$$

This transformation carries the function $u \mapsto u^2$ into the two-variable function $(u, v) \mapsto uv$, and it is devilishly effective at morphing identities with squares into new ones where the squares are replaced by products.

To see how this works, check that the four-variable identity follows from the two-variable Lagrange identity after two sequential polarizations. To keep your calculation tidy, you may want to use the shorthand

$$\begin{vmatrix} \alpha & \beta \\ \gamma & \delta \end{vmatrix} \equiv \alpha\delta - \beta\gamma \qquad (3.14)$$

and the easily verified identity that follows from the definition (3.14),

$$\begin{vmatrix} \alpha + \alpha' & \beta \\ \gamma + \gamma' & \delta \end{vmatrix} = \begin{vmatrix} \alpha & \beta \\ \gamma & \delta \end{vmatrix} + \begin{vmatrix} \alpha' & \beta \\ \gamma' & \delta \end{vmatrix}. \qquad (3.15)$$

This shorthand recalls the notation for the determinant of a two-by-two matrix, but to solve this problem one does not need to know more about determinants than the two self-evident relations (3.14) and (3.15).

Exercise 3.8 (Milne and Gauges of Proportionality)

We have seen that the form

$$Q = \frac{1}{2}\sum_{i=1}^{n}\sum_{j=1}^{n}(a_i b_j - a_j b_i)^2$$

provides a natural measure of proportionality for the pair of vectors $(a_1, a_2, \ldots, a_n)$ and $(b_1, b_2, \ldots, b_n)$, but one can think of other measures of proportionality that are just as reasonable. For example, if we restrict our attention to vectors of positive terms, then one might equally well use the self-normalized sum

$$R = \frac{1}{2}\sum_{i=1}^{n}\sum_{j=1}^{n}\frac{(a_i b_j - a_j b_i)^2}{(a_i + b_i)(a_j + b_j)}. \qquad (3.16)$$

Develop an identity containing R that will permit you to prove the inequality of E.A. Milne:

$$\left\{\sum_{j=1}^{n}(a_j + b_j)\right\}\left\{\sum_{j=1}^{n}\frac{a_j b_j}{(a_j + b_j)}\right\} \leq \left\{\sum_{j=1}^{n}a_j\right\}\left\{\sum_{j=1}^{n}b_j\right\}. \qquad (3.17)$$

Next, use your identity to show that one has equality in the bound (3.17) if and only if the vectors $(a_1, a_2, \ldots, a_n)$ and $(b_1, b_2, \ldots, b_n)$ are proportional. Incidentally, the bound (3.17) was introduced by Milne in 1925 to help explain the biases inherent in certain measurements of stellar radiation.

4
On Geometry and Sums of Squares

John von Neumann once said, "In mathematics you don't understand things, you just get used to them." The notion of n-dimensional space is now an early entrant in the mathematical curriculum, and few of us view it as particularly mysterious; nevertheless, for generations before ours this was not always the case. To be sure, our experience with the Pythagorean theorem in $\mathbb{R}^2$ and $\mathbb{R}^3$ is easily extrapolated to suggest that for two points $\mathbf{x} = (x_1, x_2, \ldots, x_d)$ and $\mathbf{y} = (y_1, y_2, \ldots, y_d)$ in $\mathbb{R}^d$ the distance $\rho(\mathbf{x}, \mathbf{y})$ between $\mathbf{x}$ and $\mathbf{y}$ should be given by

$$\rho(\mathbf{x}, \mathbf{y}) = \sqrt{(y_1 - x_1)^2 + (y_2 - x_2)^2 + \cdots + (y_d - x_d)^2}, \qquad (4.1)$$

but, despite the familiarity of this formula, it still keeps some secrets. In particular, many of us may be willing to admit to some uncertainty whether it is best viewed as a theorem or as a definition.

With proper preparation, either point of view may be supported, although the path of least resistance is surely to take the formula for $\rho(\mathbf{x}, \mathbf{y})$ as the *definition* of the Euclidean distance in $\mathbb{R}^d$. Nevertheless, there is a Faustian element to this bargain.

First, this definition makes the Pythagorean theorem into a bland triviality, and we may be saddened to see our much-proved friend treated so shabbily. Second, we need to check that this definition of distance in $\mathbb{R}^d$ meets the minimal standards that one demands of a distance function; in particular, we need to check that ρ satisfies the so-called triangle inequality, although, by a bit of luck, Cauchy's inequality will help us with this task. Third, and finally, we need to test the limits on our intuition. Our experience with $\mathbb{R}^2$ and $\mathbb{R}^3$ is a powerful guide, yet it can also mislead us, and one does well to develop a skeptical attitude about what is obvious and what is not.

Even though it may be a bit like having dessert before having dinner,

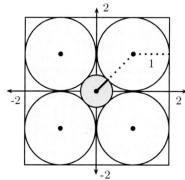

In $\mathbb{R}^2$, one places a unit circle in each quadrant of the square $[-2, 2]^2$.

A non-overlapping circle of maximal radius is then centered at the origin.

Fig. 4.1. This arrangement of $5 = 2^2 + 1$ circles in $[-2, 2]^2$ has a natural generalization to an arrangement of $2^d + 1$ spheres in $[-2, 2]^d$. This general arrangement then provokes a question which a practical person might find perplexing — or even silly. Does the central sphere stay inside the box $[-2, 2]^d$ for all values of d?

we will begin with the third task. This time the problem that guides us is framed with the help of the arrangement of circles illustrated in Figure 4.1. This simple arrangement of $5 = 2^2 + 1$ circles is not rich enough to suggest any serious questions, but it has a d-dimensional analog which puts our intuition to the test.

ON AN ARRANGEMENT IN $\mathbb{R}^d$

Consider the arrangement where for each of the 2^d points denoted by $\mathbf{e} = (e_1, e_2, \ldots, e_d)$ with $e_k = 1$ or $e_k = -1$ for all $1 \leq k \leq d$, we have a sphere $S_\mathbf{e}$ with unit radius and center $\mathbf{e}$. Each of these spheres is contained in the cube $[-2, 2]^d$ and, to complete the picture, we place a sphere $\mathcal{S}(d)$ at the origin that has the largest possible radius subject to the constraint that $\mathcal{S}(d)$ does not intersect the interior of any of the initial collection of 2^d unit spheres. We then ask ourselves a question which no normal, sensible person would ever think of asking.

Problem 4.1 (Thinking Outside the Box)

Is the central sphere $\mathcal{S}(d)$ contained in the cube $[-2, 2]^d$ for all $d \geq 2$?

Just posing this question provides a warning that we should not trust our intuition here. If we rely purely on our visual imagination, it may even seem silly to suggest that $\mathcal{S}(d)$ might somehow expand beyond the box $[-2, 2]^d$. Nevertheless, our visual imagination is largely rooted in our experience with $\mathbb{R}^2$ and $\mathbb{R}^3$, and this intuition can easily fail us in $\mathbb{R}^d$, $d \geq 4$. Instead, computation must be our guide.

Here we first note that for each of the 2^d outside spheres the corresponding center point $\mathbf{e}$ has distance $\sqrt{d}$ from the origin. Next, since each outside sphere has radius 1, we see by subtraction that the radius of the central sphere $\mathcal{S}(d)$ is equal to $\sqrt{d}-1$. Thus, we find that for $d \geq 10$ one has $\sqrt{d}-1 > 2$, and, yes, indeed, the central sphere actually extends beyond the box $[-2,2]^d$. In fact, as $d \to \infty$ the fraction of the volume of the sphere that is inside the box even goes to zero exponentially fast.

REFINING INTUITION — FACING LIMITATIONS

When one shares this example with friends, there is usually a brief moment of awe, but sooner or later someone says, "Why should we regard this as surprising? Just look how far away the point $\mathbf{e} = (e_1, e_2, \ldots, e_d)$ is from the origin! Is it really any wonder that ....". Such observations illustrate how quickly (and almost subconsciously) we refine our intuition after some experience with calculations. Nevertheless, if we accept such remarks at face value, it is easy to become overly complacent about the very real limitations on our physical intuition.

Ultimately, we may do best to take a hint from pilots who train themselves to fly safely through clouds by relying on instruments rather than physical sensations. When we work on problems in $\mathbb{R}^d$, $d \geq 4$, we benefit greatly from the analogy with $\mathbb{R}^2$ and $\mathbb{R}^3$, but at the end of the day, we must rely on computation rather than visual imagination.

MEETING THE MINIMAL REQUIREMENTS

The example of Figure 4.1 reminds us that intuition is fallible, but even our computations need guidance. One way to seek help is to force our problem into its simplest possible form, while striving to retain its essential character. Thus, a complex model is often boiled down to a simpler abstract model where we rely on a small set of rules, or axioms, to help us express the minimal demands that must be met. In this way one hopes to remove the influence of an overly active imagination, while still retaining a modicum of control.

Our next challenge is to see how the Euclidean distance (4.1) might pass through such a logical sieve. Thus, for a moment, we consider an arbitrary set S and a function $\rho: S \times S \to \mathbb{R}$ that has the four following properties:

(i) $\rho(\mathbf{x}, \mathbf{y}) \geq 0$ for all $\mathbf{x}, \mathbf{y}$ in S,
(ii) $\rho(\mathbf{x}, \mathbf{y}) = 0$ if and only if $\mathbf{x} = \mathbf{y}$,

(iii) $\rho(\mathbf{x}, \mathbf{y}) = \rho(\mathbf{y}, \mathbf{x})$ for all $\mathbf{x}, \mathbf{y}$ in S, and
(iv) $\rho(\mathbf{x}, \mathbf{y}) \leq \rho(\mathbf{x}, \mathbf{z}) + \rho(\mathbf{z}, \mathbf{y})$ for all $\mathbf{x}, \mathbf{y}$ and $\mathbf{z}$ in S.

These properties are intended to reflect the rock-bottom minimal requirements that $\rho(\cdot, \cdot)$ must meet for us to be willing to think of $\rho(\mathbf{x}, \mathbf{y})$ as the distance from $\mathbf{x}$ to $\mathbf{y}$ in S. A pair (S, ρ) with these properties is called a *metric space*, and such spaces provide the simplest possible setting for the study of problems that depend only on the notion of distance.

When we look at the Euclidean distance ρ defined by the formula (4.1), we see at a glance that properties (i)–(iii) are met. It is perhaps less evident that property (iv) is also satisfied, but the next challenge problem invites one to confirm this fact. The challenge is easily met, yet along the way we will find a simple relationship between the triangle inequality and Cauchy's inequality that puts Cauchy's inequality on a new footing. Ironically, the *axiomatic* approach to Euclidean distance adds greatly to the *intuitive* mastery of Cauchy's inequality.

Problem 4.2 (Triangle Inequality for Euclidean Distance)
Show that the function $\rho : \mathbb{R}^d \times \mathbb{R}^d \to \mathbb{R}$ defined by

$$\rho(\mathbf{x}, \mathbf{y}) = \sqrt{(y_1 - x_1)^2 + (y_2 - x_2)^2 + \cdots + (y_d - x_d)^2} \qquad (4.2)$$

satisfies the triangle inequality

$$\rho(\mathbf{x}, \mathbf{y}) \leq \rho(\mathbf{x}, \mathbf{z}) + \rho(\mathbf{z}, \mathbf{y}) \qquad \text{for all } \mathbf{x}, \mathbf{y} \text{ and } \mathbf{z} \text{ in } \mathbb{R}^d. \qquad (4.3)$$

To solve this problem, we first note from the definition (4.2) of ρ that one has the *translation property* that $\rho(\mathbf{x} + \mathbf{w}, \mathbf{y} + \mathbf{w}) = \rho(\mathbf{x}, \mathbf{y})$ for all $\mathbf{w} \in \mathbb{R}^d$; thus, to prove the triangle inequality (4.3), it suffices to show that for all $\mathbf{u}$ and $\mathbf{v}$ in $\mathbb{R}^d$ one has

$$\rho(\mathbf{0}, \mathbf{u} + \mathbf{v}) \leq \rho(\mathbf{0}, \mathbf{u}) + \rho(\mathbf{u}, \mathbf{u} + \mathbf{v}) = \rho(\mathbf{0}, \mathbf{u}) + \rho(\mathbf{0}, \mathbf{v}). \qquad (4.4)$$

By squaring this inequality and applying the definition (4.2), we see that the target inequality (4.3) is also equivalent to

$$\sum_{j=1}^{d}(u_j + v_j)^2 \leq \sum_{j=1}^{d} u_j^2 + 2\left\{\sum_{j=1}^{d} u_j^2\right\}^{1/2}\left\{\sum_{j=1}^{d} v_j^2\right\}^{1/2} + \sum_{j=1}^{d} v_j^2,$$

and this in turn may be simplified to the equivalent bound

$$\sum_{j=1}^{d} u_j v_j \leq \left\{\sum_{j=1}^{d} u_j^2\right\}^{1/2}\left\{\sum_{j=1}^{d} v_j^2\right\}^{1/2}.$$

Thus, in the end, one finds that the triangle inequality for the Euclidean distance is *equivalent* to Cauchy's inequality.

SOME NOTATION AND A MODEST GENERALIZATION

The definition (4.2) of ρ can be written quite briefly with help from the standard inner product $\langle \mathbf{u}, \mathbf{v} \rangle = u_1 v_2 + u_2 v_2 + \cdots + u_d v_d$, and, instead of (4.2), one can simply write $\rho(\mathbf{x}, \mathbf{y}) = \langle \mathbf{y} - \mathbf{x}, \mathbf{y} - \mathbf{x} \rangle^{\frac{1}{2}}$. This observation suggests a generalization of the Euclidean distance that turns out to have far reaching consequences.

To keep the logic of the generalization organized in a straight line, we begin with a formal definition. If V is a real vector space, such as $\mathbb{R}^d$, we say that the function from V to $\mathbb{R}^+$ defined by the mapping $v \mapsto \|v\|$ is a *norm* on V provided that it satisfies the following properties:

(i) $\|\mathbf{v}\| = 0$ if and only if $\mathbf{v} = \mathbf{0}$,
(ii) $\|\alpha \mathbf{v}\| = |\alpha| \|\mathbf{v}\|$ for all $\alpha \in \mathbb{R}$, and
(iii) $\|\mathbf{u} + \mathbf{v}\| \leq \|\mathbf{u}\| + \|\mathbf{v}\|$ for all $\mathbf{u}$ and $\mathbf{v}$ in V.

Also, if V is a vector space and $\|\cdot\|$ is a norm on V, then the couple $(V, \|\cdot\|)$ is called a *normed linear space*. The arguments of the preceding section can now be repeated to establish two related, but logically independent, observations:

(I). If $(V, \langle \cdot, \cdot \rangle)$ is an inner product space, then $\|\mathbf{v}\| = \langle \mathbf{v}, \mathbf{v} \rangle^{\frac{1}{2}}$ defines a norm on V. Thus, to each inner product space $(V, \langle \cdot, \cdot \rangle)$ we can associate a natural normed linear space $(V, \|\cdot\|)$.

(II). If $(V, \|\cdot\|)$ is a normed linear space, then $\rho(\mathbf{x}, \mathbf{y}) = \|\mathbf{x} - \mathbf{y}\|$ defines a metric on V. Thus, to each normed linear space we can associate a natural metric space $(V, \rho(\cdot, \cdot))$.

Here one should note that the three notions of an inner product space, a normed linear space, and a metric space are notions of strictly increasing generality. The space S with just two points $\mathbf{x}$ and $\mathbf{y}$ where ρ is defined by setting $\rho(\mathbf{x}, \mathbf{x}) = \rho(\mathbf{y}, \mathbf{y}) = 0$ and $\rho(\mathbf{x}, \mathbf{y}) = 1$ is a metric space, but it certainly is not an inner product space — the set S is not even a vector space. Later, in Chapter 9, we will also meet normed linear spaces that are not inner product spaces.

HOW MUCH INTUITION?

According to an old (and possibly apocryphal) story, during one of his lectures David Hilbert once wrote a line on the blackboard and said, "It is obvious that ...," but then Hilbert paused and thought for a moment. He then became noticeably perplexed, and he even left the room, returning only after an awkward passage of time. When Hilbert resumed his lecture, he began by saying "It is obvious that ...."

One of the tasks we assign ourselves as students of mathematics is to sort out for ourselves what is obvious and what is not. Oddly enough, this is not always an easy task. In particular, if we ask ourselves if the triangle inequality is obvious in $\mathbb{R}^d$ for $d \geq 4$, we may face a situation which is similar to the one that perplexed Hilbert.

The very young child who takes the diagonal across the park shows an intuitive understanding of the essential truth of the triangle inequality in $\mathbb{R}^2$. Moreover, anyone with some experience with $\mathbb{R}^d$ understands that if we ask a question about the relationship of three points in $\mathbb{R}^d$, $d \geq 3$, then we are "really" posing a problem in the two-dimensional plane that contains those points. These observations support the assertion that the triangle inequality in $\mathbb{R}^d$ is obvious.

The triangle inequality is indeed true in $\mathbb{R}^d$, so one cannot easily refute the claim of someone who says that it is flatly obvious. Nevertheless, algebra can be relied upon in ways that geometry cannot, and we already know from the example of Figure 4.1 that our experience with $\mathbb{R}^3$ can be misleading, or at least temporarily misleading. Sometimes questions are better than answers and, for the moment at least, we will let the issue of the obviousness of the triangle inequality remain a part of our continuing conversation. A more pressing issue is to understand the distance from a point to a line.

A CLOSEST POINT PROBLEM

For any point $\mathbf{x} \neq \mathbf{0}$ in $\mathbb{R}^d$ there is a unique line $\mathcal{L}$ through $\mathbf{x}$ and the origin $\mathbf{0} \in \mathbb{R}^d$, and one can write this line explicitly as $\mathcal{L} = \{t\mathbf{x} : t \in \mathbb{R}\}$. The closest point problem is the task of determining the point on $\mathcal{L}$ that is closest to a given point $\mathbf{v} \in \mathbb{R}^d$. By what may seem at first to be very good luck, there is an explicit formula for this closest point that one may write neatly with help from the standard the inner product $\langle \mathbf{v}, \mathbf{x} \rangle = v_1 w_1 + v_2 w_2 + \cdots + v_n w_n$.

Problem 4.3 (Projection Formula)

For each $\mathbf{v}$ and each $\mathbf{x} \neq \mathbf{0}$ in $\mathbb{R}^d$, let $P(\mathbf{v})$ denote the point on the line $\mathcal{L} = \{t\mathbf{x} : t \in \mathbb{R}\}$ that is closest to $\mathbf{v}$. Show that one has

$$P(\mathbf{v}) = \mathbf{x} \frac{\langle \mathbf{x}, \mathbf{v} \rangle}{\langle \mathbf{x}, \mathbf{x} \rangle}. \tag{4.5}$$

The point $P(\mathbf{v}) \in \mathcal{L}$ is called the *projection* of $\mathbf{v}$ on $\mathcal{L}$, and the formula (4.5) for $P(\mathbf{v})$ has many important applications in statistics and engineering, as well as in mathematics. Anyone who is already familiar with a proof of this formula should rise to this challenge by looking for a new proof. In fact, the projection formula (4.5) is wonderfully provable, and successful derivations may be obtained by calculus, by algebra, or even by direct arguments which require nothing more than a clever guess and Cauchy's inequality.

A LOGICAL CHOICE

The proof by algebra is completely elementary and relatively uncommon, so it seems like a logical choice for us. To find the value of $t \in \mathbb{R}$ that minimizes $\rho(\mathbf{v}, t\mathbf{x})$, we can just as easily try to minimize its square

$$\rho^2(\mathbf{v}, t\mathbf{x}) = \langle \mathbf{v} - t\mathbf{x}, \mathbf{v} - t\mathbf{x} \rangle,$$

which has the benefit of being a quadratic polynomial in t. If we look back on our earlier experience with such polynomials, then we will surely think of completing the square, and by doing so we find

$$\langle \mathbf{v} - t\mathbf{x}, \mathbf{v} - t\mathbf{x} \rangle = \langle \mathbf{v}, \mathbf{v} \rangle - 2t\langle \mathbf{v}, \mathbf{x} \rangle + t^2 \langle \mathbf{x}, \mathbf{x} \rangle$$

$$= \langle \mathbf{x}, \mathbf{x} \rangle \left(t^2 - 2t \frac{\langle \mathbf{v}, \mathbf{x} \rangle}{\langle \mathbf{x}, \mathbf{x} \rangle} + \frac{\langle \mathbf{v}, \mathbf{v} \rangle}{\langle \mathbf{x}, \mathbf{x} \rangle} \right)$$

$$= \langle \mathbf{x}, \mathbf{x} \rangle \left\{ \left(t - \frac{\langle \mathbf{v}, \mathbf{x} \rangle}{\langle \mathbf{x}, \mathbf{x} \rangle} \right)^2 + \frac{\langle \mathbf{v}, \mathbf{v} \rangle}{\langle \mathbf{x}, \mathbf{x} \rangle} - \frac{\langle \mathbf{v}, \mathbf{x} \rangle^2}{\langle \mathbf{x}, \mathbf{x} \rangle^2} \right\}.$$

Thus, in the end, we see that $\rho^2(\mathbf{v}, t\mathbf{x})$ has the nice representation

$$\langle \mathbf{x}, \mathbf{x} \rangle \left\{ \left(t - \frac{\langle \mathbf{v}, \mathbf{x} \rangle}{\langle \mathbf{x}, \mathbf{x} \rangle} \right)^2 + \frac{\langle \mathbf{v}, \mathbf{v} \rangle \langle \mathbf{x}, \mathbf{x} \rangle - \langle \mathbf{v}, \mathbf{x} \rangle^2}{\langle \mathbf{x}, \mathbf{x} \rangle^2} \right\}. \quad (4.6)$$

From this formula we see at a glance that $\rho(\mathbf{v}, t\mathbf{x})$ is minimized when we take $t = \langle \mathbf{v}, \mathbf{x} \rangle / \langle \mathbf{x}, \mathbf{x} \rangle$, and since this coincides exactly with the assertion of projection formula (4.5), the solution of the challenge problem is complete.

AN ACCIDENTAL COROLLARY — CAUCHY–SCHWARZ AGAIN

If we set $t = \langle \mathbf{v}, \mathbf{x} \rangle / \langle \mathbf{x}, \mathbf{x} \rangle$ in the formula (4.6), then we find that

$$\min_{t \in \mathbb{R}} \rho^2(\mathbf{v}, t\mathbf{x}) = \frac{\langle \mathbf{v}, \mathbf{v} \rangle \langle \mathbf{x}, \mathbf{x} \rangle - \langle \mathbf{v}, \mathbf{x} \rangle^2}{\langle \mathbf{x}, \mathbf{x} \rangle} \quad (4.7)$$

and, since the left-hand side is obviously nonnegative, we discover that

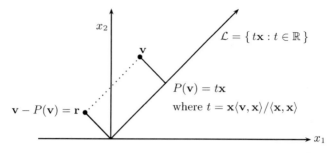

Fig. 4.2. The closest point on the line $\mathcal{L}$ to the point to $\mathbf{v} \in \mathbb{R}^d$ is the point $P(\mathbf{v})$. It is called the *projection* of $\mathbf{v}$ onto $\mathcal{L}$, and either by calculus, or by completing of the square, or by direct arguments using Cauchy's inequality, one can show that $P(\mathbf{v}) = \mathbf{x}\langle \mathbf{x}, \mathbf{v}\rangle/\langle \mathbf{x}, \mathbf{x}\rangle$. One way to characterize the projection $P(\mathbf{v})$ is that it is the unique element of $\mathcal{L}$ such that $\mathbf{r} = \mathbf{v} - P(\mathbf{v})$ is orthogonal to the vector $\mathbf{x}$ which determines the line $\mathcal{L}$.

our calculation has provided a small unanticipated bonus. The numerator on the right-hand side of the identity (4.7) must also be positive, and this observation gives us yet another proof of the Cauchy–Schwarz inequality.

There are even two further benefits to the formula (4.7). First, it gives us a geometrical interpretation of the defect $\langle \mathbf{v}, \mathbf{v}\rangle\langle \mathbf{x}, \mathbf{x}\rangle - \langle \mathbf{v}, \mathbf{x}\rangle^2$. Second, it tells us at a glance that one has $\langle \mathbf{v}, \mathbf{v}\rangle\langle \mathbf{x}, \mathbf{x}\rangle = \langle \mathbf{v}, \mathbf{x}\rangle^2$, if and only if $\mathbf{v}$ is an element of the line $\mathcal{L} = \{t\mathbf{x} : t \in \mathbb{R}\}$, which is a simple geometric interpretation of our earlier characterization of the case of equality.

How to Guess the Projection Formula

Two elements $\mathbf{x}$ and $\mathbf{y}$ of an inner product space $(V, \langle \cdot, \cdot \rangle)$ are said to be *orthogonal* if $\langle \mathbf{x}, \mathbf{y}\rangle = 0$, and one can check without difficulty that if $\langle \cdot, \cdot \rangle$ is the standard inner product on $\mathbb{R}^2$ or $\mathbb{R}^3$, then this modestly abstract notion of orthogonality corresponds to the traditional notion of orthogonality, or perpendicularity, which one meets in Euclidean geometry. If we combine this abstract definition with our intuitive understanding of $\mathbb{R}^2$, then, almost without calculation, we can derive a convincing guess for a formula for the projection $P(\mathbf{v})$.

For example, in Figure 4.2 our geometric intuition suggests that it is "obvious" (that tricky word again!) that if we want to choose t such that $P(\mathbf{v})$ is the closest point to $\mathbf{v}$ on $\mathcal{L}$, then we need to choose t so that the line from $P(\mathbf{v})$ to $\mathbf{v}$ should be orthogonal to the line $\mathcal{L}$. In

On Geometry and Sums of Squares 59

symbols, this means that we should choose t such that

$$\langle \mathbf{x}, \mathbf{v} - t\mathbf{v} \rangle = 0 \quad \text{or} \quad t = \langle \mathbf{x}, \mathbf{v} \rangle / \langle \mathbf{x}, \mathbf{x} \rangle.$$

We already know this is the value of t which yields the projection formula (4.5), so — this time at least — our intuition has given us good guidance.

If we are so inclined, we can even turn this guess into a proof. Specifically, we can use Cauchy's inequality to prove that this guess for t is actually the optimal choice. Such an argument provides us with a second, logically independent, derivation of the projection formula. This would be an instructive exercise, but, it seems better to move directly to a harder challenge.

REFLECTIONS AND PRODUCTS OF LINEAR FORMS

The projection formula and the closest point problem provide us with important new perspectives, but eventually one has to ask how these help us with our main task of discovering and proving useful inequalities. The next challenge problem clears this hurdle by suggesting an elegant bound which might be hard to discover (or to prove) without guidance from the geometry of $\mathbb{R}^n$.

Problem 4.4 (A Bound for the Product of Two Linear Forms)

Show that for all real u_j, v_j, and x_j, $1 \leq j \leq n$, one has the following upper bound for a product of two linear forms:

$$\sum_{j=1}^{n} u_j x_j \sum_{j=1}^{n} v_j x_j \leq \frac{1}{2} \Big\{ \sum_{j=1}^{n} u_j v_j + \Big(\sum_{j=1}^{n} u_j^2 \Big)^{1/2} \Big(\sum_{j=1}^{n} v_j^2 \Big)^{1/2} \Big\} \sum_{j=1}^{n} x_j^2. \quad (4.8)$$

The charm of this inequality is that it leverages the presence of two sums to obtain a bound that is sharper than the inequality which one would obtain from two applications of Cauchy's inequality to the individual multiplicands. In fact, when $\langle \mathbf{u}, \mathbf{v} \rangle \leq 0$ the new bound does better by at least a factor of one-half, and, even if the vectors $\mathbf{u} = (u_1, u_2, \ldots, u_n)$ and $\mathbf{v} = (v_1, v_2, \ldots, v_n)$ are proportional, the bound (4.8) is not worse than the one provided by Cauchy's inequality. Thus, the new inequality (4.8) provides us with a win-win situation whenever we need to estimate the product of two sums.

FOUNDATIONS FOR A PROOF

This time we will take an indirect approach to our problem and, at first, we will only try to deepen our understanding of the geometry of projection on a line. We begin by noting that Figure 4.2 strongly suggests that the projection P onto the line $\mathcal{L} = \{t\mathbf{x} : t \in \mathbb{R}\}$, must satisfy the bound

$$\|P(\mathbf{v})\| \leq \|\mathbf{v}\| \quad \text{for all } \mathbf{v} \in \mathbb{R}^d \tag{4.9}$$

and, moreover, one even expects strict inequality here unless $\mathbf{v} \in \mathcal{L}$. In fact, the proof of the bound (4.9) is quite easy since the projection formula (4.5) and Cauchy's inequality give us

$$\|P(\mathbf{v})\| = \left\|\mathbf{x}\frac{\langle \mathbf{x}, \mathbf{v}\rangle}{\langle \mathbf{x}, \mathbf{x}\rangle}\right\| = \frac{1}{\|\mathbf{x}\|}|\langle \mathbf{x}, \mathbf{v}\rangle| \leq \|\mathbf{v}\|.$$

From Projection to Reflection

We also face a similar situation when we consider the *reflection* of the point $\mathbf{v}$ through the line $\mathcal{L}$, say as illustrated by Figure 4.3. Formally, the reflection of the point $\mathbf{v}$ in the line $\mathcal{L}$ is the point $R(\mathbf{v})$ defined by the formula $R(\mathbf{v}) = 2P(\mathbf{v}) - \mathbf{v}$. In some ways, the reflection $R(\mathbf{v})$ is an even more natural object than the projection $P(\mathbf{v})$. In particular, one can guess from Figure 4.2 that the mapping $R : V \rightarrow V$ has the pleasing length preserving property

$$\|R(\mathbf{v})\| = \|\mathbf{v}\| \quad \text{for all } \mathbf{v} \in \mathbb{R}^d. \tag{4.10}$$

One can prove this identity by a direct calculation with the projection formula, but that calculation is most neatly organized if we first observe some general properties of P. In particular, we have the nice formula

$$\langle P(\mathbf{v}), P(\mathbf{v})\rangle = \left\langle \frac{\langle \mathbf{x}, \mathbf{v}\rangle \mathbf{x}}{\|\mathbf{x}\|^2}, \frac{\langle \mathbf{x}, \mathbf{v}\rangle \mathbf{x}}{\|\mathbf{x}\|^2}\right\rangle = \frac{\langle \mathbf{x}, \mathbf{v}\rangle^2}{\|\mathbf{x}\|^2},$$

while at the same time we also have

$$\langle P(\mathbf{v}), \mathbf{v}\rangle = \left\langle \frac{\langle \mathbf{x}, \mathbf{v}\rangle \mathbf{x}}{\|\mathbf{x}\|^2}, \mathbf{v}\right\rangle = \frac{\langle \mathbf{x}, \mathbf{v}\rangle^2}{\|\mathbf{x}\|^2},$$

so we may combine these observations to obtain

$$\langle P(\mathbf{v}), P(\mathbf{v})\rangle = \langle P(\mathbf{v}), \mathbf{v}\rangle.$$

This useful identity now provides a quick confirmation of the length-preserving (or *isometry*) property of the reflection R; we just expand the inner product and simplify to find

$$\|R(\mathbf{v})\|^2 = \langle 2P(\mathbf{v}) - \mathbf{v}, 2P(\mathbf{v}) - \mathbf{v}\rangle$$
$$= 4\langle P(\mathbf{v}), P(\mathbf{v})\rangle - 4\langle P(\mathbf{v}), \mathbf{v}\rangle + \langle \mathbf{v}, \mathbf{v}\rangle$$
$$= \langle \mathbf{v}, \mathbf{v}\rangle.$$

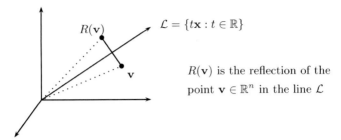

Fig. 4.3. When the point **v** is reflected in the line $\mathcal{L}$ one obtains a new point $R(\mathbf{v})$ which is the same distance from the origin as **v**. More formally, the reflection of **v** is the point $R(\mathbf{v})$ *defined* by the formula $R(\mathbf{v}) = 2P(\mathbf{v}) - \mathbf{v}$. One can then use the projection formula for P to prove that $\|R(\mathbf{v})\| = \|\mathbf{v}\|$.

RETURN TO THE CHALLENGE

The geometry of the reflection through the line $\mathcal{L} = \{t\mathbf{x} : t \in \mathbb{R}\}$ is easily understood, but sometimes the associated algebra can offer a pleasant surprise. For example, the isometry property of the reflection R and the Cauchy–Schwarz inequality can be combined to provide an almost immediate solution of our challenge problem.

From the Cauchy–Schwarz inequality and the isometry property of the reflection R we have the bound

$$\langle R(\mathbf{u}), \mathbf{v}\rangle \leq \|R(\mathbf{u})\|\|\mathbf{v}\| \leq \|\mathbf{u}\|\|\mathbf{v}\|, \tag{4.11}$$

while on the other hand, the definition of R and the projection formula give us the identity

$$\langle R(\mathbf{u}), \mathbf{v}\rangle = \langle 2P(\mathbf{u}) - \mathbf{u}, \mathbf{v}\rangle = 2\langle P(\mathbf{u}), \mathbf{v}\rangle - \langle \mathbf{u}, \mathbf{v}\rangle$$
$$= 2\left\langle \frac{\langle \mathbf{x}, \mathbf{u}\rangle \mathbf{x}}{\|\mathbf{x}\|^2}, \mathbf{v}\right\rangle - \langle \mathbf{u}, \mathbf{v}\rangle$$
$$= \frac{2}{\|x\|^2}\langle \mathbf{x}, \mathbf{u}\rangle\langle \mathbf{x}, \mathbf{v}\rangle - \langle \mathbf{u}, \mathbf{v}\rangle.$$

Thus, from Cauchy–Schwarz and the isometry bound (4.11) we have

$$\frac{2}{\|\mathbf{x}\|^2}\langle \mathbf{x}, \mathbf{u}\rangle\langle \mathbf{x}, \mathbf{v}\rangle - \langle \mathbf{u}, \mathbf{v}\rangle \leq \|\mathbf{u}\|\|\mathbf{v}\|,$$

and this may be arranged more naturally as

$$\langle \mathbf{x}, \mathbf{u}\rangle\langle \mathbf{x}, \mathbf{v}\rangle \leq \frac{1}{2}(\langle \mathbf{u}, \mathbf{v}\rangle + \|\mathbf{u}\|\|\mathbf{v}\|)\|\mathbf{x}\|^2. \tag{4.12}$$

If we now interpret these inner products as the standard inner products on $\mathbb{R}^n$, then we see that the bound (4.12) is precisely the inequality (4.8) of the challenge problem.

Thus, almost by accident, we find that the geometry of reflection has brought us to a new and informative refinement of Cauchy's inequality. Such accidents are common, and they form a thread from which Scheherazade could spin a thousand tales, all with the name *symmetry and its applications*. We will revisit this theme, but first we seek a different kind of contribution from a different kind of geometry.

THE LIGHT CONE INEQUALITY

The preceding examples suggested how Euclidean geometry helps to deepen our understanding of the theory of inequalities, but the traditional geometry of Euclid is not the only one that helps in this way. Other geometries, or geometric models, can do their part.

One especially attractive example calls on the famous space-time geometry of Einstein and Minkowski. The physical background of this model is not needed here, but, for motivation, it is useful to recall one fundamental principle of special relativity: no information of any kind can travel faster than the speed of light.

If we scale space so that the speed of light is 1, this principle tells implies that each point $\mathbf{x} = (t; x_1, x_2, \ldots, x_d)$ of time and space where one can have knowledge of an event that takes place at the origin at time 0 must satisfy the bound

$$\sqrt{x_1^2 + x_2^2 + \cdots + x_d^2} \leq t. \tag{4.13}$$

The set C of all such points in $\mathbb{R}^+ \times \mathbb{R}^d$ is called *Minkowski's light cone*, and it is illustrated in Figure 4.4.

The only further notion that we need is the *Lorentz product*, which is the bilinear form defined for pairs of elements $\mathbf{x} = (t; x_1, x_2, \ldots, x_d)$ and $\mathbf{y} = (u; y_1, y_2, \ldots, y_d)$ in the light cone C by the formula

$$[\mathbf{x}, \mathbf{y}] = tu - x_1 y_1 - x_2 y_2 - \cdots - x_d y_d. \tag{4.14}$$

This quadratic form was introduced by the Dutch physicist Hendrick Antoon Lorentz (1853–1928), who used it to simplify some of the formulas of special relativity, but for us the interesting feature of the Lorentz product is its relationship to the Cauchy–Schwarz inequality. It turns out that the Lorentz product satisfies an inequality which has a superficial resemblance to the Cauchy–Schwarz inequality, except for one remarkable twist — the inequality is *exactly reversed*!

On Geometry and Sums of Squares

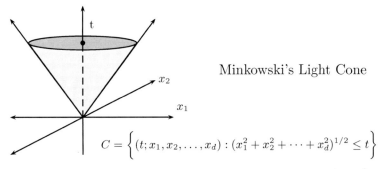

$$C = \left\{(t; x_1, x_2, \ldots, x_d) : (x_1^2 + x_2^2 + \cdots + x_d^2)^{1/2} \leq t\right\}$$

Fig. 4.4. Minkowski's light cone C is the region of space-time $\mathbb{R}^+ \times \mathbb{R}^d$ where one can have knowledge of an event that takes place at the origin at time zero. Here time is scaled so that the speed of light is equal to one.

Problem 4.5 (Light Cone Inequality)

Show that if $\mathbf{x}$ and $\mathbf{y}$ are points of $\mathbb{R}^+ \times \mathbb{R}^d$ that are elements of the light cone C defined in Figure 4.4, then the Lorentz product satisfies the inequality

$$[\mathbf{x}, \mathbf{x}]^{\frac{1}{2}} [\mathbf{y}, \mathbf{y}]^{\frac{1}{2}} \leq [\mathbf{x}, \mathbf{y}]. \tag{4.15}$$

Show, moreover, that if $\mathbf{x} = (t; x_1, x_2, \ldots, x_n)$ and $\mathbf{y} = (u; y_1, y_2, \ldots, y_n)$ then the inequality (4.15) is strict unless $ux_j = ty_j$ for all $1 \leq j \leq d$.

DEVELOPMENT OF A PLAN

If the *Cauchy–Schwarz Master Class* were to have a final exam, then the light cone inequality would provide fertile ground for the development of good problems. One can prove the light cone inequality with almost any reasonable tool — induction, the AM-GM inequality, or even a Lagrange-type identity will do the job. Here we will explore a lazy and devious route, precisely the kind favored by most mathematicians.

Since our goal is to prove a reversal of the Cauchy–Schwarz inequality, a pleasantly outrageous plan would be to look for some way to invert the famous polynomial argument of Schwarz (say, as described in Chapter 1, on page 11). In Schwarz's argument, one constructs a quadratic polynomial, makes an observation about its roots, and then draws a conclusion about the coefficients of the polynomial. That is just what we will try here — with some necessary changes. After all, we want a different conclusion about the coefficients, so we need to make a different observation about the roots.

In imitation of Schwarz's argument, we introduce the quadratic poly-

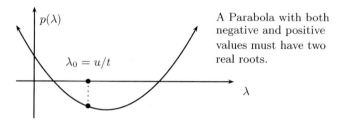

A Parabola with both negative and positive values must have two real roots.

Fig. 4.5. Schwarz's proof of the Cauchy–Schwarz inequality exploited the bound on the coefficients of a polynomial without real roots; in contrast, Minkowski's light cone inequality cone exploits the information that one gets from knowing a quadratic polynomial has two real roots.

nomial

$$p(\lambda) = [\mathbf{x} - \lambda \mathbf{y}, \mathbf{x} - \lambda \mathbf{y}] = [\mathbf{x}, \mathbf{x}] - 2\lambda[\mathbf{x}, \mathbf{y}] + \lambda^2[\mathbf{y}, \mathbf{y}] \quad (4.16)$$

$$= (t - \lambda u)^2 - \sum_{j=1}^{d}(x_j - \lambda y_j)^2, \quad (4.17)$$

and we immediately address ourselves to understanding its roots. To side-step trivialities, we first note that if $t = 0$ then our assumption that $\mathbf{x} = (t; x_1, x_2, \ldots, x_d) \in C$ tells us that $\mathbf{x} = \mathbf{0}$. In this case, the light cone inequality (4.15) is trivially true, so, without loss of generality, we can assume that $t \neq 0$.

Next, for space-time vectors $\mathbf{x}$ and $\mathbf{y}$ in C one sees from Cauchy's inequality and the definition of the light cone that the spatial components $(x_1, x_2, \ldots, x_n)$ and $(y_1, y_2, \ldots, y_n)$ must satisfy the bound

$$\sum_{k=1}^{n} x_k y_k \leq \left(\sum_{k=1}^{n} x_k^2\right)^{\frac{1}{2}} \left(\sum_{k=1}^{n} y_k^2\right)^{\frac{1}{2}} \leq tu.$$

In the language of the Lorentz product, this says $[\mathbf{x}, \mathbf{y}] \geq 0$, and as a consequence we see that the light cone inequality is trivially true whenever $[\mathbf{x}, \mathbf{x}] = 0$ or $[\mathbf{y}, \mathbf{y}] = 0$. Thus, without loss of generality, we can assume *both* of these Lorentz products are nonzero.

Now, we are ready for the main argument. For $u \neq 0$, we may then take $\lambda_0 = t/u$, and the first sum of the expanded polynomial (4.17) vanishes. We then see that either (i) $ux_j = ty_j$ for all $1 \leq j \leq d$ or else we have (ii) $p(\lambda_0) < 0$. In the first situation, we have the case of equality which was suggested by the challenge problem, so to complete the solution we just need to confirm that in the second situation we have required strict inequality.

Since we have assumed that $[\mathbf{y},\mathbf{y}] > 0$, we see from the product form (4.16) that $p(\lambda) \to \infty$ as $\lambda \to \infty$ or $\lambda \to -\infty$ and we know that $p(\lambda_0) < 0$ so the equation $p(\lambda) = A\lambda^2 + 2B\lambda + C = 0$ must have two distinct real roots. The binomial formula for the quadratic equation then tells us that $AC < B$. When we identify the coefficients of $p(\lambda)$ from its product form (4.16) we find $A = [\mathbf{x},\mathbf{x}]$, $B = [\mathbf{x},\mathbf{y}]$, and $C = [\mathbf{y},\mathbf{y}]$, so $AC < B$ gives us the strict inequality $[\mathbf{x},\mathbf{x}][\mathbf{y},\mathbf{y}] < [\mathbf{x},\mathbf{y}]^2$, which we hoped to show.

COMPLEX INNER PRODUCT SPACES

If V is a complex vector space, such as $\mathbb{C}^d$ or the set of complex valued continuous functions on $[0,1]$, then we say that a function on $V \times V$ defined by the mapping $(\mathbf{a},\mathbf{b}) \mapsto \langle \mathbf{a},\mathbf{b} \rangle \in \mathbb{C}$ is an *complex inner product* and we say that $(V, \langle \cdot, \cdot \rangle)$ is a *complex inner product space* provided that the pair $(V, \langle \cdot, \cdot \rangle)$ has five basic properties. The first four of these perfectly parallel those required of a real inner product space:

(i) $\langle \mathbf{v}, \mathbf{v} \rangle \geq 0$ for all $\mathbf{v} \in V$,
(ii) $\langle \mathbf{v}, \mathbf{v} \rangle = 0$ if and only if $\mathbf{v} = 0$
(iii) $\langle \alpha \mathbf{v}, \mathbf{w} \rangle = \alpha \langle \mathbf{v}, \mathbf{w} \rangle$ for all $\alpha \in \mathbb{C}$ and $\mathbf{v}, \mathbf{w} \in V$,
(iv) $\langle \mathbf{u}, \mathbf{v} + \mathbf{w} \rangle = \langle \mathbf{u}, \mathbf{v} \rangle + \langle \mathbf{u}, \mathbf{w} \rangle$ for all $\mathbf{u}, \mathbf{v}$ and $\mathbf{w} \in V$,

but the fifth property requires a modest change; specifically, for a complex inner product space we assume that

(v) $\langle \mathbf{v}, \mathbf{w} \rangle = \overline{\langle \mathbf{w}, \mathbf{v} \rangle}$ for all $\mathbf{v}, \mathbf{w} \in V$.

Problem 4.6 (Cauchy–Schwarz for a Complex Inner Product)

Show that in a complex inner product space $(V, \langle \cdot, \cdot \rangle)$ one has

$$|\langle \mathbf{v}, \mathbf{w} \rangle| \leq \langle \mathbf{v}, \mathbf{v} \rangle^{\frac{1}{2}} \langle \mathbf{w}, \mathbf{w} \rangle^{\frac{1}{2}}. \tag{4.18}$$

Furthermore, show that $\mathbf{v} \neq 0$ then one has equality in the bound (4.18) if and only if $\mathbf{w} = \lambda \mathbf{v}$ for some $\lambda \in \mathbb{C}$.

A NATURAL PLAN AND A NEW OBSTACLE

A natural plan for proving the Cauchy–Schwarz inequality for a complex inner product space is to mimic the proof for a real inner product space while paying attention to any changes which may be required by the new "property (v)." Thus, we compute

$$0 \le \langle \mathbf{v}-\mathbf{w}, \mathbf{v}-\mathbf{w}\rangle = \langle \mathbf{v}, \mathbf{v}\rangle + \langle \mathbf{w}, \mathbf{w}\rangle - \langle \mathbf{v}, \mathbf{w}\rangle - \langle \mathbf{w}, \mathbf{v}\rangle$$
$$= \langle \mathbf{v}, \mathbf{v}\rangle + \langle \mathbf{w}, \mathbf{w}\rangle - \{\langle \mathbf{v}, \mathbf{w}\rangle + \overline{\langle \mathbf{v}, \mathbf{w}\rangle}\}$$
$$= \langle \mathbf{v}, \mathbf{v}\rangle + \langle \mathbf{w}, \mathbf{w}\rangle - 2\mathrm{Re}\,\langle \mathbf{v}, \mathbf{w}\rangle,$$

and we deduce that

$$\mathrm{Re}\,\langle \mathbf{v}, \mathbf{w}\rangle \le \frac{1}{2}\langle \mathbf{v}, \mathbf{v}\rangle + \frac{1}{2}\langle \mathbf{w}, \mathbf{w}\rangle, \qquad (4.19)$$

where we have strict inequality unless $\mathbf{v} = \mathbf{w}$.

The additive bound (4.19) must be converted to one that is multiplicative. If we call on the familiar normalization method and introduce

$$\hat{\mathbf{v}} = \mathbf{v}/\langle \mathbf{v}, \mathbf{v}\rangle^{\frac{1}{2}} \quad \text{and} \quad \hat{\mathbf{w}} = \mathbf{w}/\langle \mathbf{w}, \mathbf{w}\rangle^{\frac{1}{2}},$$

then arithmetic brings us quickly to the bound

$$\mathrm{Re}\,\langle \mathbf{v}, \mathbf{w}\rangle \le \langle \mathbf{v}, \mathbf{v}\rangle^{\frac{1}{2}} \langle \mathbf{w}, \mathbf{w}\rangle^{\frac{1}{2}}. \qquad (4.20)$$

Unfortunately, this starts to look worrisome. We hoped to obtain a bound on $|\langle \mathbf{v}, \mathbf{w}\rangle|$ but we have only found bound on $\mathrm{Re}\,\langle \mathbf{v}, \mathbf{w}\rangle$, a term which may be arbitrarily smaller than $|\langle \mathbf{v}, \mathbf{w}\rangle|$. Is it possible that this approach has failed?

SAVED BY A SELF-IMPROVEMENT

The saving grace of inequality (4.20) is that it is of the self-improving kind. If we exploit its generality appropriately, we can derive an apparently stronger inequality.

If we write $\langle \mathbf{v}, \mathbf{w}\rangle = \rho e^{i\theta}$ with $\rho > 0$ and if we set $\tilde{\mathbf{v}} = e^{-i\theta}\mathbf{v}$, then the properties of the complex inner product give us the identities

$$\langle \tilde{\mathbf{v}}, \tilde{\mathbf{v}}\rangle = \langle \mathbf{v}, \mathbf{v}\rangle \quad \text{and} \quad \langle \tilde{\mathbf{v}}, \mathbf{w}\rangle = \mathrm{Re}\,\langle \tilde{\mathbf{v}}, \mathbf{w}\rangle = |\langle \mathbf{v}, \mathbf{w}\rangle|,$$

so the real part bound (4.20) for the pair $\tilde{\mathbf{v}}$ and $\mathbf{w}$ gives us

$$|\langle \mathbf{v}, \mathbf{w}\rangle| = \mathrm{Re}\,\langle \tilde{\mathbf{v}}, \mathbf{w}\rangle \le \langle \tilde{\mathbf{v}}, \tilde{\mathbf{v}}\rangle^{\frac{1}{2}} \langle \mathbf{w}, \mathbf{w}\rangle^{\frac{1}{2}} = \langle \mathbf{v}, \mathbf{v}\rangle^{\frac{1}{2}} \langle \mathbf{w}, \mathbf{w}\rangle^{\frac{1}{2}}.$$

The outside terms yield the complex Cauchy–Schwarz inequality in the precisely the form we expected, so the bound (4.20) was strong enough after all.

THE TRICK OF "MAKING IT REAL"

In this argument, we faced an inequality which was made more complicated because of the presence of a real part. This is a common difficulty, and it is often addressed by the trick used here: one pre-multiplies by a well-chosen complex number in order to guarantee that some critical quantity will be real. This is one of the most widely used maneuvers in the theory of complex inequalities, and it should never be far out of mind.

Finally, to complete the solution of the challenge problem, we should confirm the alleged necessary and sufficient conditions for equality. Here it is honestly easy to retrace the steps of our argument to confirm the stated conditions, but, as we will discuss later (page 138), such backtrack arguments are not always trouble free. For the standard complex inner product one also has another option which is perhaps more satisfying; one can simply use the complex Lagrange identity (4.23) as suggested by Exercise 4.4.

EXERCISES

Exercise 4.1 (Triangle Inequality "Pot Shots")

The triangle inequality in $\mathbb{R}^d$ may seem obvious, but some of its consequences can be puzzling when they are presented out of context. Here, the next three exercises are not at all hard, but you might ask yourself, "Would these have been so easy yesterday?"

(a) Show for nonnegative x, y, z that
$$(x+y+z)\sqrt{2} \leq \sqrt{x^2+y^2} + \sqrt{y^2+z^2} + \sqrt{x^2+z^2}.$$

(b) Show for $0 < x \leq y \leq z$ that
$$\sqrt{y^2+z^2} \leq x\sqrt{2} + \sqrt{(y-x)^2 + (z-x)^2}.$$

(c) Show for positive x, y, z that
$$2\sqrt{3} \leq \sqrt{x^2+y^2+z^2} + \sqrt{x^{-2}+y^{-2}+z^{-2}}.$$

This list can be continued almost without limit, yet there is really only one theme: any time you see a sum of square roots in an inequality, you should give at least a moment's thought to the possibility that the triangle inequality may help.

Exercise 4.2 (The Geometry of "Steepest Ascent")

If $f : \mathbb{R}^n \to \mathbb{R}$ is a differentiable function, then one often hears that the gradient
$$\nabla f(\mathbf{x}) = \left(\frac{\partial f}{\partial x_1}, \frac{\partial f}{\partial x_2}, \ldots, \frac{\partial f}{\partial x_n}\right)$$

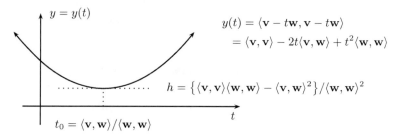

Fig. 4.6. By calculus, or by completing the square, one finds that the quadratic polynomial $P(t) = \langle \mathbf{v} - t\mathbf{w}, \mathbf{v} - t\mathbf{w} \rangle$ takes it minimum at $t_0 = \langle \mathbf{v}, \mathbf{w} \rangle / \langle \mathbf{w}, \mathbf{w} \rangle$. The nonnegativity of h is enough to prove Cauchy's inequality for the nth time, but geometry adds details which can be critical.

points in the direction of *steepest ascent* of f provided that $\nabla f \neq 0$. In longhand, this says for any unit vector $\mathbf{u}$ one has the bound

$$\left. \frac{d}{dt} f(\mathbf{x} + t\mathbf{u}) \right|_{t=0} \leq \left. \frac{d}{dt} f(\mathbf{x} + t\mathbf{v}) \right|_{t=0} \qquad (4.21)$$

where $\mathbf{v} = \nabla f(\mathbf{x}) / \|\nabla f(\mathbf{x})\|$. Prove this inequality and show that it is strict unless $\mathbf{u} = \mathbf{v}$.

Exercise 4.3 (Cauchy via Another Identity)

Lagrange's identity is not the only formula that gives an instant proof of Cauchy's inequality. Check that in any real inner product space the difference $\langle \mathbf{v}, \mathbf{v} \rangle \langle \mathbf{w}, \mathbf{w} \rangle - \langle \mathbf{v}, \mathbf{w} \rangle^2$ can be written as

$$\langle \mathbf{w}, \mathbf{w} \rangle \left\{ \left\langle \mathbf{v} - \frac{\langle \mathbf{w}, \mathbf{v} \rangle}{\langle \mathbf{w}, \mathbf{w} \rangle} \mathbf{w}, \; \mathbf{v} - \frac{\langle \mathbf{w}, \mathbf{v} \rangle}{\langle \mathbf{w}, \mathbf{w} \rangle} \mathbf{w} \right\rangle \right\}, \qquad (4.22)$$

and explain why this also implies the general Cauchy–Schwarz inequality.

Incidentally, one does not need a flash of algebraic insight to discover the representation (4.22). As Figure 4.6 suggests, this formula cannot remain hidden for long once we ask ourselves about minimization of the polynomial $P(t) = \langle \mathbf{v} - t\mathbf{w}, \mathbf{v} - t\mathbf{w} \rangle$.

Exercise 4.4 (Lagrange's Identity for Complex Numbers)

Prove that for complex a_k and b_k, $1 \leq k \leq n$, one has

$$\sum_{k=1}^{n} |a_k|^2 \sum_{k=1}^{n} |b_k|^2 - \left| \sum_{k=1}^{n} a_k b_k \right|^2 = \sum\sum_{1 \leq j < k \leq n} |\bar{a}_j b_k - a_k \bar{b}_j|^2, \qquad (4.23)$$

and show that this identity yields the complex Cauchy inequality as

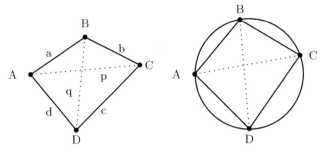

Fig. 4.7. Ptolemy's inequality and the condition for equality.

well as the necessary and sufficient conditions for equality. Here one should note that this identity *does not* follow from direct substitution of complex numbers into the Lagrange's identity for real numbers; those pesky absolute values get in the way. A slightly more sophisticated approach is required.

Exercise 4.5 (A Vector-Scalar Melange)

Consider real weights $p_j > 0$, $j = 1, 2, \ldots, n$, arbitrary real numbers α_j, $j = 1, 2, \ldots, n$, and an inner product space $(V, \langle \cdot, \cdot \rangle)$. Find an analog of Lagrange's identity which suffices to prove that one has the inequality

$$\left\| \sum_{j=1}^{n} p_j \alpha_j \mathbf{x}_j \right\|^2 \leq \sum_{j=1}^{n} p_j \alpha_j^2 \sum_{k=1}^{n} p_k \|\mathbf{x}_k\|^2 \qquad (4.24)$$

for all $\mathbf{x}_k$, $1 \leq k \leq n$, in V. Also, check that your identity implies that equality holds if and only if we have $\alpha_j \mathbf{x}_k = \alpha_k \mathbf{x}_j$ for all $1 \leq j, k \leq n$.

Exercise 4.6 (Ptolemy's Inequality)

Ptolemy may be best known for founding a theory of planetary motion which was overturned by Copernicus, but parts of Ptolemy's legacy have stood the test of time. Among these, Ptolemy has a namesake inequality which even today is a workhorse of the theory of geometric inequalities. Ptolemy's inequality asserts that in a convex quadrilateral "the product of the diagonals is bounded by the sum of the products of the opposite sides," or, in the notation of Figure 4.7,

$$pq \leq ac + bd. \qquad (4.25)$$

Prove this inequality and show that equality holds if and only if the four vertices A, B, C, D are all on the circumference of a circle.

Exercise 4.7 (Representations of Complex Inner Products)

(a) If $\langle \cdot, \cdot \rangle$ is a complex inner product and if $\alpha \in \mathbb{C}$ and $\alpha^N = 1$ but $\alpha^2 \neq 1$, then show that one has the representation

$$\langle x, y \rangle = \frac{1}{N} \sum_{n=0}^{N-1} \|x + \alpha^n y\|^2 \alpha^n \tag{4.26}$$

where, as usual, $\|w\| = \langle w, w \rangle^{1/2}$.

(b) Similarly show that for any complex inner product one has

$$\langle x, y \rangle = \frac{1}{2\pi} \int_{-\pi}^{\pi} \|x + e^{i\theta} y\|^2 e^{i\theta} \, d\theta. \tag{4.27}$$

One benefit of identities such as these is that they may help us convert facts for $\|\cdot\|$ into facts for $\langle \cdot, \cdot \rangle$ or vice-versa. One can say that these are "just" variants of the polarization identity, but there are times when they are just the variant one needs.

Exercise 4.8 (A Concrete Model of an Abstract Space)

If $\mathbf{x}_1, \mathbf{x}_2, \ldots, \mathbf{x}_n$ are linearly independent elements of the (real or complex) inner product space $(V, \langle \cdot, \cdot \rangle)$, we define a new sequence $\mathbf{e}_1, \mathbf{e}_2, \ldots, \mathbf{e}_n$ by setting $\mathbf{e}_1 = \mathbf{x}_1 / \|\mathbf{x}_1\|$ and by applying the two-part recursion

$$\mathbf{z}_k = \mathbf{x}_k - \sum_{j=1}^{k-1} \langle \mathbf{x}_k, \mathbf{e}_j \rangle \mathbf{e}_j \quad \text{and} \quad \mathbf{e}_k = \frac{\mathbf{z}_k}{\|\mathbf{z}_k\|} \tag{4.28}$$

for $k = 2, 3, \ldots, n$. This algorithm is known as the *Gram–Schmidt process*, and it provides a systematic tool for reducing questions in an inner product space to questions for real or complex numbers. In this exercise we develop the most basic properties of this process, and in the next four exercises we show how these properties are used in practice.

(a) Show that $\{\mathbf{e}_k : 1 \leq k \leq n\}$ is an *orthonormal sequence* in the sense that for all $1 \leq j, k \leq n$ one has

$$\langle \mathbf{e}_j, \mathbf{e}_k \rangle = \begin{cases} 1 & \text{if } j = k \\ 0 & \text{if } j \neq k. \end{cases}$$

(b) Show that $\{\mathbf{x}_k : 1 \leq k \leq n\}$ and $\{\mathbf{e}_k : 1 \leq k \leq n\}$ satisfy the triangular system of linear relations

$$\mathbf{x}_1 = \langle \mathbf{x}_1, \mathbf{e}_1 \rangle \mathbf{e}_1$$
$$\mathbf{x}_2 = \langle \mathbf{x}_2, \mathbf{e}_1 \rangle \mathbf{e}_1 + \langle \mathbf{x}_2, \mathbf{e}_2 \rangle \mathbf{e}_2$$
$$\cdots = \cdots$$
$$\mathbf{x}_n = \langle \mathbf{x}_n, \mathbf{e}_1 \rangle \mathbf{e}_1 + \langle \mathbf{x}_n, \mathbf{e}_2 \rangle \mathbf{e}_2 + \cdots + \langle \mathbf{x}_n, \mathbf{e}_n \rangle \mathbf{e}_n.$$

Exercise 4.9 (Gram–Schmidt Implies Cauchy–Schwarz)

Apply the Gram–Schmidt process to the two term sequence $\{\mathbf{x}, \mathbf{y}\}$ and show that it reduces the inequality $|\langle \mathbf{x}, \mathbf{y} \rangle| \leq \langle \mathbf{x}, \mathbf{x} \rangle^{\frac{1}{2}} \langle \mathbf{y}, \mathbf{y} \rangle^{\frac{1}{2}}$ to a bound that is obvious. Thus, the Gram–Schmidt process gives an automatic proof of the Cauchy–Schwarz inequality.

Exercise 4.10 (Gram–Schmidt Implies Bessel)

If $\{\mathbf{y}_k : 1 \leq k \leq n\}$ is an orthonormal sequence from a (real or complex) inner product space $(V, \langle \cdot, \cdot \rangle)$, then Bessel's inequality asserts that

$$\sum_{k=1}^{n} |\langle \mathbf{x}, \mathbf{y}_k \rangle|^2 \leq \langle \mathbf{x}, \mathbf{x} \rangle \qquad \text{for all } \mathbf{x} \in V. \tag{4.29}$$

Show that the Gram–Schmidt process yields a semi-automatic proof of Bessel's inequality. Incidentally, one should also note that the case $n = 1$ of Bessel's inequality is equivalent to the Cauchy–Schwarz inequality.

Exercise 4.11 (Gram–Schmidt and Products of Linear Forms)

Use the Gram–Schmidt process for the three-term sequence $\{\mathbf{x}, \mathbf{y}, \mathbf{z}\}$ to show that in a real inner product space one has

$$\langle \mathbf{x}, \mathbf{y} \rangle \langle \mathbf{x}, \mathbf{z} \rangle \leq \frac{1}{2} (\langle \mathbf{y}, \mathbf{z} \rangle + \|\mathbf{y}\| \|\mathbf{z}\|) \|\mathbf{x}\|^2, \tag{4.30}$$

a bound which we used earlier (page 61) to illustrate the use of isometries and projections.

Exercise 4.12 (A Gram–Schmidt Finale)

Show that if $\mathbf{x}, \mathbf{y}, \mathbf{z}$ are elements of a (real or complex) inner product space V and if $\|\mathbf{x}\| = \|\mathbf{y}\| = \|\mathbf{z}\| = 1$, then one has the inequality

$$|\langle \mathbf{x}, \mathbf{x} \rangle \langle \mathbf{y}, \mathbf{z} \rangle - \langle \mathbf{x}, \mathbf{y} \rangle \langle \mathbf{x}, \mathbf{z} \rangle|$$
$$\leq \{\langle \mathbf{x}, \mathbf{x} \rangle^2 - |\langle \mathbf{x}, \mathbf{y} \rangle|^2\}\{\langle \mathbf{x}, \mathbf{x} \rangle^2 - |\langle \mathbf{x}, \mathbf{z} \rangle|^2\} \tag{4.31}$$

and the inequality

$$\langle \mathbf{x}, \mathbf{x} \rangle^2 \left(|\langle \mathbf{y}, \mathbf{z} \rangle|^2 + |\langle \mathbf{y}, \mathbf{x} \rangle|^2 + |\langle \mathbf{x}, \mathbf{z} \rangle|^2 \right)$$
$$\leq \langle \mathbf{x}, \mathbf{x} \rangle^4 + \langle \mathbf{x}, \mathbf{x} \rangle \langle \mathbf{z}, \mathbf{y} \rangle \langle \mathbf{y}, \mathbf{x} \rangle \langle \mathbf{x}, \mathbf{z} \rangle$$
$$+ \langle \mathbf{x}, \mathbf{x} \rangle \langle \mathbf{y}, \mathbf{z} \rangle \langle \mathbf{x}, \mathbf{y} \rangle \langle \mathbf{z}, \mathbf{x} \rangle. \qquad (4.32)$$

At first glance, these bounds may seem intimidating, but after one uses the Gram–Schmidt process to strip away the inner products, they are just like the kind of bounds we have met many times before.

Exercise 4.13 (Equivalence of Isometry and Orthonormality)

This exercise shows how an important algebraic identity can be proved with help from the condition for equality in the Cauchy–Schwarz bound. The task is to show that if the $n \times n$ matrix A preserves the Euclidean length of each $\mathbf{v}$ in $\mathbb{R}^n$ then its columns are orthonormal. In the useful shorthand of matrix algebra, one needs to show

$$\|A\mathbf{v}\| = \|\mathbf{v}\| \text{ for all } \mathbf{v} \in \mathbb{R}^d \iff A^T A = I,$$

where I is the identity matrix, A^T is the transpose of A, and $\|\mathbf{v}\|$ is the Euclidean length of $\mathbf{v}$.

As a hint, one might first show that $\|A^T \mathbf{v}\| \leq \|\mathbf{v}\|$; that is, one might show that the transpose A^T does not increase length. One can then argue that if Cauchy–Schwarz inequality is applied to the inner product $\langle \mathbf{v}, A^T A \mathbf{v} \rangle$ then *equality* actually holds.

5
Consequences of Order

One of the natural questions that accompanies any inequality is the possibility that it admits a converse of one sort or another. When we pose this question for Cauchy's inequality, we find a challenge problem that is definitely worth our attention. It not only leads to results that are useful in their own right, but it also puts us on the path of one of the most fundamental principles in the theory of inequalities — the systematic exploitation of order relationships.

Problem 5.1 (The Hunt for a Cauchy Converse)
Determine the circumstances which suffice for nonnegative real numbers a_k, b_k, $k = 1, 2, \ldots, n$ to satisfy an inequality of the type

$$\left(\sum_{k=1}^{n} a_k^2\right)^{\frac{1}{2}} \left(\sum_{k=1}^{n} b_k^2\right)^{\frac{1}{2}} \leq \rho \sum_{k=1}^{n} a_k b_k \tag{5.1}$$

for a given constant ρ.

ORIENTATION

Part of the challenge here is that the problem is not fully framed — there are circumstances and conditions that remain to be determined. Nevertheless, uncertainty is an inevitable part of research, and practice with modestly ambiguous problems can be particularly valuable.

In such situations, one almost always begins with some experimentation, and since the case $n = 1$ is trivial, the simplest case worth study is given by taking the vectors $(1, a)$ and $(1, b)$ with $a > 0$ and $b > 0$. In this case, the two sides of the conjectured Cauchy converse (5.1) relate the quantities

$$(1 + a^2)^{\frac{1}{2}}(1 + b^2)^{\frac{1}{2}} \quad \text{and} \quad 1 + ab,$$

and this calculation already suggests a useful inference. If a and b are chosen so that the product ab is held constant while $a \to \infty$, then one finds that the right-hand expression is bounded, but the left-hand expression is unbounded. This observation shows in essence that for a given fixed value of $\rho \geq 1$ the conjecture (5.1) cannot hold unless the ratios a_k/b_k are required to be bounded from above and below.

Thus, we come to a more refined point of view, and we see that it is natural to conjecture that a bound of the type (5.1) will hold provided that the summands satisfy the ratio constraint

$$m \leq \frac{a_k}{b_k} \leq M \qquad \text{for all } k = 1, 2, \ldots n, \qquad (5.2)$$

for some constants $0 < m \leq M < \infty$. In this new interpretation of the conjecture (5.1), one naturally permits ρ to depend on the values of m and M, though we would hope to show that ρ can be chosen so that it does not have any further dependence on the individual summands a_k and b_k. Now, the puzzle is to find a way to exploit the betweenness bounds (5.2).

EXPLOITATION OF BETWEENNESS

When we look at our *unknown* (the conjectured inequality) and then look at the *given* (the betweenness bounds), we may have the lucky idea of hunting for clues in our earlier proofs of Cauchy's inequality. In particular, if we recall the proof that took $(a-b)^2 \geq 0$ as its departure point, we might start to suspect that an analogous idea could help here. Is there some way to obtain a useful quadratic bound from the betweenness relation (5.2)?

Once the question is put so bluntly, one does not need long to notice that the two-sided bound (5.2) gives us a cheap quadratic bound

$$\left(M - \frac{a_k}{b_k}\right)\left(\frac{a_k}{b_k} - m\right) \geq 0. \qquad (5.3)$$

Although one cannot tell immediately if this observation will help, the analogy with the earlier success of the trivial bound $(a-b)^2 \geq 0$ provides ground for optimism.

At a minimum, we should have the confidence needed to unwrap the bound (5.3) to find the equivalent inequality

$$a_k^2 + (mM)\, b_k^2 \leq (m+M)\, a_k b_k \qquad \text{for all } k = 1, 2, \ldots, n. \qquad (5.4)$$

Now we seem to be in luck; we have found a bound on a sum of squares by a product, and this is precisely what a converse to Cauchy's inequality

Consequences of Order

requires. The eventual role to be played by M and m is still uncertain, but the scent of progress is in the air.

The bounds (5.4) call out to be summed over $1 \leq k \leq n$, and, upon summing, the factors mM and $m+M$ come out neatly to give us

$$\sum_{k=1}^{n} a_k^2 + (mM) \sum_{k=1}^{n} b_k^2 \leq (m+M) \sum_{k=1}^{n} a_k b_k, \quad (5.5)$$

which is a fine additive bound. Thus, we face a problem of a kind we have met before — we need to convert an additive bound to one that is multiplicative.

Passage to a Product

If we cling to our earlier pattern, we might now be tempted to introduce normalized variables $\hat{a}_k$ and $\hat{b}_k$, but this time normalization runs into trouble. The problem is that the inequality (5.5) may be applied to $\hat{a}_k$ and $\hat{b}_k$ only if they satisfy the ratio bound $m \leq \hat{a}_k/\hat{b}_k \leq M$, and these constraints rule out the natural candidates for the normalizations $\hat{a}_k$ and $\hat{b}_k$. We need a new idea for passing to a product.

Conceivably, one might get stuck here, but help is close at hand provided that we pause to ask clearly what is needed — which is just a lower bound for a sum of two expressions by a product of their square roots. Once this is said, one can hardly fail to think of using the AM-GM inequality, and when it is applied to the additive bound (5.5), one finds

$$\left(\sum_{k=1}^{n} a_k^2\right)^{\frac{1}{2}} \left(mM \sum_{k=1}^{n} b_k^2\right)^{\frac{1}{2}} \leq \frac{1}{2}\left\{\sum_{k=1}^{n} a_k^2 + (mM)\sum_{k=1}^{n} b_k^2\right\}$$

$$\leq \frac{1}{2}\left\{(m+M)\sum_{k=1}^{n} a_k b_k\right\}.$$

Now, with just a little rearranging, we come to the inequality that completes our quest. Thus, if we set

$$A = (m+M)/2 \quad \text{and} \quad G = \sqrt{mM}, \quad (5.6)$$

then, for all nonnegative a_k, b_k, $k = 1, 2, \ldots, n$ with

$$0 < m \leq a_k/b_k \leq M < \infty,$$

we find the we have established the bound

$$\left(\sum_{k=1}^{n} a_k^2\right)^{\frac{1}{2}} \left(\sum_{k=1}^{n} b_k^2\right)^{\frac{1}{2}} \leq \frac{A}{G} \sum_{k=1}^{n} a_k b_k; \quad (5.7)$$

thus, in the end, one sees that there is indeed a natural converse to Cauchy's inequality.

ON THE CONVERSION OF INFORMATION

When one looks back on the proof of the converse Cauchy inequality (5.7), one may be struck by how quickly progress followed once the two *order relationships*, $m \leq a_k/b_k$ and $a_k/b_k \leq M$, were put together to build the simple *quadratic inequality* $(M - a_k/b_k)(a_k/b_k - m) \geq 0$. In the context of a single example, this could just be a lucky accident, but something deeper is afoot.

In fact, the device of order-to-quadratic conversion is remarkably versatile tool with a wide range of applications. The next few challenge problems illustrate some of these that are of independent interest.

MONOTONICITY AND CHEBYSHEV'S "ORDER INEQUALITY"

One way to put a large collection of order relationships at your fingertips is to focus your attention on monotone sequences and monotone functions. This suggestion is so natural that it might not stir high hopes, but in fact it does lead to an important result with many applications, especially in probability and statistics.

The result is due to Pafnuty Lvovich Chebyshev (1821–1894) who apparently had his first exposure to probability theory from our earlier acquaintance Victor Yacovlevich Bunyakovsky. Probability theory was one of those hot new mathematical topics which Bunyakovsky brought back to St. Petersburg when he returned from his student days studying with Cauchy in Paris. Another topic was the theory of complex variables which we will engage a bit later.

Problem 5.2 (Chebyshev's Order Inequality)

Suppose that $f : \mathbb{R} \to \mathbb{R}$ *and* $g : \mathbb{R} \to \mathbb{R}$ *are nondecreasing and suppose* $p_j \geq 0$, $j = 1, 2, \ldots, n$, *satisfy* $p_1 + p_2 + \cdots + p_n = 1$. *Show that for any nondecreasing sequence* $x_1 \leq x_2 \leq \cdots \leq x_n$ *one has the inequality*

$$\left\{\sum_{k=1}^n f(x_k) p_k\right\}\left\{\sum_{k=1}^n g(x_k) p_k\right\} \leq \sum_{k=1}^n f(x_k) g(x_k) p_k. \qquad (5.8)$$

CONNECTIONS TO PROBABILITY AND STATISTICS

The inequality (5.8) is easily understood without relying on its connection to probability theory, and it has many applications in other areas of mathematics. Nevertheless, the probabilistic interpretation of the bound

(5.8) is particularly compelling. In the language of probability, it says that if X is a random variable for which one has $P(X = x_k) = p_k$ for $k = 1, 2, \ldots, n$ then

$$E[f(X)]E[g(X)] \leq E[f(X)g(X)], \tag{5.9}$$

where, as usual, P stands for probability and E stands for the mathematical expectation. In other words, if random variables Y and Z may be written as nondecreasing functions of a single random variable X, then Y and Z must be nonnegatively correlated. Without Chebyshev's inequality, the intuition that is commonly attached to the statistical notion of correlation would stand on shaky ground.

Incidentally, there is another inequality due to Chebyshev that is even more important in probability theory; it tells us that for any random variable X with a finite mean $\mu = E(X)$ one has the bound

$$P(|X - \mu| \geq \lambda) \leq \frac{1}{\lambda^2} E(|X - \mu|^2). \tag{5.10}$$

The proof of this bound is almost trivial, especially with the hint offered in Exercise 5.11, but it is such a day-to-day workhorse in probability theory that Chebyshev's *order* (5.9) inequality is often jokingly called Chebyshev's *other* inequality.

A Proof from Our Pocket

Chebyshev's inequality (5.8) is quadratic, and the hypotheses provide order information, so, even if one were to meet Chebyshev's inequality (5.8) in a dark alley, the *order-to-quadratic conversion* is likely to come to mind. Here the monotonicity of f and g give us the quadratic bound

$$0 \leq \{f(x_k) - f(x_j)\}\{g(x_k) - g(x_j)\},$$

and this may be expanded in turn to give

$$f(x_k)g(x_j) + f(x_j)g(x_k) \leq f(x_j)g(x_j) + f(x_k)g(x_k). \tag{5.11}$$

From this point, we only need to bring the p_j's into the picture and meekly agree to take whatever arithmetic gives us.

Thus, when we multiply the bound (5.11) by $p_j p_k$ and sum over $1 \leq j \leq n$ and $1 \leq k \leq n$, we find that the left-hand sum gives us

$$\sum_{j,k=1}^{n} \{f(x_k)g(x_j) + f(x_j)g(x_k)\} p_j p_k = 2 \left\{ \sum_{k=1}^{n} f(x_k) p_k \right\} \left\{ \sum_{k=1}^{n} g(x_k) p_k \right\},$$

78   *Consequences of Order*

while the right-hand sum gives us

$$\sum_{j,k=1}^{n} \{f(x_j)g(x_j) + f(x_k)g(x_k)\}p_j p_k = 2\bigg\{\sum_{k=1}^{n} f(x_k)g(x_k)p_k\bigg\}.$$

Thus, the bound between the summands (5.11) does indeed yield the proof of Chebyshev's inequality.

ORDER, FACILITY, AND SUBTLETY

The proof of Chebyshev's inequality leads us to a couple of observations. First, there are occasions when the application of the order-to-quadratic conversion is an automatic, straightforward affair. Even so, the conversion has led to some remarkable results, including the versatile *rearrangement inequality* which is developed in our next challenge problem. The rearrangement inequality is not much harder to prove than Chebyshev's inequality, but some of its consequences are simply stunning. Here, and subsequently, we let $[n]$ denote the set $\{1, 2, \ldots, n\}$, and we recall that a permutation of $[n]$ is just a one-to-one mapping from $[n]$ into $[n]$.

Problem 5.3 (The Rearrangement Inequality)

Show that for each pair of ordered real sequences

$$-\infty < a_1 \leq a_2 \leq \cdots \leq a_n < \infty \quad \text{and} \quad -\infty < b_1 \leq b_2 \leq \cdots \leq b_n < \infty$$

and for each permutation $\sigma : [n] \to [n]$, *one has*

$$\sum_{k=1}^{n} a_k b_{n-k+1} \leq \sum_{k=1}^{n} a_k b_{\sigma(k)} \leq \sum_{k=1}^{n} a_k b_k. \qquad (5.12)$$

AUTOMATIC — BUT STILL EFFECTIVE

This problem offers us a hypothesis that provides order relations and asks us for a conclusion that is quadratic. This familiar combination may tempt one to just to dive in, but sometimes it pays to be patient. After all, the statement of the rearrangement inequality is a bit involved, and one probably does well to first consider the simplest case $n = 2$.

In this case, the order-to-quadratic conversion reminds us that

$$a_1 \leq a_2 \quad \text{and} \quad b_1 \leq b_2 \quad \text{imply} \quad 0 \leq (a_2 - a_1)(b_2 - b_1),$$

and when this is unwrapped, we find

$$a_1 b_2 + a_2 b_1 \leq a_1 b_1 + a_2 b_2,$$

which is precisely the rearrangement inequality (5.12) for $n = 2$. Nothing could be easier than this warm-up case; the issue now is to see if a similar idea can be used to deal with the more general sums

$$S(\sigma) = \sum_{k=1}^{n} a_k b_{\sigma(k)}.$$

INVERSIONS AND THEIR REMOVAL

If σ is not the identity permutation, then there must exist some pair $j < k$ such that $\sigma(k) < \sigma(j)$. Such a pair is called an *inversion*, and the observation that one draws from the case $n = 2$ is that if we switch the values of $\sigma(k)$ and $\sigma(j)$, then the value of the associated sum will increase — or, at least not decrease. To make this idea formal, we first introduce a new permutation τ by the recipe

$$\tau(i) = \begin{cases} \sigma(i) & \text{if } i \neq j \text{ and } i \neq k \\ \sigma(j) & \text{if } i = k \\ \sigma(k) & \text{if } i = j \end{cases} \quad (5.13)$$

which is illustrated in Figure 5.1. By the definition of τ and by factorization, we then find

$$\begin{aligned} S(\tau) - S(\sigma) &= a_j b_{\tau(j)} + a_k b_{\tau(k)} - a_j b_{\sigma(j)} - a_k b_{\sigma(k)} \\ &= a_j b_{\tau(j)} + a_k b_{\tau(k)} - a_j b_{\tau(k)} - a_k b_{\tau(j)} \\ &= (a_k - a_j)(b_{\tau(k)} - b_{\tau(j)}) \geq 0. \end{aligned}$$

Thus, the transformation $\sigma \mapsto \tau$ achieves two goals; first, it increases S, so $S(\sigma) \leq S(\tau)$, and second, the number of inversions of τ is forced to be strictly fewer than the number of inversions of the permutation σ.

REPEATING THE PROCESS — CLOSING THE LOOP

A permutation has at most $n(n-1)/2$ inversions and only the identity permutation has no inversions, so there exists a finite sequence of inversion removing transformations that move in sequence from σ to the identity. If we denote these by $\sigma = \sigma_0, \sigma_1, \ldots, \sigma_m$ where σ_m is the identity and $m \leq n(n-1)/2$, then, by applying the bound $S(\sigma_{j-1}) \leq S(\sigma_j)$ for $j = 1, 2, \ldots, m$, we find

$$S(\sigma) \leq \sum_{k=1}^{n} a_k b_k.$$

This completes the proof of the upper half of the rearrangement inequality (5.12).

80 Consequences of Order

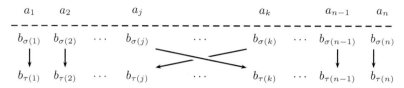

Fig. 5.1. An interchange operation converts the permutation σ to a permutation τ. By design, the new permutation τ has fewer inversions than σ; by calculation, one also finds that $S(\sigma) \leq S(\tau)$.

The easy way to get the lower half is then to notice that it is an immediate consequence of the upper half. Thus, if we consider $b'_1 = -b_n, b'_2 = -b_{n-1}, \ldots, b'_n = -b_1$ we see that

$$b'_1 \leq b'_2 \leq \cdots \leq b'_n$$

and, by the upper half of the rearrangement inequality (5.12) applied to the sequence $b'_1, b'_2, \ldots, b'_n$ we get the lower half of the inequality (5.12) for the sequence $b_1, b_2, \ldots, b_n$.

LOOKING BACK — TESTING NEW PROBES

The statement of the rearrangement inequality is exceptionally natural, and it does not provide us with any obvious loose ends. We might look back on it many times and never think of any useful variations of either its statement or its proof. Nevertheless, such variations can always be found; one just needs to use the right probes.

Obviously, no single probe, or even any set of probes, can lead with certainty to a useful variation of a given result, but there are a few generic questions that are almost always worth our time. One of the best of these asks: "Is there a *nonlinear* version of this result?"

Here, to make sense of this question, we first need to notice that the rearrangement inequality is a statement about sums of linear functions of the ordered n-tuples

$$\{b_{n-k+1}\}_{1 \leq k \leq n}, \quad \{b_{\sigma(k)}\}_{1 \leq k \leq n} \quad \text{and} \quad \{b_k\}_{1 \leq k \leq n},$$

where the "linear functions" are simply the n mappings given by

$$x \mapsto a_k x \quad k = 1, 2, \ldots, n.$$

Such simple linear maps are usually not worth naming, but here we have a higher purpose in mind. In particular, with this identification behind us, we may not need long to think of some ways that the monotonicity condition $a_k \leq a_{k+1}$ might be re-expressed.

Several variations of the rearrangement inequality may come to mind, and our next challenge problem explores one of the simplest of these. It was first studied by A. Vince, and it has several informative consequences.

Problem 5.4 (A Nonlinear Rearrangement Inequality)
Let $f_1, f_2, \ldots, f_n$ be functions from the interval I into $\mathbb{R}$ such that

$$f_{k+1}(x) - f_k(x) \quad \text{is nondecreasing for all } 1 \leq k \leq n. \tag{5.14}$$

Let $b_1 \leq b_2 \leq \cdots \leq b_n$ be an ordered sequence of elements of I, and show that for each permutation $\sigma : [n] \to [n]$, one has the bound

$$\sum_{k=1}^{n} f_k(b_{n-k+1}) \leq \sum_{k=1}^{n} f_k(b_{\sigma(k)}) \leq \sum_{k=1}^{n} f_k(b_k). \tag{5.15}$$

TESTING THE WATERS

This problem is intended to generalize the rearrangement inequality, and we see immediately that it does when we identify $f_k(x)$ with the map $x \mapsto a_k x$. To be sure, there are far more interesting nonlinear examples which one can find after even a little experimentation.

For instance, one might take $a_1 \leq a_2 \leq \cdots \leq a_n$ and consider the functions $x \mapsto \log(a_k + x)$. Here one finds

$$\log(a_{k+1} + x) - \log(a_k + x) = \log\left(\frac{(a_{k+1} + x)}{(a_k + x)}\right),$$

and if we set $r(x) = (a_{k+1} + x)/(a_k + x)$, then direct calculation gives

$$r'(x) = \frac{a_k - a_{k+1}}{(a_k + x)^2} \leq 0,$$

so, if we take

$$f_k(x) = -\log(a_k + x) \quad \text{for } k = 1, 2, \ldots, n,$$

then condition (5.14) is satisfied. Thus, by Vince's inequality and exponentiation one finds that for each permutation $\sigma : [n] \to [n]$ that

$$\prod_{k=1}^{n}(a_k + b_k) \leq \prod_{k=1}^{n}(a_k + b_{\sigma(k)}) \leq \prod_{k=1}^{n}(a_k + b_{n-k+1}). \tag{5.16}$$

This interesting product bound (5.16) shows that there is power in Vince's inequality, though in this particular case the bound was known earlier. Still, we see that a proof of Vince's inequality will be worth our time — even if only because of the corollary (5.16).

Recycling an Algorithmic Proof

If we generalize our earlier sums and write

$$S(\sigma) = \sum_{k=1}^{n} f_k(b_{\sigma(k)}),$$

then we already know from the definition (5.13) and discussion of the inversion decreasing transformation $\sigma \mapsto \tau$ that we only need to show

$$S(\sigma) \leq S(\tau).$$

Now, almost as before, we calculate the difference

$$\begin{aligned}S(\tau) - S(\sigma) &= f_j(b_{\tau(j)}) + f_k(b_{\tau(k)}) - f_j(b_{\sigma(j)}) - f_k(b_{\sigma(k)}) \\ &= f_j(b_{\tau(j)}) + f_k(b_{\tau(k)}) - f_j(b_{\tau(k)}) - f_k(b_{\tau(j)}) \\ &= \{f_k(b_{\tau(k)}) - f_j(b_{\tau(k)})\} - \{f_k(b_{\tau(j)}) - f_j(b_{\tau(j)})\} \geq 0,\end{aligned}$$

and this time the last inequality comes from $b_{\tau(j)} \leq b_{\tau(k)}$ and our hypothesis that $f_k(x) - f_j(x)$ is a nondecreasing function of $x \in I$. From this relation, one then sees that no further change is needed in our earlier arguments, and the proof of the nonlinear version of the rearrangement inequality is complete.

Exercises

Exercise 5.1 (Baseball and Cauchy's Third Inequality)

In the remarkable *Note II* of 1821 where Cauchy proved both his namesake inequality and the fundamental AM-GM bound, one finds a third inequality which is not as notable nor as deep but which is still handy from time to time. The inequality asserts that for any positive real numbers $h_1, h_2, \ldots, h_n$ and $b_1, b_2, \ldots, b_n$ one has the ratio bounds

$$m = \min_{1 \leq j \leq n} \frac{h_j}{b_j} \leq \frac{h_1 + h_2 + \cdots + h_n}{b_1 + b_2 + \cdots + b_n} \leq \max_{1 \leq j \leq n} \frac{h_j}{b_j} = M. \qquad (5.17)$$

Sports enthusiasts may imagine, as Cauchy never would, that b_j denotes the number of times a baseball player j goes to bat, and h_j denotes the number of times he gets a hit. The inequality confirms the intuitive fact that the batting average of a team is never worse than that of its worst hitter and never better than that of its best hitter.

Prove the inequality (5.17) and put it to honest mathematical use by proving that for any polynomial $P(x) = c_0 + c_1 x + c_2 x^2 + \cdots + c_n x^n$ with positive coefficients one has the monotonicity relation

$$0 < x \leq y \quad \Longrightarrow \quad \left(\frac{x}{y}\right)^n \leq \frac{P(x)}{P(y)} \leq 1.$$

Exercise 5.2 (Betweenness and an Inductive Proof of AM-GM)

One can build an inductive proof of the basic AM-GM inequality (2.3) by exploiting the conversion of an order relation to a quadratic bound. To get started, first consider $0 < a_1 \leq a_2 \leq \cdots \leq a_n$, set $A = (a_1 + a_2 + \cdots + a_n)/n$, and then show that one has

$$a_1 a_n / A \leq a_1 + a_n - A.$$

Now, complete the induction step of the AM-GM proof by considering the $n - 1$ element set $S = \{a_2, a_3, \ldots, a_{n-1}\} \cup \{a_1 + a_n - A\}$.

Exercise 5.3 (Cauchy–Schwarz and the Cross-Term Defect)

If u and v are elements of the real inner product space V for which on has the upper bounds

$$\langle \mathbf{u}, \mathbf{u} \rangle \leq A^2 \quad \text{and} \quad \langle \mathbf{v}, \mathbf{v} \rangle \leq B^2,$$

then Cauchy's inequality tells us $\langle \mathbf{u}, \mathbf{v} \rangle \leq AB$. Show that one then also has a lower bound on the cross-term difference $AB - \langle \mathbf{u}, \mathbf{v} \rangle$, namely,

$$\left\{ A^2 - \langle \mathbf{u}, \mathbf{u} \rangle \right\}^{\frac{1}{2}} \left\{ B^2 - \langle \mathbf{v}, \mathbf{v} \rangle \right\}^{\frac{1}{2}} \leq AB - \langle \mathbf{u}, \mathbf{v} \rangle. \tag{5.18}$$

Exercise 5.4 (A Remarkable Inequality of I. Schur)

Show that for all values of $x, y, z \geq 0$, one has for all $\alpha \geq 0$ that

$$x^\alpha (x-y)(x-z) + y^\alpha (y-x)(y-z) + z^\alpha (z-x)(z-y) \geq 0. \tag{5.19}$$

Moreover, show that one has equality here if and only if one has *either* $x = y = x$ *or* two of the variables are equal and the third is zero.

Schur's inequality can sometimes saves the day in problems where the AM-GM inequality looks like the natural tool, yet it comes up short. Sometimes the two-pronged condition for equality also provides a clue that Schur's inequality may be of help.

Exercise 5.5 (The Pólya–Szegő Converse Restructured)

The converse Cauchy inequality (5.7) is expressed with the aid of bounds on the ratios a_k/b_k, but for many applications it is useful to know that one also has a natural converse under the more straightforward hypothesis that

$$0 < a \leq a_k \leq A \quad \text{and} \quad 0 < b \leq b_k \leq B \qquad \text{for all } k = 1, 2, \ldots, n.$$

Use the Cauchy converse (5.7) to prove that in this case one has

$$\left\{ \sum_{k=1}^{n} a_k^2 \sum_{k=1}^{n} b_k^2 \right\} \bigg/ \left\{ \sum_{k=1}^{n} a_k b_k \right\}^2 \leq \frac{1}{4} \left\{ \sqrt{\frac{AB}{ab}} + \sqrt{\frac{ab}{AB}} \right\}^2.$$

Exercise 5.6 (A Competition Perennial)

Show that if $a > 0$, $b > 0$, and $c > 0$ then one has the elegant symmetric bound

$$\frac{3}{2} \leq \frac{a}{b+c} + \frac{b}{a+c} + \frac{c}{a+b}. \qquad (5.20)$$

This is known as Nesbitt's inequality, and along with several natural variations, it has served a remarkable number of mathematical competitions, from Moscow in 1962 to the Canadian Maritimes in 2002.

Exercise 5.7 (Rearrangement, Cyclic Shifts, and the AM-GM)

Skillful use of the rearrangement inequality often calls for one to exploit symmetry and to look for clever specializations of the resulting bounds. This problem outlines a proof of the AM-GM inequality that nicely illustrates these steps.

(a) Show that for positive c_k, $k = 1, 2, \ldots, n$ one has

$$n \leq \frac{c_1}{c_n} + \frac{c_2}{c_1} + \frac{c_3}{c_2} + \cdots + \frac{c_n}{c_{n-1}}.$$

(b) Specialize the result of part (a) to show that for all positive x_k, $k = 1, 2, \ldots, n$, one has the rational bound

$$n \leq \frac{x_1}{x_1 x_2 \cdots x_n} + x_2 + x_3 + \cdots + x_n.$$

(c) Specialize a third time to show that for $\rho > 0$ one also has

$$n \leq \frac{\rho x_1}{\rho^n x_1 x_2 \cdots x_n} + \rho x_2 + \rho x_3 + \cdots + \rho x_n,$$

and finally indicate how the right choice of ρ now yields the AM-GM inequality (2.3).

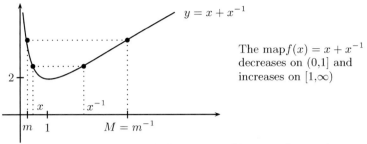

Fig. 5.2. One key to the proof of Kantorovich's inequality is the geometry of the map $x \mapsto x + x^{-1}$; another key is that a multiplicative inequality is sometimes proved most easily by first establishing an appropriate additive inequality. To say much more would risk giving away the game.

Exercise 5.8 (Kantorovich's Inequality for Reciprocals)

Show that if $0 < m = x_1 \leq x_2 \leq \cdots \leq x_n = M < \infty$ then for nonnegative weights with $p_1 + p_2 + \cdots + p_n = 1$ one has

$$\left\{\sum_{j=1}^n p_j x_j\right\}\left\{\sum_{j=1}^n p_j \frac{1}{x_j}\right\} \leq \frac{\mu^2}{\gamma^2} \qquad (5.21)$$

where $\mu = (m + M)/2$ and $\gamma = \sqrt{mM}$. This bound provides a natural complement to the elementary inequality of Exercise 1.2 (page 12), but it also has important applications in numerical analysis where, for example, it has been used to estimated the rate of convergence of the method of steepest ascent. To get started with the proof, one might note that by homogeneity it suffices to consider the case when $\gamma = 1$; the geometry of Figure 5.2 then tells a powerful tale.

Exercise 5.9 (Monotonicity Method)

Suppose $a_k > 0$ and $b_k > 0$ for $k = 1, 2, \ldots, n$ and for fixed $\theta \in \mathbb{R}$ consider the function

$$f_\theta(x) = \left\{\sum_{j=1}^n a_j^{\theta+x} b_j^{\theta-x}\right\}\left\{\sum_{j=1}^n a_j^{\theta-x} b_j^{\theta+x}\right\}, \qquad x \in \mathbb{R}.$$

If we set $\theta = 1$, we see that $f_1(0)^{1/2}$ gives us the left side of Cauchy's inequality while $f_1(1)^{1/2}$ gives us the right side. Show that $f_\theta(x)$ is a monotone increasing of x on $[0, 1]$, a fact which gives us a parametric family of inequalities containing Cauchy's inequality as a special case.

Exercise 5.10 (A Proto-Muirhead Inequality)

If the nonnegative real numbers a_1, a_2, b_1, and b_2 satisfy

$$\max\{a_1, a_2\} \geq \max\{b_1, b_2\} \quad \text{and} \quad a_1 + a_2 = b_1 + b_2,$$

then for nonnegative x and y, one has

$$x^{b_1} y^{b_2} + x^{b_2} y^{b_1} \leq x^{a_1} y^{a_2} + x^{a_2} y^{a_1}. \tag{5.22}$$

Prove this assertion by considering an appropriate factorization of the difference of the two sides.

Exercise 5.11 (Chebyshev's Inequality for Tail Probabilities)

One of the most basic properties of the mathematical expectation $E(\cdot)$ that one meets in probability theory is that for any random variables X and Y with finite expectations the relationship $X \leq Y$ implies that $E(X) \leq E(Y)$. Use this fact to show that for any random variable Z with finite mean $\mu = E(Z)$ one has the inequality

$$P(|Z - \mu| \geq \lambda) \leq \frac{1}{\lambda^2} E(|Z - \mu|^2). \tag{5.23}$$

This bound provides one concrete expression of the notion that a random variable is not likely to be too far away from its mean, and it is surely the most used of the several inequalities that carry Chebyshev's name.

6
Convexity — The Third Pillar

There are three great pillars of the theory of inequalities: positivity, monotonicity, and convexity. The notions of positivity and monotonicity are so intrinsic to the subject that they serve us steadily without ever calling attention to themselves, but convexity is different. Convexity expresses a second order effect, and for it to provide assistance we almost always need to make some deliberate preparations.

To begin, we first recall that a function $f : [a, b] \to \mathbb{R}$ is said to be *convex* provided that for all $x, y \in [a, b]$ and all $0 \leq p \leq 1$ one has

$$f(px + (1-p)y) \leq pf(x) + (1-p)f(y). \tag{6.1}$$

With nothing more than this definition and the intuition offered by the first frame of Figure 6.1, we can set a challenge problem which creates a fundamental link between the notion of convexity and the theory of inequalities.

Problem 6.1 (Jensen's Inequality)
Suppose that $f : [a, b] \to \mathbb{R}$ is a convex function and suppose that the nonnegative real numbers p_j, $j = 1, 2, \ldots, n$ satisfy

$$p_1 + p_2 + \cdots + p_n = 1.$$

Show that for all $x_j \in [a, b]$, $j = 1, 2, \ldots, n$ one has

$$f\left(\sum_{j=1}^{n} p_j x_j\right) \leq \sum_{j=1}^{n} p_j f(x_j). \tag{6.2}$$

When $n = 2$ we see that Jensen's inequality (6.2) is nothing more than the definition of convexity, so our instincts may suggest that we look for a proof by induction. Such an approach calls for one to relate averages of size $n-1$ to averages of size n, and this can be achieved several ways.

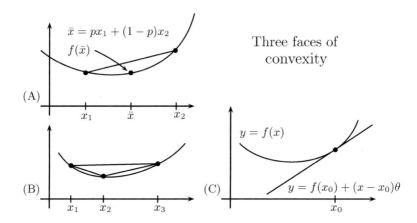

Fig. 6.1. By definition, a function f is convex provided that it satisfies the condition (6.1) which is illustrated in frame (A), but a convex function may be characterized in several other ways. For example, frame (B) illustrates that a function is convex if and only if its sequential secants have increasing slopes, and frame (C) illustrates that a function is convex if and only if for each point p on its graph there is line through p that lies below the graph. None of these criteria requires that f be differentiable.

One natural idea is simply to pull out the last summand and to renormalize the sum that is left behind. More precisely, we first note that there is no loss of generality if we assume $p_n > 0$ and, in this case, we can write

$$\sum_{j=1}^{n} p_j x_j = p_n x_n + (1 - p_n) \sum_{j=1}^{n-1} \frac{p_j}{1 - p_n} x_j.$$

Now, from this representation, the definition of convexity, and the induction hypothesis — all applied in that order — we see that

$$f\left(\sum_{j=1}^{n} p_j x_j\right) \leq p_n f(x_n) + (1 - p_n) f\left(\sum_{j=1}^{n-1} \frac{p_j}{1 - p_n} x_j\right)$$

$$\leq p_n f(x_n) + (1 - p_n) \sum_{j=1}^{n-1} \frac{p_j}{1 - p_n} f(x_j)$$

$$= \sum_{j=1}^{n} p_j f(x_j).$$

Convexity — The Third Pillar

This bound completes the induction step and thus completes the solution to one of the easiest — but most useful — of all our challenge problems.

THE CASE OF EQUALITY

We will find many applications of Jensen's inequality, and some of the most engaging of these will depend on understanding the conditions where one has equality. Here it is useful to restrict attention to those functions $f : [a, b] \to \mathbb{R}$ such that for all $x, y \in [a, b]$ and all $0 < p < 1$ and $x \neq y$ one has the strict inequality

$$f(px + (1-p)y) < pf(x) + (1-p)f(y). \tag{6.3}$$

Such functions are said to be *strictly convex*, and they help us frame the next challenge problem.

Problem 6.2 (The Case of Equality in Jensen's Inequality)
Suppose that $f : [a, b] \to \mathbb{R}$ is strictly convex and show that if

$$f\left(\sum_{j=1}^{n} p_j x_j\right) = \sum_{j=1}^{n} p_j f(x_j) \tag{6.4}$$

where the positive reals p_j, $j = 1, 2, \ldots, n$ have sum $p_1 + p_2 + \cdots + p_n = 1$, then one must have

$$x_1 = x_2 = \cdots = x_n. \tag{6.5}$$

Once more, our task is easy, but, as with Jensen's inequality, the importance of the result justifies its role as a challenge problem. For many inequalities one discovers when equality can hold by taking the proof of the inequality and running it backwards. This approach works perfectly well with Jensen's inequality, but logic of the argument still deserves some attention.

First, if the conclusion (6.5) does not hold, then the set

$$S = \left\{ j : x_j \neq \max_{1 \leq k \leq n} x_k \right\}$$

is a proper subset of $\{1, 2, \ldots, n\}$, and we will argue that this leads one to a contradiction. To see why this is so, we first set

$$p = \sum_{j \in S} p_j, \quad x = \sum_{j \in S} \frac{p_j}{p} x_j, \quad \text{and} \quad y = \sum_{j \notin S} \frac{p_j}{1-p} x_j,$$

from which we note that the *strict* convexity of f implies

$$f\left(\sum_{j=1}^{n} p_j x_j\right) = f(px + (1-p)y) < pf(x) + (1-p)f(y). \qquad (6.6)$$

Moreover, by the plain vanilla convexity of f applied separately at x and y, we also have the inequality

$$pf(x)+(1-p)f(y) \le p\sum_{j\in S}\frac{p_j}{p}f(x_j)+(1-p)\sum_{j\notin S}\frac{p_j}{1-p}f(x_j) = \sum_{j=1}^{n} p_j f(x_j).$$

Finally, from this bound and the strict inequality (6.6), we find

$$f\left(\sum_{j=1}^{n} p_j x_j\right) < \sum_{j=1}^{n} p_j f(x_j),$$

and since this inequality contradicts the assumption (6.4), the solution of the challenge problem is complete.

THE DIFFERENTIAL CRITERION FOR CONVEXITY

A key benefit of Jensen's inequality is its generality, but before Jensen's inequality can be put to work in a concrete problem, one needs to establish the convexity of the relevant function. On some occasions this can be achieved by direct application of the definition (6.1), but more commonly, convexity is established by applying the differential criterion provided by the next challenge problem.

Problem 6.3 (Differential Criterion for Convexity)
Show that if $f : (a,b) \to \mathbb{R}$ is twice differentiable, then

$$f''(x) \ge 0 \text{ for all } x \in (a,b) \text{ implies } f(\cdot) \text{ is convex on } (a,b),$$

and, in parallel, show that

$$f''(x) > 0 \text{ for all } x \in (a,b) \text{ implies } f(\cdot) \text{ is strictly convex on } (a,b).$$

If one simply visualizes the meaning of the condition $f''(x) \ge 0$, then this problem may seem rather obvious. Nevertheless, if one wants a complete proof, rather than an intuitive sketch, then the problem is not as straightforward as the graphs of Figure 6.1 might suggest.

Here, since we need to relate the function f to its derivatives, it is perhaps most natural to begin with the representation of f provided by the fundamental theorem of calculus. Specifically, if we fix a value

$x_0 \in [a, b]$, then we have the representation

$$f(x) = f(x_0) + \int_{x_0}^{x} f'(u)\, du \qquad \text{for all } x \in [a, b], \tag{6.7}$$

and once this formula is written down, we may not need long to think of exploiting the hypothesis $f''(\cdot) \geq 0$ by noting that it implies that the integrand $f'(\cdot)$ is nondecreasing. In fact, our hypothesis contains no further information, so the representation (6.7), the monotonicity of $f'(\cdot)$, and honest arithmetic *must* carry us the rest of the way.

To forge ahead, we take $a \leq x < y \leq b$ and $0 < p < 1$ and we also set $q = 1 - p$, so by applying the representation (6.7) to x, y, and $x_0 = px + qy$ we see $\Delta = pf(x) + qf(y) - f(px + qy)$ may be written as

$$\Delta = q \int_{px+qy}^{y} f'(u)\, du - p \int_{x}^{px+qy} f'(u)\, du. \tag{6.8}$$

For $u \in [x, px + qy]$ one has $f'(u) \leq f'(px + qy)$, so we have the bound

$$p \int_{x}^{px+qy} f'(u)\, du \leq qp(y-x)f'(px+qy), \tag{6.9}$$

while for $u \in [px + qy, y]$ one has $f'(u) \geq f'(px + qy)$, so we have the matching bound

$$q \int_{px+qy}^{y} f'(u)\, du \geq qp(y-x)f'(px+qy). \tag{6.10}$$

Therefore, from the integral representation (6.8) for Δ and the two monotonicity estimates (6.9) and (6.10), we find $\Delta \geq 0$, just as we needed to complete the solution of the first half of the problem.

For the second half of the theorem, we only need to note that if $f''(x) > 0$ for all $x \in (a, b)$, then both of the inequalities (6.9) and (6.10) are strict. Thus, the representation (6.8) for Δ gives us $\Delta > 0$, and we have the strict convexity of f.

Before leaving this challenge problem, we should note that there is an alternative way to proceed that is also quite instructive. In particular, one can rely on Rolle's theorem to help estimate Δ by comparison to an appropriate polynomial; this solution is outlined in Exercise 6.10.

THE AM-GM INEQUALITY AND THE SPECIAL NATURE OF $x \mapsto e^x$

The derivative criterion tells us that the map $x \mapsto e^x$ is convex, so Jensen's inequality tells us that for all real $y_1, y_2, \ldots, y_n$ and all positive

p_j, $j = 1, 2, \ldots, n$ with $p_1 + p_2 + \cdots + p_n = 1$, one has

$$\exp\left(\sum_{j=1}^n p_j y_j\right) \leq \sum_{j=1}^n p_j e^{y_j}.$$

Now, when we set $x_j = e^{y_j}$, we then find the familiar relation

$$\prod_{j=1}^n x_j^{p_j} \leq \sum_{j=1}^n p_j x_j.$$

Thus, with lightning speed and crystal clear logic, Jensen's inequality leads one to the general AM-GM bound.

Finally, this view of the AM-GM inequality as a special instance of Jensen's inequality for the function $x \mapsto e^x$ puts the AM-GM inequality in a unique light — one that may reveal the ultimate source of its vitality. Quite possibly, the pervasive value of the AM-GM bound throughout the theory of inequalities is simply one more reflection of the fundamental role of the exponential function as an isomorphism between two most important groups in mathematics: addition on the real line and multiplication on the positive real line.

How to Use Convexity in a Typical Problem

Many of the familiar functions of trigonometry and geometry have easily established convexity properties, and, more often than not, this convexity has useful consequences. The next challenge problem comes with no hint of convexity in its statement, but, if one is sensitive to the way Jensen's inequality helps us understand averages, then the required convexity is not hard to find.

Problem 6.4 (On the Maximum of the Product of Two Edges)

In an equilateral triangle with area A, the product of any two sides is equal to $(4/\sqrt{3})A$. Show that this represents the extreme case in the sense that for a triangle with area A there must exist two sides the lengths of which have a product that is at least as large as $(4/\sqrt{3})A$.

To get started we need formulas which relate edge lengths to areas, and, in the traditional notation of Figure 6.2, there are three equally viable formulas:

$$A = \frac{1}{2}ab\sin\gamma = \frac{1}{2}ac\sin\beta = \frac{1}{2}bc\sin\alpha.$$

Convexity — The Third Pillar

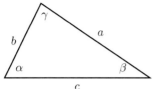

The area A of the generic triangle has three basic representations:
$$A = \tfrac{1}{2}ab\sin\gamma = \tfrac{1}{2}ac\sin\beta = \tfrac{1}{2}bc\sin\alpha$$

Fig. 6.2. All of the trigonometric functions are convex (or concave) if their arguments are restricted to an appropriate domain, and, as a consequence, there are many interesting geometric consequences of Jensen's inequality.

Now, if we average these representations, then we find that

$$\frac{1}{3}(ab + ac + bc) = (2A)\frac{1}{3}\left\{\frac{1}{\sin\alpha} + \frac{1}{\sin\beta} + \frac{1}{\sin\gamma}\right\}, \qquad (6.11)$$

and this is a formula that almost begs us to ask about the convexity of $1/\sin x$. The plot of $x \mapsto 1/\sin x$ for $x \in (0,\pi)$ certainly looks convex, and our suspicions can be confirmed by calculating the second derivative,

$$\left(\frac{1}{\sin x}\right)'' = \frac{1}{\sin x} + 2\frac{\cos^2 x}{\sin^3 x} > 0 \qquad \text{for all } x \in (0,\pi). \qquad (6.12)$$

Therefore, since we have $(\alpha + \beta + \gamma)/3 = \pi/3$, we find from Jensen's inequality that

$$\frac{1}{3}\left\{\frac{1}{\sin\alpha} + \frac{1}{\sin\beta} + \frac{1}{\sin\gamma}\right\} \geq \frac{1}{\sin\pi/3} = \frac{2}{\sqrt{3}},$$

so, by inequality (6.11), we do obtain the conjectured bound

$$\max(ab, ac, bc) \geq \frac{1}{3}(ab + ac + bc) \geq \frac{4}{\sqrt{3}}A. \qquad (6.13)$$

CONNECTIONS AND REFINEMENTS

This challenge problem is closely related to a well-known inequality of Weitzenböck which asserts that in any triangle one has

$$a^2 + b^2 + c^2 \geq \frac{4}{\sqrt{3}}A. \qquad (6.14)$$

In fact, to pass from the bound (6.13) to Weitzenböck's inequality one only has to recall that

$$ab + ac + bc \leq a^2 + b^2 + c^2,$$

which is a familiar fact that one can obtain in at least three ways —

Cauchy's inequality, the AM-GM bound, or the rearrangement inequality will all do the trick with equal grace.

Weitzenböck's inequality turns out to have many instructive proofs — Engel (1998) gives eleven! It also has several informative refinements, one of which is developed in Exercise 6.9 with help from the convexity of the map $x \mapsto \tan x$ on $[0, \pi/2]$.

How to Do Better Much of the Time

There are some mathematical methods which one might call *generic improvers*; broadly speaking, these are methods that can be used in a semi-automatic way to generalize an identity, refine an inequality, or otherwise improve a given result. A classic example which we saw earlier is the polarization device (see page 49) which often enables one to convert an identity for squares into a more general identity for products.

The next challenge problem provides an example of a different sort. It suggests how one might think about sharpening almost any result that is obtained via Jensen's inequality.

Problem 6.5 (Hölder's Defect Formula)

If $f : [a, b] \to \mathbb{R}$ is twice differentiable and if we have the bounds

$$0 \leq m \leq f''(x) \leq M \quad \text{for all } x \in [a, b], \tag{6.15}$$

then for any real values $a \leq x_1 \leq x_2 \leq \cdots \leq x_n \leq b$ and any nonnegative reals p_k, $k = 1, 2, \ldots, n$ with $p_1 + p_2 + \cdots + p_n = 1$, there exists a real value $\mu \in [m, M]$ for which one has the formula

$$\sum_{k=1}^{n} p_k f(x_k) - f\left(\sum_{k=1}^{n} p_k x_k\right) = \frac{1}{4} \mu \sum_{j=1}^{n} \sum_{k=1}^{n} p_j p_k (x_j - x_k)^2. \tag{6.16}$$

Context and a Plan

This result is from the same famous 1885 paper of Otto Ludwig Hölder (1859-1937) in which one finds his proof of the inequality that has come to be know universally as "Hölder's inequality." The defect formula (6.16) is much less well known, but it is nevertheless valuable. It provides a perfectly natural measure of the difference between the two sides of Jensen's inequality, and it tells us how to beat the plain vanilla version of Jensen's inequality whenever we can check the additional hypothesis (6.15). More often than not, the extra precision does not justify the added complexity, but it is a safe bet that some good problems are waiting to be cracked with just this refinement.

Hölder's defect formula (6.16) also deepens one's understanding of the relationship of convex functions to the simpler affine or quadratic functions. For example, if the difference $M - m$ is small, the bound (6.16) tells us that f behaves rather like a quadratic function on $[a, b]$. Moreover, in the extreme case when $m = M$, one finds that f is exactly quadratic, say $f(x) = \alpha + \beta x + \gamma x^2$ with $m = M = \mu = 2\gamma$, and the defect formula (6.16) reduces to a simple quadratic identity.

Similarly, if M is small, say $0 \leq M \leq \epsilon$, then the bound (6.16) tells us that f behaves rather like an affine function $f(x) = \alpha + \beta x$. For an exactly affine function, the left-hand side of the bound (6.16) is identically equal to zero, but in general the bound (6.16) asserts a more subtle relation. More precisely, it tells us that the left-hand side is a small multiple of a measure of the extent to which the values x_j, $j = 1, 2, \ldots, n$ are diffused throughout the interval $[a, b]$.

CONSIDERATION OF THE CONDITION

This challenge problem leads us quite naturally to an intermediate question: How can we use the fact that $0 \leq m \leq f''(x) \leq M$? Once this question is asked, one may not need long to observe that the two closely related functions

$$g(x) = \frac{1}{2}Mx^2 - f(x) \quad \text{and} \quad h(x) = f(x) - \frac{1}{2}mx^2$$

are again convex. In turn, this observation almost begs us to ask what Jensen's inequality says for these functions.

For $g(x)$, Jensen's inequality gives us the bound

$$\frac{1}{2}M\bar{x}^2 - f(\bar{x}) \leq \sum_{k=1}^{n} p_k \left\{ \frac{1}{2}Mx_k^2 - f(x_k) \right\}$$

where we have set $\bar{x} = p_1 x_1 + p_2 x_2 + \cdots + p_n x_n$, and this bound is easily rearranged to yield

$$\left\{ \sum_{k=1}^{n} p_k f(x_k) \right\} - f(\bar{x}) \leq \frac{1}{2}M \left\{ \left(\sum_{k=1}^{n} p_k x_k^2 \right) - \bar{x}^2 \right\} = \frac{1}{2}M \sum_{k=1}^{n} p_k (x_k - \bar{x})^2.$$

The perfectly analogous computation for $h(x)$ gives us a lower bound

$$\left\{ \sum_{k=1}^{n} p_k f(x_k) \right\} - f(\bar{x}) \geq \frac{1}{2}m \sum_{k=1}^{n} p_k (x_k - \bar{x})^2,$$

and these upper and lower bounds almost complete the proof of the assertion (6.16). The only missing element is the identity

$$\sum_{k=1}^{n} p_k(x_k - \bar{x})^2 = \frac{1}{2}\sum_{j=1}^{n}\sum_{k=1}^{n} p_j p_k (x_j - x_k)^2$$

which is easily checked by algebraic expansion and the definition of $\bar{x}$.

PREVAILING AFTER A NEAR FAILURE

Convexity and Jensen's inequality provide straightforward solutions to many problems. Nevertheless, they will sometimes run into a unexpected roadblock. Our next challenge comes from the famous problem section of the *American Mathematical Monthly*, and it provides a classic example of this phenomenon.

At first the problem looks invitingly easy, but, soon enough, it presents difficulties. Fortunately, these turn out to be of a generous kind. After we deepen our understanding of convex functions, we find that Jensen's inequality does indeed prevail.

Problem 6.6 (AMM 2002, Proposed by M. Mazur)

Show that if a, b, and c, are positive real numbers for which one has the lower bound $abc \geq 2^9$, then

$$\frac{1}{\sqrt{1 + (abc)^{1/3}}} \leq \frac{1}{3}\left\{\frac{1}{\sqrt{1+a}} + \frac{1}{\sqrt{1+b}} + \frac{1}{\sqrt{1+c}}\right\}. \tag{6.17}$$

The average on the right-hand side suggests that Jensen's inequality might prove useful, while the geometric mean on the left-hand side suggests that the exponential function will have a role. With more exploration — and some luck — one may not need long to guess that the function

$$f(x) = \frac{1}{\sqrt{1 + e^x}}$$

might help bring Jensen's inequality properly into play. In fact, once this function is written down, one may check almost without calculation that the proposed inequality (6.17) is equivalent to the assertion that

$$f\left(\frac{x+y+z}{3}\right) \leq \frac{1}{3}\{f(x) + f(y) + f(z)\} \tag{6.18}$$

for all real x, y, and z such that $\exp(x + y + z) \geq 2^9$.

To see if Jensen's inequality may be applied, we need to assess the

Convexity — The Third Pillar 97

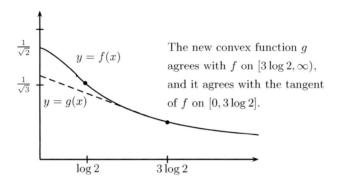

Fig. 6.3. Effective use of Jensen's inequality calls for one to find a function that is convex on all of $[0, \infty)$ and that is never larger than f. (Note: To make the concavity of f on $[0, \log 2)$ visible, the graph is not drawn to scale.)

convexity properties of f, so we just differentiate twice to find

$$f'(x) = -\frac{e^x}{2(1+e^x)^{3/2}}$$

and

$$f''(x) = -\frac{1}{2}(1+e^x)^{-3/2}e^x + \frac{3}{4}(1+e^x)^{-5/2}e^{2x}.$$

The second formula tells us that $f''(x) \geq 0$ if and only if we have $e^x \geq 2$, so by Jensen's inequality one finds that the target inequality (6.17) holds provided that *each* of the terms a, b, and c is at least as large as 2.

DIFFICULTIES, EXPLORATIONS, AND POSSIBILITIES

The difficulty we face here is that the hypothesis of Problem 6.6 only tells us that product abc is at least as large as 2^9; we are not given any bounds on the individual terms except that $a > 0$, $b > 0$, and $c > 0$. Thus, Jensen's inequality cannot complete the proof all by itself, and we must seek help from some other resources.

There are many ideas one might try, but before going too far, one should surely consider the graph of $f(x)$. What one finds from the plot in Figure 6.3 is that the $f(x)$ looks remarkably convex over the interval $[0, 10]$ despite the fact that calculation that shows $f(x)$ is concave on $[0, \log 2]$ and convex on $[\log 2, \infty)$. Thus, our plot holds out new hope; perhaps some small modification of f might have the convexity that we need to solve our problem.

THE IDEA OF A CONVEX MINORANT

When we think about the way we hoped to use f with Jensen's inequality, we soon realize that we can make our task a little bit easier. Suppose, for example, that we can find a convex function $g : [0, \infty) \to \mathbb{R}$ such that we have both the condition

$$g(x) \leq f(x) \qquad \text{for all } x \in [0, \infty) \qquad (6.19)$$

and the complementary condition

$$g(x) = f(x) \qquad \text{for all } x \geq 3\log 2. \qquad (6.20)$$

For such a function, Jensen's inequality would tell us that for $x, y,$ and z with $\exp(x + y + z) \geq 2^9$ we have the bound

$$\begin{aligned} f\left(\frac{x+y+z}{3}\right) &= g\left(\frac{x+y+z}{3}\right) \\ &\leq \frac{1}{3}\Big\{g(x) + g(y) + g(z)\Big\} \\ &\leq \frac{1}{3}\Big\{f(x) + f(y) + f(z)\Big\}. \end{aligned}$$

The first and last terms of this bound recover the inequality (6.18) so the solution of the challenge problem would be complete except for one small detail — we still need to show that there is a convex g on $[0, \infty)$ such that $g(x) \leq f(x)$ for $x \in [0, 3\log 2]$ and $f(x) = g(x)$ for all $x \geq 3\log 2$.

CONSTRUCTION OF THE CONVEX MINORANT

One way to construct a convex function g with the minorization properties describe above is to just take $g(x) = f(x)$ for $x \geq 3\log 2$ and to define $g(x)$ on $[0, 3\log 2]$ by linear extrapolation. Thus, for $x \in [0, 3\log 2]$, we take

$$\begin{aligned} g(x) &= f(3\log 2) + (x - 3\log 2)f'(3\log 2) \\ &= \frac{1}{3} + (3\log 2 - x)(4/27). \end{aligned}$$

Three simple observations now suffice to show that $g(x) \leq f(x)$ for all $x \geq 0$. First, for $x \geq 3\log 2$, we have $g(x) = f(x)$ by definition. Second, for $\log 2 \leq x \leq 3\log 2$ we have $g(x) \leq f(x)$ because in this range $g(x)$ has the value of a tangent line to $f(x)$ and by convexity of f on $\log 2 \leq x \leq 3\log 2$ the tangent line is below f. Third, in the critical region $0 \leq x \leq \log 2$ we have $g(x) \leq f(x)$ because (i) f is concave, (ii)

Convexity — The Third Pillar

g is linear, and (iii) f is larger than g at the end points of the interval $[0, \log 2]$. More precisely, at the first end point one has

$$g(0) = 0.641 \cdots \leq f(0) = \frac{1}{\sqrt{2}} = 0.707\ldots,$$

while at the second end point one has

$$g(\log 2) = 0.538 \cdots \leq f(\log 2) = \frac{1}{\sqrt{3}} = 0.577\ldots.$$

Thus, the convex function g is indeed a minorant of f which agrees with f on $[3 \log 2, \infty)$, so the solution to the challenge problem is complete.

JENSEN'S INEQUALITY IN PERSPECTIVE

Jensen's inequality may lack the primordial nature of either Cauchy's inequality or the AM-GM inequality, but, if one were forced to pick a single result on which to build a theory of mathematical inequalities, Jensen's inequality would be an excellent choice. It can be used as a starting point for the proofs of almost all of the results we have seen so far, and, even then, it is far from exhausted.

EXERCISES

Exercise 6.1 (A Renaissance Inequality)

The Renaissance mathematician Pietro Mengoli (1625–1686) only needed simple algebra to prove the pleasing symmetric inequality

$$\frac{1}{x-1} + \frac{1}{x} + \frac{1}{x+1} > \frac{3}{x} \qquad \text{for all } x > 1, \qquad (6.21)$$

yet he achieved a modest claim on intellectual immortality when he used it to give one of the earliest proofs of the divergence of the harmonic series,

$$H_n = 1 + \frac{1}{2} + \frac{1}{3} + \cdots + \frac{1}{n} \quad \Longrightarrow \quad \lim_{n \to \infty} H_n = \infty. \qquad (6.22)$$

Rediscover Mengoli's algebraic proof of the inequality (6.21) and check that it also follows from Jensen's inequality. Further, show, as Mengoli did, that the inequality (6.21) implies the divergence of H_n.

Exercise 6.2 (A Perfect Cube and a Triple Product)

Show that if $x, y, z > 0$ and $x + y + z = 1$ then one has

$$64 \leq \left(1 + \frac{1}{x}\right)\left(1 + \frac{1}{y}\right)\left(1 + \frac{1}{z}\right).$$

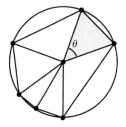

An inscribed polygon can be decomposed into triangles like the shaded one which has area $\frac{1}{2}\sin\theta$.

Fig. 6.4. If a convex polygon with n sides is inscribed the unit circle, our visual imagination suggests that the area is maximized only by a regular polygon. This conjecture can be proved by methods which would have been familiar to Euclid, but a modern proof by convexity is easier.

Exercise 6.3 (Area Inequality for n-gons)

Figure 6.4 suggests that among all convex n-sided convex polygons that one can inscribed in a circle, only the regular n-gon has maximal area. Can Jensen's inequality be used to confirm this suggestion?

Exercise 6.4 (Investment Inequalities)

If $0 < r_k < \infty$, and if our investment of one dollar in year k grows to $1 + r_k$ dollars at the end of the year, we call r_k the return on our investment in year k. Show that the value $V = (1+r_1)(1+r_2)\cdots(1+r_n)$ of our investments after n years must satisfy the bounds

$$(1 + r_G)^n \leq \prod_{k=1}^{n}(1 + r_k) \leq (1 + r_A)^n, \qquad (6.23)$$

where $r_G = (r_1 r_2 \cdots r_n)^{1/n}$ and $r_A = (r_1 + r_2 + \cdots + r_n)/n$. Also explain why this bound might be viewed as a *refinement* of the AM-GM inequality.

Exercise 6.5 (Superadditivity of the Geometric Means)

We have seen before in Exercise 2.11 that for nonnegative a_j and b_j, $j = 1, 2, \ldots, n$ one has superadditivity of the geometric mean:

$$(a_1 a_2 \cdots a_n)^{1/n} + (b_1 b_2 \cdots b_n)^{1/n} \leq \{(a_1 + b_1)(a_2 + b_2) \cdots (a_n + b_n)\}^{1/n}.$$

Does this also follow from Jensen's inequality?

Convexity — The Third Pillar 101

Exercise 6.6 (Cauchy's Technique and Jensen's Inequality)

In 1906, J.L.W.V. Jensen wrote an article that was inspired by the proof given by Cauchy's for the AM-GM inequality, and, in an effort to get to the heart of Cauchy's argument, Jensen introduced the class of functions that satisfy the inequality

$$f\left(\frac{x+y}{2}\right) \leq \frac{f(x)+f(y)}{2} \quad \text{for all } x,y \in [a,b]. \tag{6.24}$$

Such functions are now called J-convex functions, and, as we note below in Exercise 6.7, they are just slightly more general than the convex functions defined by condition (6.1).

For a moment, step into Jensen's shoes and show how one can modify Cauchy's leap-forward fall-back induction (page 20) to prove that for all J-convex functions one has

$$f\left(\frac{1}{n}\sum_{k=1}^{n} x_k\right) \leq \frac{1}{n}\sum_{k=1}^{n} f(x_k) \quad \text{for all } \{x_k : 1 \leq k \leq n\} \subset [a,b]. \tag{6.25}$$

Here one might note that near the end of his 1906 article, Jensen expressed the bold view that perhaps someday the class of convex function might seen to be as fundamental as the class of positive functions or the class of increasing functions. If one allows for the mild shift from the specific notion of J-convexity to the more modern interpretation of convexity (6.1), then Jensen's view turned out to be quite prescient.

Exercise 6.7 (Convexity and J-Convexity)

Show that if $f : [a,b] \to \mathbb{R}$ is continuous and J-convex, then f must be convex in the modern sense expressed by the condition (6.1). As a curiosity, we should note that there do exist J-convex functions that are not convex in the modern sense. Nevertheless, such functions are wildly discontinuous, and they are quite unlikely to turn up unless they are explicitly invited.

Exercise 6.8 (A "One-liner" That Could Have Taken All Day)

Show that for all $0 \leq x,y,z \leq 1$, one has the bound

$$L(x,y,z) = \frac{x^2}{1+y} + \frac{y^2}{1+z} + \frac{z^2}{1+x+y} + x^2(y^2-1)(z^2-1) \leq 2.$$

Placed suggestively in a chapter on convexity, this problem is not much more than a one-liner, but in a less informative location, it might send one down a long trail of fruitless algebra.

Exercise 6.9 (Hadwiger–Finsler Inequality)

For any triangle with the traditional labelling of Figure 6.2, the law of cosines tells us that $a^2 = b^2 + c^2 - 2bc\cos\alpha$. Show that this law implies the area formula

$$a^2 = (b-c)^2 + 4A\tan(\alpha/2),$$

then show how Jensen's inequality implies that in any triangle one has

$$a^2 + b^2 + c^2 \geq (a-b)^2 + (b-c)^2 + (c-a)^2 + 4\sqrt{3}A.$$

This bound is known as the Hadwiger–Finsler inequality, and it provides one of the nicest refinements of Weitzenböck's inequality.

Exercise 6.10 (The f'' Criterion and Rolle's Theorem)

We saw earlier (page 90) that the fundamental theorem of calculus implies that if one has $f''(x) \geq 0$ for all $x \in [a, b]$, then f is convex on $[a, b]$. This exercise sketches how one can also prove this important fact by estimating the difference $f(px_1+qx_2)-pf(x_1)-qf(x_2)$ by comparison with an appropriate polynomial.

(a) Take $0 < p < 1$, $q = 1 - p$ and set $\mu = px_1 + qx_2$ where $x_1 < x_2$. Find the unique quadratic polynomial $Q(x)$ such that

$$Q(x_1) = f(x_1), \quad Q(x_2) = f(x_2), \quad \text{and} \quad Q(\mu) = f(\mu).$$

(b) Use the fact that $\Delta(x) = f(x) - Q(x)$ has three distinct zeros in $[a, b]$ to show that there is an x^* such that $\Delta''(x^*) = 0$.

(c) Finally, explain how $f''(x) \geq 0$ for all $x \in [a, b]$ and $\Delta''(x^*) = 0$ imply that $f(px_1 + qx_2) - pf(x_1) - qf(x_2) \geq 0$.

Exercise 6.11 (Transformation to Achieve Convexity)

Show that for positive a, b, and c such that $a + b + c = abc$ one has

$$\frac{1}{\sqrt{1+a^2}} + \frac{1}{\sqrt{1+b^2}} + \frac{1}{\sqrt{1+c^2}} \leq \frac{3}{2}.$$

This problem from the 1998 Korean National Olympiad is not easy, even with the hint provided by the exercise's title. Someone who is lucky may draw a link between the hypothesis $a + b + c = abc$ and the reasonably well-known fact that in a triangle labeled as in Figure 6.2 one has

$$\tan(\alpha) + \tan(\beta) + \tan(\gamma) = \tan(\alpha)\tan(\beta)\tan(\gamma).$$

This identity is easily checked by applying the addition formula for the tangent to the sum $\gamma = \pi - (\alpha + \beta)$, but it is surely easier to remember than to discover on the spot.

A point z outside of a closed bounded set H determines a natural "viewing angle" 2ψ.

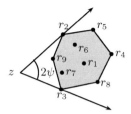

Fig. 6.5. The viewing angle 2ψ of the convex hull of the set of roots $r_1, r_2, \ldots, r_n$ of $P(z)$ determines the parameter ψ that one finds in Wilf's quantitative refinement of the Gauss–Lucas Theorem.

Exercise 6.12 (The Gauss–Lucas Theorem)

Show that for any complex polynomial $P(z) = a_0 + a_1 z + \cdots + a_n z^n$, the roots of the derivative $P'(z)$ are contained in the convex hull H of the roots of $P(z)$.

Exercise 6.13 (Wilf's Inequality)

Show that if H is the convex hull of the roots of the complex polynomial $P = a_0 + a_1 z + \cdots + a_n z^n$, then one has

$$\left|\frac{a_n}{P(z)}\right|^{1/n} \leq \frac{1}{n \cos \psi} \left|\frac{P'(z)}{P(z)}\right| \qquad \text{for all } z \notin H, \tag{6.26}$$

where the angle ψ is defined by Figure 6.5. This inequality provides both a new proof and a quantitative refinement of the classic Gauss–Lucas Theorem of Exercise 6.12.

Exercise 6.14 (A Polynomial Lower Bound)

Given that the zeros of the polynomial $P(z) = a_n z^n + \cdots + a_1 z + a_0$ are contained in the unit disc $U = \{z : |z| \leq 1\}$, show that one has

$$n|a_n|^{1/n} |P(z)|^{(n-1)/n} \sqrt{1 - |z|^{-2}} \leq |P'(z)| \qquad \text{for all } z \notin U. \tag{6.27}$$

Exercise 6.15 (A Complex Mean Product Theorem)

Show that if $0 < r < 1$ and if the complex numbers $z_1, z_2, \ldots, z_n$ are in the disk $D = \{z : |z| \leq r\}$, then there exists a $z_0 \in D$ such that

$$\prod_{j=1}^{n} (1 + z_j) = (1 + z_0)^n. \tag{6.28}$$

Exercise 6.16 (Shapiro's Cyclic Sum Inequality)

Show that for positive a_1, a_2, a_3, and a_4, one has the bound

$$2 \leq \frac{a_1}{a_2 + a_3} + \frac{a_2}{a_3 + a_4} + \frac{a_3}{a_4 + a_1} + \frac{a_4}{a_1 + a_2}. \tag{6.29}$$

Incidentally, the review of Bushell (1994) provides a great deal of information about the inequalities of the form

$$n/2 \leq \frac{x_1}{x_2 + x_3} + \frac{x_2}{x_3 + x_4} + \cdots + \frac{x_{n-1}}{x_n + x_1} + \frac{x_n}{x_1 + x_2}.$$

This bound is known to fail for $n \geq 25$, yet the precise set of n for which it is valid has not yet been determined.

Exercise 6.17 (The Three Chord Lemma)

Show that if $f : [a, b] \to \mathbb{R}$ is convex and $a < x < b$, then one has

$$\frac{f(x) - f(a)}{x - a} \leq \frac{f(b) - f(a)}{b - a} \leq \frac{f(b) - f(x)}{b - x}. \tag{6.30}$$

As the next two exercises suggest, this bound is the key to some of the most basic regularity properties of convex functions.

Exercise 6.18 (Near Differentiability of Convex Functions)

Use the Three Chord Lemma to show that for convex $f : [a, b] \to \mathbb{R}$ and $a < x < b$ one has the existence of the finite limits

$$f'_+(x) \stackrel{\text{def}}{=} \lim_{h \downarrow 0} \frac{f(x + h) - f(x)}{h} \quad \text{and} \quad f'_-(x) \stackrel{\text{def}}{=} \lim_{h \downarrow 0} \frac{f(x - h) - f(x)}{h}.$$

Exercise 6.19 (Ratio Bounds and Linear Minorants)

For convex $f : [a, b] \to \mathbb{R}$ and $a < x < y < b$, show that one has

$$f'_-(x) \leq f'_+(x) \leq \frac{f(y) - f(x)}{y - x} \leq f'_-(y) \leq f'_+(y). \tag{6.31}$$

In particular, note that for each $\theta \in [f'_-(x), f'_+(x)]$ one has the bound

$$f(y) \geq f(x) + (y - x)\theta \qquad \text{for all } y \in [a, b]. \tag{6.32}$$

The linear lower bound (6.32) is more effective that its simplicity would suggest, and it has some notable consequences. In the next chapter we will find that it yields and exceptionally efficient proof of Jensen's inequality.

7
Integral Intermezzo

The most fundamental inequalities are those for finite sums, but there can be no doubt that inequalities for integrals also deserve a fair share of our attention. Integrals are pervasive throughout science and engineering, and they also have some mathematical advantages over sums. For example, integrals can be cut up into as many pieces as we like, and integration by parts is almost always more graceful than summation by parts. Moreover, any integral may be reshaped into countless alternative forms by applying the change-of-variables formula.

Each of these themes contributes to the theory of integral inequalities. These themes are also well illustrated by our favorite device — concrete challenge problems which have a personality of their own.

Problem 7.1 (A Continuum of Compromise)
Show that for an integrable $f : \mathbb{R} \to \mathbb{R}$, one has the bound

$$\int_{-\infty}^{\infty} |f(x)|\,dx \leq 8^{\frac{1}{2}} \left(\int_{-\infty}^{\infty} |xf(x)|^2\,dx \right)^{\frac{1}{4}} \left(\int_{-\infty}^{\infty} |f(x)|^2\,dx \right)^{\frac{1}{4}}. \quad (7.1)$$

A QUICK ORIENTATION AND A QUALITATIVE PLAN

The one-fourth powers on the right side may seem strange, but they are made more reasonable if one notes that each side of the inequality is homogenous of order one in f; that is, if f is replaced by λf where λ is a positive constant, then each side is multiplied by λ. This observation makes the inequality somewhat less strange, but one may still be stuck for a good idea.

We faced such a predicament earlier where we found that one often does well to first consider a simpler *qualitative* challenge. Here the natural candidate is to try to show that the left side is finite whenever both integrals on the right are finite.

Once we ask this question, we are not likely to need long to think of looking for separate bounds for the integral of $|f(x)|$ on the interval $T = (-t, t)$ and its complement T^c. If we also ask ourselves how we might introduce the term $|xf(x)|$, then we are almost forced to think of using the splitting trick on the set T^c. Pursuing this thought, we then find for all $t > 0$ that we have the bound

$$\int_{-\infty}^{\infty} |f(x)|\, dx = \int_T |f(x)|\, dx + \int_{T^c} \frac{1}{|x|} |xf(x)|\, dx$$

$$\leq (2t)^{\frac{1}{2}} \left(\int_T |f(x)|^2\, dx \right)^{\frac{1}{2}} + \left(\frac{2}{t} \right)^{\frac{1}{2}} \left(\int_{T^c} |xf(x)|^2\, dx \right)^{\frac{1}{2}}, \quad (7.2)$$

where in the second line we just applied Schwarz's inequality twice.

This bound is not the one we hoped to prove, but it makes the same qualitative case. Specifically, it confirms that the integral of $|f(x)|$ is finite when the bounding terms of the inequality (7.1) are finite. We now need to pass from our additive bound to one that is multiplicative, and we also need to exploit our free parameter t.

We have no specific knowledge about the integrals over T and T^c, so there is almost no alternative to using the crude bound

$$\int_T |f(x)|^2\, dx \leq \int_{\mathbb{R}} |f(x)|^2\, dx \stackrel{\text{def}}{=} A$$

and its cousin

$$\int_{T^c} |xf(x)|^2\, dx \leq \int_{\mathbb{R}} |xf(x)|^2\, dx \stackrel{\text{def}}{=} B.$$

The sum (7.2) is therefore bounded above by $\phi(t) \stackrel{\text{def}}{=} 2^{\frac{1}{2}} t^{\frac{1}{2}} A^{\frac{1}{2}} + 2^{\frac{1}{2}} t^{-\frac{1}{2}} B^{\frac{1}{2}}$, and we can use calculus to minimize $\phi(t)$. Since $\phi(t) \to \infty$ as $t \to 0$ or $t \to \infty$ and since $\phi'(t) = 0$ has the unique root $t_0 = B^{\frac{1}{2}}/A^{\frac{1}{2}}$, we find $\min_{t:t>0} \phi(t) = \phi(t_0) = 8^{\frac{1}{2}} A^{\frac{1}{4}} B^{\frac{1}{4}}$, and this gives us precisely the bound proposed by the challenge problem.

Dissections and Benefits of the Continuum

The inequality (7.1) came to us with only a faint hint that one might do well to cut the target integral into the piece over $T = (-t, t)$ and the piece over T^c, yet once this dissection was performed, the solution came to us quickly. The impact of dissection is usually less dramatic, but on a qualitative level at least, dissection can be counted upon as one of the most effective devices we have for estimation of integrals.

Integral Intermezzo 107

Here our use of a flexible, parameter-driven, dissection also helped us to take advantage the intrinsic richness of the continuum. Without a pause, we were led to the problem of minimizing $\phi(t)$, and this turned out to be a simple calculus exercise. It is far less common for a discrete problem to crack so easily; even if one finds the analogs of t and $\phi(t)$, the odds are high that the resulting discrete minimization problem will be a messy one.

BEATING SCHWARZ BY TAKING A DETOUR

Many problems of mathematical analysis call for a bound that beats the one which we get from an immediate application of Schwarz's inequality. Such a refinement may require a subtle investigation, but sometimes the critical improvement only calls for one to exercise some creative self-restraint. A useful motto to keep in mind is "Transform-Schwarz-Invert," but to say any more might give away the solution to the next challenge problem.

Problem 7.2 (Doing Better Than Schwarz)
Show that if $f : [0, \infty) \to [0, \infty)$ is a continuous, nonincreasing function which is differentiable on $(0, \infty)$, then for any pair of parameters $0 < \alpha, \beta < \infty$, the integral

$$I = \int_0^\infty x^{\alpha+\beta} f(x) \, dx \qquad (7.3)$$

satisfies the bound

$$I^2 \leq \left\{ 1 - \left(\frac{\alpha - \beta}{\alpha + \beta + 1} \right)^2 \right\} \int_0^\infty x^{2\alpha} f(x) \, dx \int_0^\infty x^{2\beta} f(x) \, dx. \qquad (7.4)$$

What makes this inequality instructive is that the direct application of Schwarz's inequality to the splitting

$$x^{\alpha+\beta} f(x) = x^\alpha \sqrt{f(x)} \; x^\beta \sqrt{f(x)}$$

would give one a weaker inequality where the first factor on the right-hand side of the bound (7.4) would be replaced by 1. The essence of the challenge is therefore to beat the naive immediate application of Schwarz's inequality.

TAKING THE HINT

If we want to apply the pattern of "Transform-Schwarz-Invert," we need to think of ways we might transform the integral (7.3), and, from the specified hypotheses, the natural transformation is simply integration by parts. To explore the feasibility of this idea we first note that by the continuity of f we have $x^{\gamma+1}f(x) \to 0$ as $x \to 0$, so integration by parts provides the nice formula

$$\int_0^\infty x^\gamma f(x)\, dx = \frac{1}{1+\gamma} \int_0^\infty x^{\gamma+1} |f'(x)|\, dx, \tag{7.5}$$

provided that we also have

$$x^{\gamma+1} f(x) \to 0 \qquad \text{as } x \to \infty. \tag{7.6}$$

Before we worry about checking this limit (7.6), we should first see if the formula (7.5) actually helps.

If we first apply the formula (7.5) to the integral I of the challenge problem, we have $\gamma = \alpha + \beta$ and

$$(\alpha + \beta + 1)I = \int_0^\infty x^{\alpha+\beta+1} |f'(x)|\, dx.$$

Thus, if we then apply Schwarz's inequality to the splitting

$$x^{\alpha+\beta+1}|f'(x)| = \{x^{(2\alpha+1)/2}|f'(x)|^{1/2}\}\{x^{(2\beta+1)/2}|f'(x)|^{1/2}\}$$

we find the nice intermediate bound

$$(1+\alpha+\beta)^2 I^2 \leq \int_0^\infty x^{2\alpha+1}|f'(x)|\, dx \int_0^\infty x^{2\beta+1}|f'(x)|\, dx.$$

Now we see how we can *invert*; we just apply integration by parts (7.5) to each of the last two integrals to obtain

$$I^2 \leq \frac{(2\alpha+1)(2\beta+1)}{(\alpha+\beta+1)^2} \int_0^\infty x^{2\alpha} f(x)\, dx \int_0^\infty x^{2\beta} f(x)\, dx.$$

Here, at last, we find after just a little algebraic manipulation of the first factor that we do indeed have the inequality of the challenge problem.

Our solution is therefore complete except for one small point; we still need to check that our three applications of the integration by parts formula (7.5) were justified. For this it suffices to show that we have the limit (7.6) when γ equals 2α, 2β, or $\alpha + \beta$, and it clearly suffices to check the limit for the largest of these, which we can take to be 2α. Moreover, we can assume that in addition to the hypotheses of the challenge problem that we also have the condition

$$\int_0^\infty x^{2\alpha} f(x)\, dx < \infty, \tag{7.7}$$

since otherwise our target inequality (7.4) is trivial.

Integral Intermezzo 109

A POINTWISE INFERENCE

These considerations present an amusing intermediate problem; we need to prove a pointwise condition (7.6) with an integral hypothesis (7.7). It is useful to note that such an inference would be impossible here without the additional information that f is monotone decreasing.

We need to bring the value of f at a fixed point into clear view, and here it is surely useful to note that for any $0 \leq t < \infty$ we have

$$\int_0^t x^{2\alpha} f(x)\,dx = \frac{f(t)t^{2\alpha+1}}{2\alpha+1} - \frac{1}{2\alpha+1}\int_0^t x^{2\alpha+1} f'(x)\,dx$$

$$= \frac{f(t)t^{2\alpha+1}}{2\alpha+1} + \frac{1}{2\alpha+1}\int_0^t x^{2\alpha+1} |f'(x)|\,dx \qquad (7.8)$$

$$\geq \frac{1}{2\alpha+1}\int_0^t x^{2\alpha+1} |f'(x)|\,dx.$$

By the hypothesis (7.7) the first integral has a finite limit as $t \to \infty$, so the last integral also has a finite limit as $t \to \infty$. From the identity (7.8) we see that $f(t)t^{2\alpha+1}/(2\alpha+1)$ is the difference of these integrals, so we find that there exists a constant $0 \leq c < \infty$ such that

$$\lim_{t\to\infty} t^{2\alpha+1} f(t) = c. \qquad (7.9)$$

Now, if $c > 0$, then there is a T such that $t^{2\alpha+1} f(t) \geq c/2$ for $t \geq T$, and in this case one would have

$$\int_0^\infty x^{2\alpha} f(x)\,dx \geq \int_T^\infty \frac{c}{2x}\,dx = \infty. \qquad (7.10)$$

Since this bound contradicts our assumption (7.7), we find that $c = 0$, and this fact confirms that our three applications of the integration by parts formula (7.5) were justified.

ANOTHER POINTWISE CHALLENGE

In the course of the preceding challenge problem, we noted that the monotonicity assumption on f was essential, yet one can easily miss the point in the proof where that hypothesis was applied. It came in quietly on the line (7.8) where the integration by parts formula was restructured to express $f(t)t^{2\alpha+1}$ as the difference of two integrals with finite limits.

One of the recurring challenges of mathematical analysis is the extraction of local, pointwise information about a function from aggregate information which is typically expressed with the help of integrals. If one does not know something about the way or the rate at which the function changes, the task is usually impossible. In some cases one can succeed with just aggregate information about the rate of change. The next challenge problem provides an instructive example.

Problem 7.3 (A Pointwise Bound)

Show that if $f : [0, \infty) \to \mathbb{R}$ satisfies the two integral bounds

$$\int_0^\infty x^2 |f(x)|^2 \, dx < \infty \quad \text{and} \quad \int_0^\infty |f'(x)|^2 \, dx < \infty,$$

then for all $x > 0$ one has the inequality

$$|f(x)|^2 \leq \frac{4}{x} \left\{ \int_x^\infty t^2 |f(t)|^2 \, dt \right\}^{1/2} \left\{ \int_x^\infty |f'(t)|^2 \, dt \right\}^{1/2} \tag{7.11}$$

and, consequently, $\sqrt{x} |f(x)| \to \infty$ as $x \to \infty$.

ORIENTATION AND A PLAN

In this problem, as in many others, we must find a way to get started even though we do not have a clear idea how we might eventually reach our goal. Our only guide here is that we know we must relate f' to f, and thus we may suspect that the fundamental theorem of calculus will somehow help.

This is *The Cauchy-Schwarz Master Class*, so here one may not need long to think of applying the 1-trick and Schwarz's inequality to get the bound

$$|f(x+t) - f(x)| = \left| \int_x^{x+t} f'(u) \, du \right| \leq t^{1/2} \left\{ \int_x^{x+t} |f'(u)|^2 \, du \right\}^{1/2}.$$

In fact, this estimate gives us both an upper bound

$$|f(x+t)| \leq |f(x)| + t^{1/2} \left\{ \int_x^\infty |f'(u)|^2 \, du \right\}^{1/2} \tag{7.12}$$

and a lower bound

$$|f(x+t)| \geq |f(x)| - t^{1/2} \left\{ \int_x^\infty |f'(u)|^2 \, du \right\}^{1/2}, \tag{7.13}$$

and each of these offers a sense of progress. After all, we needed to find roles for both of the integrals

$$F^2(x) \stackrel{\text{def}}{=} \int_x^\infty u^2 |f(u)|^2 \, du \quad \text{and} \quad D^2(x) \stackrel{\text{def}}{=} \int_x^\infty |f'(u)|^2 \, du,$$

and now we at least see how $D(x)$ can play a part.

When we look for a way to relate $F(x)$ and $D(x)$, it is reasonable to

Integral Intermezzo 111

think of using $D(x)$ and our bounds (7.12) and (7.13) to build upper and lower estimates for $F(x)$. To be sure, it is not clear that such estimates will help us with our challenge problem, but there is also not much else we can do.

After some exploration, one does discover that it is the trickier lower estimate which brings home the prize. To see how this goes, we first note that for any value of $0 \leq h$ such that $h^{\frac{1}{2}} \leq f(x)/D(x)$ one has

$$F^2(x) \geq \int_0^h u^2 |f(u)|^2 \, du = \int_0^h (x+t)^2 |f(x+t)|^2 \, dt$$
$$\geq \int_0^h (x+t)^2 |f(x) - t^{\frac{1}{2}} D(x)|^2 \, dt$$
$$\geq hx^2 \{f(x) - h^{\frac{1}{2}} D(x)\}^2,$$

or, a bit more simply, we have

$$F(x) \geq h^{\frac{1}{2}} x \{f(x) - h^{\frac{1}{2}} D(x)\}.$$

To maximize this lower bound we take $h^{\frac{1}{2}} = f(x)/\{2D(x)\}$, and we find

$$F(x) \geq \frac{xf^2(x)}{4D(x)} \quad \text{or} \quad xf^2(x) \leq 4F(x)D(x),$$

just as we were challenged to show.

PERSPECTIVE ON LOCALIZATION

The two preceding problems required us to extract pointwise estimates from integral estimates, and this is often a subtle task. More commonly one faces the simpler challenge of converting an estimate for one type of integral into an estimate for another type of integral. We usually do not have derivatives at our disposal, yet we may still be able to exploit local estimates for global purposes.

Problem 7.4 (A Divergent Integral)
Given $f : [1, \infty) \to (0, \infty)$ *and a constant* $c > 0$, *show that if*

$$\int_1^t f(x) \, dx \leq ct^2 \quad \text{for all } 1 \leq t < \infty \quad \text{then} \quad \int_1^\infty \frac{1}{f(x)} \, dx = \infty.$$

AN IDEA THAT DOES NOT QUITE WORK

Given our experiences with sums of reciprocals (e.g., Exercise 1.2, page 12), it is natural to think of applying Schwarz's inequality to the splitting $1 = \sqrt{f(x)} \cdot \{1/\sqrt{f(x)}\}$. This suggestion leads us to

$$(t-1)^2 = \left(\int_1^t 1\,dx\right)^2 \leq \int_1^t f(x)\,dx \int_1^t \frac{1}{f(x)}\,dx, \qquad (7.14)$$

so, by our hypothesis we find

$$c^{-1} t^{-2} (t-1)^2 \leq \int_1^t \frac{1}{f(x)}\,dx,$$

and when we let $t \to \infty$ we find the bound

$$c^{-1} \leq \int_1^\infty \frac{1}{f(x)}\,dx. \qquad (7.15)$$

Since we were challenged to show that the last integral is infinite, we have fallen short of our goal. Once more we need to find some way to *sharpen Schwarz*.

FOCUSING WHERE ONE DOES WELL

When Schwarz's inequality disappoints us, we often do well to ask how our situation differs from the case when Schwarz's inequality is at its best. Here we applied Schwarz's inequality to the product of $\phi(x) = f(x)$ and $\psi(x) = 1/f(x)$, and we know that Schwarz's inequality is sharp if and only if $\phi(x)$ and $\psi(x)$ are proportional. Since $f(x)$ and $1/f(x)$ are far from proportional on the infinite interval $[0, \infty)$, we get a mild hint: perhaps we can do better if we restrict our application of Schwarz's inequality to the corresponding integrals over appropriately chosen finite intervals $[A, B]$.

When we repeat our earlier calculation for a generic interval $[A, B]$ with $1 \leq A < B$, we find

$$(B - A)^2 \leq \int_A^B f(x)\,dx \int_A^B \frac{1}{f(x)}\,dx, \qquad (7.16)$$

and, now, we cannot do much better in our estimate of the first integral than to exploit our hypothesis via the crude bound

$$\int_A^B f(x)\,dx < \int_1^B f(x)\,dx \leq cB^2,$$

after which inequality (7.16) gives us

$$\frac{(B-A)^2}{cB^2} \leq \int_A^B \frac{1}{f(x)}\,dx. \qquad (7.17)$$

The issue now is to see if perhaps the flexibility of the parameters A and B can be of help.

Integral Intermezzo 113

This turns out to be a fruitful idea. If we take $A = 2^j$ and $B = 2^{j+1}$, then for all $0 \le j < \infty$ we have
$$\frac{1}{4c} \le \int_{2^j}^{2^{j+1}} \frac{1}{f(x)}\,dx,$$
and if we sum these estimates over $0 \le j < k$ we find
$$\frac{k}{4c} \le \int_1^{2^k} \frac{1}{f(x)}\,dx \le \int_1^\infty \frac{1}{f(x)}\,dx. \qquad (7.18)$$
Since k is arbitrary, the last inequality does indeed complete the solution to our fourth challenge problem.

A FINAL PROBLEM: JENSEN'S INEQUALITY FOR INTEGRALS

The last challenge problem could be put simply: "Prove an integral version of Jensen's inequality." Naturally, we can also take this opportunity to add something extra to the pot.

Problem 7.5 (Jensen's Inequality: An Integral Version)
Show that for each interval $I \subset \mathbb{R}$ and each convex $\Phi : I \to \mathbb{R}$, one has the bound
$$\Phi\left(\int_D h(x)w(x)\,dx\right) \le \int_D \Phi\bigl(h(x)\bigr)\,w(x)\,dx, \qquad (7.19)$$
for each $h : D \to I$ and each weight function $w : D \to [0, \infty)$ such that
$$\int_D w(x)\,dx = 1.$$

THE OPPORTUNITY TO TAKE A GEOMETRIC PATH

We could prove the conjectured inequality (7.19) by working our way up from Jensen's inequality for finite sums, but it is probably more instructive to take a hint from Figure 7.1. If we compare the figure to our target inequality and if we ask ourselves about reasonable choices for μ, one candidate which is sure to make our list is
$$\mu = \int_D h(x)w(x)\,dx;$$
after all, $\Phi(\mu)$ is already present in the inequality (7.19).

Noting that the parameter t is still at our disposal, we now see that $\Phi(h(x))$ may be brought into action if we set $t = h(x)$. If θ denotes the slope of the support line pictured in Figure 7.1, then we have the bound
$$\Phi(\mu) + (h(x) - \mu)\theta \le \Phi(h(x)) \qquad \text{for all } x \in D. \qquad (7.20)$$

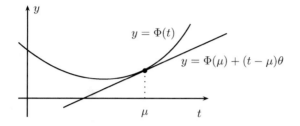

The linear lower bound is often more powerful than one might guess.

Fig. 7.1. For each point $p = (\mu, \Phi(\mu))$ on the graph of a convex function Φ, there is a line through p which never goes above the graph of Φ. If Φ is differentiable, the slope θ of this line is $\Phi'(\mu)$, and if Φ is not differentiable, then according to Exercise 6.19 one can take θ to be any point in the interval $[\Phi'_-(\mu), \Phi'_+(\mu)]$ determined by the left and right derivatives.

If we multiply the bound (7.20) by the weight factor $w(x)$ and integrate, then the conjectured bound (7.19) falls straight into our hands because of the relation

$$\int_D (h(x) - \mu) w(x) \theta \, dx = \theta \left\{ \int_D h(x) w(x) \, dx - \mu \right\} = 0.$$

PERSPECTIVES AND COROLLARIES

Many integral inequalities can be proved by a two-step pattern where one proves a pointwise inequality and then one integrates. As the proof of Jensen's inequality suggests, this pattern is particularly effective when the pointwise bound contains a nontrivial term which has integral zero.

There are many corollaries of the continuous version of Jensen's inequality, but probably none of these is more important than the one we obtain by taking $\Phi(x) = e^x$ and by replacing $h(x)$ by $\log h(x)$. In this case, we find the bound

$$\exp\left(\int_D \log\{h(x)\} w(x) \, dx \right) \leq \int_D h(x) w(x) \, dx, \qquad (7.21)$$

which is the natural integral analogue of the arithmetic-geometric mean inequality.

To make the connection explicit, one can set $h(x) = a_k > 0$ on $[k-1, k)$ and set $w(x) = p_k \geq 0$ on $[k-1, k)$ for $1 \leq k \leq n$. One then finds that for $p_1 + p_2 + \cdots + p_n = 1$ the bound (7.21) reduces to exactly to the classic AM-GM bound,

$$\prod_{k=1}^{n} a_k^{p_k} \leq \sum_{k=1}^{n} p_k a_k. \qquad (7.22)$$

Incidentally, the integral analog (7.21) of the AM-GM inequality (7.22) has a long and somewhat muddy history. Apparently, the inequality was first recorded (for $w(x) \equiv 1$) by none other than V. Y. Bunyakovsky. It even appears in the famous *Mémoire* 1859 where Bunyakovsky introduced his integral analog of Cauchy's inequality. Nevertheless, in this case, Bunyakovsky's contribution seems to have been forgotten even by the experts.

EXERCISES

Exercise 7.1 (Integration of a Well-Chosen Pointwise Bound)
Many significant integral inequalities can be proved by integration of an appropriately constructed pointwise bound. For example, the integral version (7.19) of Jensen's inequality was proved this way.

For a more flexible example, show that there is a pointwise integration proof of Schwarz's inequality which flows directly from the symmetrizing substitutions

$$u \mapsto f(x)g(y) \quad \text{and} \quad v \mapsto f(y)g(x)$$

and familiar bound $2uv \leq u^2 + v^2$.

Exercise 7.2 (A Centered Version of Schwarz's Inequality)
If $w(x) \geq 0$ for all $x \in \mathbb{R}$ and if the integral w over $\mathbb{R}$ is equal to 1, then the *weighted average* of a (suitably integrable) function $f : \mathbb{R} \to \mathbb{R}$ is defined by the formula

$$A(f) = \int_{-\infty}^{\infty} f(x)w(x)\,dx.$$

Show that for functions f and g, one has the following bound on the average of their product,

$$\{A(fg) - A(f)A(g)\}^2 \leq \{A(f^2) - A^2(f)\}\{A(g^2) - A^2(g)\},$$

provided that all of the indicated integrals are well defined.

This inequality, like other variations of the Cauchy and Schwarz inequalities, owes its usefulness to its ability to help us convert information on two individual functions to information about their product. Here we see that the average of the product, $A(fg)$, cannot differ too greatly from the product of the averages, $A(f)A(g)$, provided that the *variance* terms, $A(f^2) - A^2(f)$ and $A(g^2) - A^2(g)$, are not too large.

Exercise 7.3 (A Tail and Smoothness Bound)
Show that if $f : \mathbb{R} \to \mathbb{R}$ has a continuous derivative then
$$\int_{-\infty}^{\infty} |f(x)|^2 \, dx \leq 2 \left(\int_{-\infty}^{\infty} x^2 |f(x)|^2 \, dx \right)^{\frac{1}{2}} \left(\int_{-\infty}^{\infty} |f'(x)|^2 \, dx \right)^{\frac{1}{2}}.$$

Exercise 7.4 (Reciprocal on a Square)
Show that for $a \geq 0$ and $b \geq 0$ one has the bound
$$\frac{1}{a+b+1} < \int_a^{a+1} \int_b^{b+1} \frac{dx \, dy}{x+y},$$
which is a modest — but useful — improvement on the naive lower bound $1/(a+b+2)$ which one gets by minimizing the integrand.

Exercise 7.5 (Estimates via Integral Representations)
The complicated formula for the derivative
$$\frac{d^4}{dx^4} \frac{\sin t}{t} = \frac{\sin t}{t} + \frac{2 \cos t}{t^2} - \frac{12 \sin t}{t^3} - \frac{24 \cos t}{t^4} + \frac{25 \sin t}{t^5}$$
may make one doubt the possibility of proving a simple bound such as
$$\left| \frac{d^4}{dx^4} \frac{\sin t}{t} \right| \leq \frac{1}{5} \quad \text{for all } t \in \mathbb{R}. \tag{7.23}$$
Nevertheless, this bound and its generalization for the n-fold derivative are decidedly easy if one thinks of using the integral representation
$$\frac{\sin t}{t} = \int_0^1 \cos(st) \, ds. \tag{7.24}$$
Show how the representation (7.24) may be used to prove the bound (7.23), and give at least one further example of a problem where an analogous integral representation may be used in this way. The moral of this story is that many apparently subtle quantities can be estimated efficiently if they can first be represented as integrals.

Integral Intermezzo 117

Exercise 7.6 (Confirmation by Improvement)

Confirm your mastery of the fourth challenge problem (page 111) by showing that you can get the same conclusion from a weaker hypothesis. For example, show that if there is a constant $0 < c < \infty$ such that the function $f : [1, \infty) \to (0, \infty)$ satisfies the bound

$$\int_1^t f(x)\,dx \leq ct^2 \log t, \qquad (7.25)$$

then one still has divergence of the reciprocal integral

$$\int_1^\infty \frac{1}{f(x)}\,dx = \infty.$$

Exercise 7.7 (Triangle Lower Bound)

Suppose the function $f : [0, \infty) \to [0, \infty)$ is convex on $[T, \infty)$ and show that for all $t \geq T$ one has

$$\frac{1}{2} f^2(t)/|f'(t)| \leq \int_t^\infty f(u)\,du. \qquad (7.26)$$

This is called the *triangle lower bound*, and it is often applied in probability theory. For example, if we take $f(u) = e^{-u^2/2}/\sqrt{2\pi}$ then it gives the lower bound

$$\frac{e^{-t^2/2}}{2t\sqrt{2\pi}} \leq \frac{1}{\sqrt{2\pi}} \int_t^\infty e^{-u^2/2}\,du \qquad \text{for } t \geq 1,$$

although one can do a little better in this specific case.

Exercise 7.8 (The Slip-in Trick: Two Examples)

(a) Show that for all $n = 1, 2, \ldots$ one has the lower bound

$$I_n = \int_0^{\pi/2} (1 + \cos t)^n\,dt \geq \frac{2^{n+1} - 1}{n+1}.$$

(b) Show that for all $x > 0$ one has the upper bound

$$I'_n = \int_x^\infty e^{-u^2/2}\,du \leq \frac{1}{x} e^{-x^2/2}.$$

No one should pass up this problem. The "slip-in trick" is one of the most versatile tools we have for the estimation of integrals and sums; to be unfamiliar with it would be to suffer an unnecessary handicap.

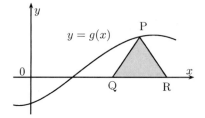

A favorite example of J.E. Littlewood which illustrates the legitimacy of pictorial arguments

Fig. 7.2. Consider a function $g(x)$ for which $|g'(x)| \leq B$, so g cannot change too rapidly. If $g(x_0) = P > 0$ for some x_0, then there is a certain triangle which must lie under the graph of g. This observation reveals an important relation between g, g', and the integral of g.

Exercise 7.9 (Littlewood's Middle Derivative Squeeze)

Show that if $f : [0, \infty) \to \mathbb{R}$ is twice differentiable and if $|f''(x)|$ is bounded, then

$$\lim_{x \to \infty} f(x) = 0 \quad \text{implies} \quad \lim_{x \to \infty} f'(x) = 0.$$

In his *Miscellany*, J.E. Littlewood suggests that "pictorial arguments, while not so purely conventional, can be quite legitimate." The result of this exercise is his leading example, and the picture he offered is essentially that of Figure 7.2.

Exercise 7.10 (Monotonicity and Integral Estimates)

Although the point was not stressed in this chapter, many of the most useful day-to-day estimates of integrals are found with help from monotonicity. Gain some practical experience by proving that

$$\int_x^1 \log(1+t) \frac{dt}{t} < (2 \log 2) \frac{1-x}{1+x} \quad \text{for all } 0 < x < 1$$

and by showing that $2 \log 2$ cannot be replaced by a smaller constant. Incidentally, this particular inequality is one we will see again when it helps us with Exercise 11.6.

Exercise 7.11 (A Continuous Carleman-Type Inequality)

Given an integrable $f : [a, b] \to [0, \infty)$ and an integrable weight function $w : [a, b] \to [0, \infty)$ with integral 1 on $[a, b]$, show that one has

$$\exp \int_a^b \{\log f(x)\} w(x) \, dx \leq e \int_a^b f(x) w(x) \, dx. \qquad (7.27)$$

Exercise 7.12 (Grüss's Inequality — Integrals of Products)
Suppose that $-\infty < \alpha \leq A < \infty$ and $-\infty < \beta \leq B < \infty$ and suppose that functions f and g satisfy the bounds

$$\alpha \leq f(x) \leq A \quad \text{and} \quad \beta \leq g(x) \leq B \qquad \text{for all } 0 \leq x \leq 1.$$

Show that one has the bound

$$\left| \int_0^1 f(x)g(x)\,dx - \int_0^1 f(x)\,dx \int_0^1 g(x)\,dx \right| \leq \frac{1}{4}(A-\alpha)(B-\beta),$$

and show by example that the factor of $1/4$ cannot be replaced by a smaller constant.

8
The Ladder of Power Means

The quantities that provide the upper bound in Cauchy's inequality are special cases of the general means

$$M_t = M_t[\mathbf{x}; \mathbf{p}] \equiv \left\{ \sum_{k=1}^{n} p_k x_k^t \right\}^{1/t} \quad (8.1)$$

where $\mathbf{p} = (p_1, p_2, \ldots, p_n)$ is a vector of positive weights with total mass of $p_1+p_2+\cdots+p_n = 1$ and $\mathbf{x} = (x_1, x_2, \ldots, x_n)$ is a vector of nonnegative real numbers. Here the parameter t can be taken to be any real value, and one can even take $t = -\infty$ or $t = \infty$, although in these cases and the case $t = 0$ the general formula (8.1) requires some reinterpretation. The proper definition of the power mean M_0 is motivated by the natural desire to make the map $t \mapsto M_t$ a continuous function on all of $\mathbb{R}$. The first challenge problem suggests how this can be achieved, and it also adds a new layer of intuition to our understanding of the geometric mean.

Problem 8.1 (The Geometric Mean as a Limit)
For nonnegative real numbers x_k, $k = 1, 2, \ldots, n$, and nonnegative weights p_k, $k = 1, 2, \ldots, n$ with total mass $p_1 + p_2 + \cdots + p_n = 1$, one has the limit

$$\lim_{t \to 0} \left\{ \sum_{k=1}^{n} p_k x_k^t \right\}^{1/t} = \prod_{k=1}^{n} x_k^{p_k}. \quad (8.2)$$

APPROXIMATE EQUALITIES AND LANDAU'S NOTATION

The solution of this challenge problem is explained most simply with the help of Landau's little o and big O notation. In this useful shorthand, the statement $\lim_{t \to 0} f(t)/g(t) = 0$ is abbreviated simply by writing

The Ladder of Power Means

$f(t) = o(g(t))$ as $t \to 0$, and, analogously, the statement that the ratio $f(t)/g(t)$ is bounded in some neighborhood of 0 is abbreviated by writing $f(t) = O(g(t))$ as $t \to 0$. By hiding details that are irrelevant, this notation often allows one to render a mathematical inequality in a form that gets most quickly to its essential message.

For example, it is easy to check that for all $x > -1$ one has a natural two-sided estimate for $\log(1+x)$,

$$\frac{x}{1+x} \leq \int_1^{1+x} \frac{du}{u} = \log(1+x) \leq x,$$

yet, for many purposes, these bounds are more efficiently summarized by the simpler statement

$$\log(1+x) = x + O(x^2) \qquad \text{as } x \to 0. \tag{8.3}$$

Similarly, one can check that for all $|x| \leq 1$ one has the bound

$$1 + x \leq e^x = \sum_{j=0}^{\infty} \frac{x^j}{j!} \leq 1 + x + x^2 \sum_{j=2}^{\infty} \frac{x^{j-2}}{j!} \leq 1 + x + ex^2,$$

though, again, for many calculations we only need to know that these bounds give us the relation

$$e^x = 1 + x + O(x^2) \qquad \text{as } x \to 0. \tag{8.4}$$

Landau's notation and the big-O relations (8.3) and (8.4) for the logarithm and the exponential now help us calculate quite smoothly that as $t \to 0$ one has

$$\log\left\{\left(\sum_{k=1}^n p_k x_k^t\right)^{1/t}\right\} = \frac{1}{t}\log\left\{\sum_{k=1}^n p_k e^{t\log x_k}\right\}$$

$$= \frac{1}{t}\log\left\{\sum_{k=1}^n p_k\left(1 + t\log x_k + O(t^2)\right)\right\}$$

$$= \frac{1}{t}\log\left\{1 + t\sum_{k=1}^n p_k \log x_k + O(t^2)\right\}$$

$$= \sum_{k=1}^n p_k \log x_k + O(t).$$

This big-O identity is even a bit stronger than one needs to confirm the limit (8.2), so the solution of the challenge problem is complete.

A Corollary

The formula (8.2) provides a general representation of the geometric mean as a limit of a sum, and it is worth noting that for two summands it simply says that

$$\lim_{p\to\infty} \left\{ \theta a^{1/p} + (1-\theta) b^{1/p} \right\}^p = a^\theta b^{1-\theta}, \tag{8.5}$$

all nonnegative a, b, and $\theta \in [0,1]$. This formula and its more complicated cousin (8.2) give us a general way to convert information for a sum into information for a product.

Later we will draw some interesting inferences from this observation, but first we need to develop an important relation between the power means and the geometric mean. We will do this by a method that is often useful as an exploratory tool in the search for new inequalities.

Siegel's Method of Halves

Carl Ludwig Siegel (1896–1981) observed in his lectures on the geometry of numbers that the limit representation (8.2) for the geometric mean can be used to prove an elegant refinement of the AM-GM inequality. The proof calls on nothing more than Cauchy's inequality and the limit characterization of the geometric mean, yet it illustrates a sly strategy which opens many doors.

Problem 8.2 (Power Mean Bound for the Geometric Mean)

Follow in Siegel's footsteps and prove that for any nonnegative weights p_k, $k=1,2,\ldots,n$ with total mass $p_1 + p_2 + \cdots + p_n = 1$ and for any nonnegative real numbers x_k, $k=1,2,\ldots,n$, one has the bound

$$\prod_{k=1}^n x_k^{p_k} \leq \left\{ \sum_{k=1}^n p_k x_k^t \right\}^{1/t} \quad \text{for all } t > 0. \tag{8.6}$$

As the section title hints, one way to approach such a bound is to consider what happens when t is halved (or doubled). Specifically, one might first aim for an inequality such as

$$M_t \leq M_{2t} \quad \text{for all } t > 0, \tag{8.7}$$

and afterwards one can then look for a way to draw the connection to the limit (8.2).

As usual, Cauchy's inequality is our compass, and again it points us to the splitting trick. If we write $p_k x_k^t = p_k^{\frac{1}{2}} p_k^{\frac{1}{2}} x_k^t$ we find

$$M_t^t = \sum_{k=1}^n p_k x_k^t = \sum_{k=1}^n p_k^{1/2} p_k^{1/2} x_k^t$$

$$\leq \left(\sum_{k=1}^n p_k\right)^{\frac{1}{2}} \left(\sum_{k=1}^n p_k x_k^{2t}\right)^{\frac{1}{2}} = M_{2t}^t,$$

and now when we take the tth root of both sides, we have before us the conjectured doubling formula (8.7).

To complete the solution of the challenge problem, we can simply iterate the process of taking halves, so, after j steps, we find for all real $t > 0$ that

$$M_{t/2^j} \leq M_{t/2^{j-1}} \leq \cdots \leq M_{t/2} \leq M_t. \tag{8.8}$$

Now, from the limit representation of the geometric mean (8.2) we have

$$\lim_{j \to \infty} M_{t/2^j} = M_0 = \prod_{k=1}^n x_k^{p_k},$$

so from the halving bound (8.8) we find that for all $t \geq 0$ one has

$$\prod_{k=1}^n x_k^{p_k} = M_0 \leq M_t = \left\{\sum_{k=1}^n p_k x_k^t\right\}^{1/t} \quad \text{for all } t > 0. \tag{8.9}$$

MONOTONICITY OF THE MEANS

Siegel's doubling relation (8.7) and the plot given in Figure 8.1 of the two-term power mean $(px^t + qy^t)^{1/t}$ provide us with big hints about the quantitative and qualitative features of the general mean M_t. Perhaps the most basic among these is the monotonicity of the map $t \mapsto M_t$ which we address in the next challenge problem.

Problem 8.3 (Power Mean Inequality)

Consider positive weights p_k, $k = 1, 2, \ldots, n$ which have total mass $p_1 + p_2 + \cdots + p_n = 1$, and show that for nonnegative real numbers x_k, $k = 1, 2, \ldots, n$, the mapping $t \mapsto M_t$ is a nondecreasing function on all of $\mathbb{R}$. That is, show that for all $-\infty < s < t < \infty$ one has

$$\left\{\sum_{k=1}^n p_k x_k^s\right\}^{1/s} \leq \left\{\sum_{k=1}^n p_k x_k^t\right\}^{1/t}. \tag{8.10}$$

Finally, show that then one has equality in the bound (8.10) if and only if $x_1 = x_2 = \cdots = x_n$.

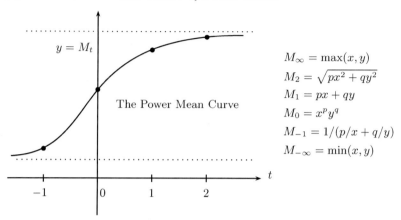

Fig. 8.1. If $x > 0$, $y > 0$, $0 < p < 1$ and $q = 1 - p$, then a qualitative plot of $M_t = (px^t + qy^t)^{1/t}$ for $-\infty < t < \infty$ suggests several basic relationships between the power means. Perhaps the most productive of these is simply the fact that M_t is a monotone increasing function of the power t, but all of the elements of the diagram have their day.

THE FUNDAMENTAL SITUATION: $0 < s < t$

One is not likely to need long to note the resemblance of our target inequality (8.10) to the bound one obtains from Jensen's inequality for the map $x \mapsto x^p$ with $p > 1$,

$$\left\{\sum_{k=1}^{n} p_k x_k\right\}^p \leq \sum_{k=1}^{n} p_k x_k^p.$$

In particular, if we assume $0 < s < t$ then the substitutions $y_k^s = x_k$ and $p = t/s > 1$ give us

$$\left\{\sum_{k=1}^{n} p_k y_k^s\right\}^{t/s} \leq \sum_{k=1}^{n} p_k y_k^t, \tag{8.11}$$

so taking the tth root gives us the power mean inequality (8.10) in the most basic case. Moreover, the strict convexity of $x \mapsto x^p$ for $p > 1$ tells us that if $p_k > 0$ for all $k = 1, 2, \ldots, n$, then we have equality in the bound (8.11) if and only if $x_1 = x_2 = \cdots = x_n$.

THE REST OF THE CASES

There is something aesthetically unattractive about breaking a problem into a collection of special cases, but sometimes such decompositions are unavoidable. Here, as Figure 8.2 suggests, there are two further cases to consider. The most pressing of these is Case II where $s < t < 0$, and

The Ladder of Power Means

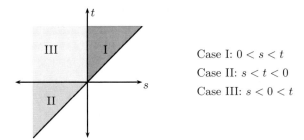

Case I: $0 < s < t$
Case II: $s < t < 0$
Case III: $s < 0 < t$

Fig. 8.2. The power mean inequality deals with all $-\infty < s < t < \infty$ and Jensen's inequality deals directly with Case I and indirectly with Case II. Case III has two halves $s = 0 < t$ and $s < t = 0$ which are consequences of the geometric mean power mean bound (8.6).

we cover it by applying the result of Case I. Since $-t > 0$ is smaller than $-s > 0$, the bound of Case I gives us

$$\left\{\sum_{k=1}^{n} p_k x_k^{-t}\right\}^{-1/t} \leq \left\{\sum_{k=1}^{n} p_k x_k^{-s}\right\}^{-1/s}.$$

Now, when we take reciprocals we find

$$\left\{\sum_{k=1}^{n} p_k x_k^{-s}\right\}^{1/s} \leq \left\{\sum_{k=1}^{n} p_k x_k^{-t}\right\}^{1/t},$$

so when we substitute $x_k = y_k^{-1}$, we get the power mean inequality for $s < t < 0$.

Case III of Figure 8.2 is the easiest of the three. By the PM-GM inequality (8.6) for x_k^{-t}, $1 \leq k \leq n$, and the power $0 \leq -s$, we find after taking reciprocals that

$$\left\{\sum_{k=1}^{n} p_k x_k^s\right\}^{1/s} \leq \prod_{k=1}^{n} x_k^{p_k} \quad \text{for all } s < 0. \tag{8.12}$$

Together with the basic bound (8.6) for $0 < t$, this completes the proof of Case III.

All that remains now is to acknowledge that the three cases still leave some small cracks unfilled; specifically, the boundary situations $0 = s < t$ and $s < t = 0$ have been omitted from the three cases of Figure 8.2. Fortunately, these situations were already covered by the bounds (8.6) and (8.12), so the solution of the challenge problem really is complete.

In retrospect, Cases II and III resolved themselves more easily than one might have guessed. There is even some charm in the way the geometric mean resolved the relation between the power means with positive and negative powers. Perhaps we can be encouraged by this experience the next time we are forced to face a case-by-case argument.

SOME SPECIAL MEANS

We have already seen that some of the power means deserve special attention, and, after $t = 2$, $t = 1$, and $t = 0$, the cases most worthy of note are $t = -1$ and the limit values one obtains by taking $t \to \infty$ or by taking $t \to -\infty$. When $t = -1$, the mean M_{-1} is called the *harmonic mean* and in longhand it is given by

$$M_{-1} = M_{-1}[\mathbf{x}; \mathbf{p}] = \frac{1}{p_1/x_1 + p_2/x_2 + \cdots + p_n/x_n}.$$

From the power mean inequality (8.10) we know that M_{-1} provides a lower bound on the geometric mean, and, *a fortiori*, one has a bound on the arithmetic mean. Specifically, we have the harmonic mean-geometric mean inequality (or the HM-GM inequality)

$$\frac{1}{p_1/x_1 + p_2/x_2 + \cdots + p_n/x_n} \leq x_1^{p_1} x_2^{p_2} \cdots x_n^{p_n} \qquad (8.13)$$

and, as a corollary, one also has the harmonic mean-arithmetic mean inequality (or the HM-AM inequality)

$$\frac{1}{p_1/x_1 + p_2/x_2 + \cdots + p_n/x_n} \leq p_1 x_1 + p_2 x_2 + \cdots + p_n x_n. \qquad (8.14)$$

Sometimes these inequalities come into play just as they are written, but perhaps more often we use them "upside down" where they give us useful lower bounds for the weighted sums of reciprocals:

$$\frac{1}{x_1^{p_1} x_2^{p_2} \cdots x_n^{p_n}} \leq \frac{p_1}{x_1} + \frac{p_2}{x_2} + \cdots + \frac{p_n}{x_n}, \qquad (8.15)$$

$$\frac{1}{p_1 x_1 + p_2 x_2 + \cdots + p_n x_n} \leq \frac{p_1}{x_1} + \frac{p_2}{x_2} + \cdots + \frac{p_n}{x_n}. \qquad (8.16)$$

GOING TO EXTREMES

The last of the power means to require special handling are those for the extreme values $t = -\infty$ and $t = \infty$ where the appropriate definitions are given by

$$M_{-\infty}[\mathbf{x}; \mathbf{p}] \equiv \min_k x_k \quad \text{and} \quad M_{\infty}[\mathbf{x}; \mathbf{p}] \equiv \max_k x_k. \qquad (8.17)$$

The Ladder of Power Means

With this interpretation one has all of the properties that Figure 8.1 suggests. In particular, one has the obvious (but useful) bounds

$$M_{-\infty}[\mathbf{x}; \mathbf{p}] \leq M_t[\mathbf{x}; \mathbf{p}] \leq M_\infty[\mathbf{x}; \mathbf{p}] \qquad \text{for all } t \in \mathbb{R},$$

and one also has the two continuity relations

$$\lim_{t \to \infty} M_t[\mathbf{x}; \mathbf{p}] = M_\infty[\mathbf{x}; \mathbf{p}] \quad \text{and} \quad \lim_{t \to -\infty} M_t[\mathbf{x}; \mathbf{p}] = M_{-\infty}[\mathbf{x}; \mathbf{p}].$$

To check these limits, we first note that for all $t > 0$ and all $1 \leq k \leq n$ we have the elementary bounds

$$p_k x_k^t \leq M_t^t[\mathbf{x}; \mathbf{p}] \leq M_\infty^t[\mathbf{x}; \mathbf{p}],$$

and, since $p_k > 0$ we have $p_k^{1/t} \to 1$ as $t \to \infty$, so we can take roots and let $t \to \infty$ to deduce that for all $1 \leq k \leq n$ we have

$$x_k \leq \liminf_{t \to \infty} M_t[\mathbf{x}; \mathbf{p}] \leq \limsup_{t \to \infty} M_t[\mathbf{x}; \mathbf{p}] \leq M_\infty[\mathbf{x}; \mathbf{p}].$$

Since $\max_k x_k = M_\infty[\mathbf{x}; \mathbf{p}]$, we have the same bound on both the extreme left and extreme right, so in the end we see

$$\lim_{t \to \infty} M_t[\mathbf{x}; \mathbf{p}] = M_\infty[\mathbf{x}; \mathbf{p}].$$

This confirms the first continuity relation, and in view of the general identity $M_{-t}(x_1, x_2, \ldots, x_n; \mathbf{p}) = M_t^{-1}(1/x_1, 1/x_2, \ldots, 1/x_n; \mathbf{p})$, the second continuity relation follows from the first.

THE INTEGRAL ANALOGS

The integral analogs of the power means are also important, and their relationships follows in lock-step with those one finds for sums. To make this notion precise, we take $D \subset \mathbb{R}$ and we consider a weight function $w : D \to [0, \infty)$ which satisfies

$$\int_D w(x)\, dx = 1 \quad \text{and} \quad w(x) > 0 \qquad \text{for all } x \in D,$$

then for $f : D \to [0, \infty]$ and $t \in (-\infty, 0) \cup (0, \infty)$ we define the tth *power mean* of f by the formula

$$M_t = M_t[f; w] \equiv \left\{ \int_D f^t(x) w(x)\, dx \right\}^{1/t}. \qquad (8.18)$$

As in the discrete case, the mean M_0 requires special attention, and for the integral mean the appropriate definition requires one to set

$$M_0[f; w] \equiv \exp\left(\int_D \left\{ \log f(x) \right\} w(x)\, dx \right). \qquad (8.19)$$

Despite the differences in the two forms (8.18) and (8.19), the definition (8.19) should not come as a surprise. After all, we found earlier (page 114) that the formula (8.19) is the natural integral analog of the geometric mean of f with respect to the weight function w.

Given the definitions (8.18) and (8.19), one now has the perfect analog of the discrete power mean inequality; specifically, one has

$$M_s[f;w] \leq M_t[f;w] \quad \text{for all } -\infty < s < t < \infty. \tag{8.20}$$

Moreover, for well-behaved f, say, those that are continuous, one has equality in the bound (8.20) if and only if f is constant on D.

We have already invested considerable effort on the discrete power mean inequality (8.10), so we will not take the time here to work out a proof of the continuous analog (8.20), even though such a proof provides worthwhile exercise that every reader is encouraged to pursue. Instead, we take up a problem which shows as well as any other just how effective the basic bound $M_0[f;w] \leq M_1[f;w]$ is. In fact, we will only use the simplest case when $D = [0,1]$ and $w(x) = 1$ for all $x \in D$.

CARLEMAN'S INEQUALITY AND THE CONTINUOUS AM-GM BOUND

In Chapter 2 we used Pólya's proof of Carleman's geometric mean bound,

$$\sum_{k=1}^{\infty}(a_1 a_2 \cdots a_k)^{1/k} \leq e \sum_{k=1}^{\infty} a_k, \tag{8.21}$$

as a vehicle to help illustrate the value of restructuring a problem so that the AM-GM inequality could be used where it is most efficient. Pólya's proof is an inspirational classic, but if one is specifically curious about Carleman's inequality, then there are several natural questions that Pólya's analysis leaves unanswered.

One feature of Pólya's proof that many people find perplexing is that it somehow manages to provide an effective estimate of the total of all the summands $(a_1 a_2 \cdots a_k)^{1/k}$ without providing a compelling estimate for the individual summands when they are viewed one at a time. The next challenge problem solves part of this mystery by showing that there is indeed a bound for the individual summands which is good enough so that it can be summed to obtain Carleman's inequality.

The Ladder of Power Means

Problem 8.4 (Termwise Bounds for Carleman's Summands)
Show that for positive real numbers a_k, $k = 1, 2, \ldots$, one has

$$(a_1 a_2 \cdots a_n)^{1/n} \leq \frac{e}{2n^2} \sum_{k=1}^{n} (2k-1) a_k \quad \text{for } n = 1, 2, \ldots, \quad (8.22)$$

and then show that these bounds can be summed to prove the classical Carleman inequality (8.21).

A REASONABLE FIRST STEP

The unspoken hint of our problem's location suggests that one should look for a role for the integral analogs of the power means. Since we need to estimate the terms $(a_1 a_2 \cdots a_n)^{1/n}$ it also seems reasonable to consider the integrand $f : [0, \infty) \to \mathbb{R}$ where we take $f(x)$ to be equal to a_k on the interval $(k-1, k]$ for $1 \leq k < \infty$. This choice makes it easy for us to put the left side of the target inequality (8.22) into an integral form:

$$\left\{ \prod_{k=1}^{n} a_k \right\}^{1/n} = \exp\left\{ \frac{1}{n} \sum_{k=1}^{n} \log a_k \right\}$$
$$= \exp\left\{ \frac{1}{n} \int_0^n \log f(x)\, dx \right\}$$
$$= \exp\left\{ \int_0^1 \log f(ny)\, dy \right\}. \quad (8.23)$$

This striking representation for the geometric mean almost begs us to apply continuous version of the AM-GM inequality.

Unfortunately, if we were to acquiesce, we would find ourselves embarrassed; the immediate application of the continuous AM-GM inequality to the formula (8.23) returns us unceremoniously back at the classical discrete AM-GM inequality. For the moment, it may seem that the nice representation (8.23) really accomplishes nothing, and we may even be tempted to abandon this whole line of investigation. Here, and at similar moments, one should take care not to desert a natural plan too quickly.

A DEEPER LOOK

The naive application of the AM-GM bound leaves us empty handed, but surely there is something more that we can do. At a minimum, we can review some of Pólya's questions and, as we work down the list, we may be struck by the one that asks, "Is it possible to satisfy the condition?"

Here the notion of condition and conclusion are intertwined, but ultimately we need a bound like the one given by the right side of our target inequality (8.22). Once this is said, we will surely ask ourselves where the constant factor e is to be found. Such a factor is not in the formula (8.23) as it stands, but perhaps we can put it there.

This question requires exploration, but if one thinks how e might be expressed in a form that is analogous to the right side of the formula (8.23), then sooner or later one is likely to have the lucky thought of replacing $f(ny)$ by y. One would then notice that

$$e = \exp\left\{-\int_0^1 \log y \, dy\right\}, \qquad (8.24)$$

and this identity puts us back on the scent. We just need to slip $\log y$ into the integrand and return to our original plan. Specifically, we find

$$\exp\left\{\int_0^1 \log f(ny)\, dy\right\} = \exp\left\{\int_0^1 \log\{yf(ny)\} - \log y \, dy\right\}$$

$$= e \exp\left\{\int_0^1 \log\{yf(ny)\}\, dy\right\}$$

$$\leq e \int_0^1 yf(ny)\, dy, \qquad (8.25)$$

where in the last step we finally get to apply the integral version of the AM-GM inequality.

Two Final Steps

Now, for the function f defined by setting $f(x) = a_k$ for $x \in (k-1, k]$, we have the elementary identity

$$\int_0^1 yf(ny)\, dy = \sum_{k=1}^n \int_{(k-1)/n}^{k/n} ya_k \, dy = \frac{1}{2n^2}\sum_{k=1}^n (2k-1)a_k, \qquad (8.26)$$

so, in view of the general bound (8.25) and the identity (8.23), the proof of the first inequality (8.22) of the challenge problem is complete.

All that remains is for us to add up the termwise bounds (8.22) and check that the sum yields the classical form of Carleman's inequality (8.21). This is easy enough, but some care is still needed to squeeze out

The Ladder of Power Means

exactly the right final bound. Specifically, we note that

$$\sum_{n=1}^{\infty} \frac{1}{n^2} \sum_{k=1}^{n} (2k-1)a_k = \sum_{k=1}^{\infty} (2k-1)a_k \sum_{n=k}^{\infty} \frac{1}{n^2}$$

$$\leq \sum_{k=1}^{\infty} (2k-1)a_k \sum_{n=k}^{\infty} \left\{ \frac{1}{n-\frac{1}{2}} - \frac{1}{n+\frac{1}{2}} \right\}$$

$$= \sum_{k=1}^{\infty} \frac{2k-1}{k-\frac{1}{2}} a_k = 2 \sum_{k=1}^{\infty} a_k$$

and, when we insert this bound in the identity (8.26), we see that the estimate (8.25) does indeed complete the proof of Carleman's inequality.

EXERCISES

Exercise 8.1 (Power Means in Disguise)

To use the power mean inequality effectively one must be able to pick power means out of a crowd, and this exercise provides some practice. Prove that for positive x, y, and z, one has

$$\frac{9}{2(x+y+z)} \leq \frac{1}{x+y} + \frac{1}{x+z} + \frac{1}{y+z} \quad (8.27)$$

and prove that for $p \geq 1$ one also has

$$\frac{1}{2} 3^{2-p} (x+y+z)^{p-1} \leq \frac{x^p}{y+z} + \frac{y^p}{x+z} + \frac{z^p}{x+y}. \quad (8.28)$$

Incidentally, one might note that for $p = 1$ the second bound reduces to the much-proved Nesbitt inequality of Exercise 5.6.

Exercise 8.2 (Harmonic Means and Recognizable Sums)

Suppose $x_1, x_2, \ldots, x_n$ are positive and let S denote their sum. Show that we have the bound

$$\frac{n^2}{(2n-1)} \leq \frac{S}{2S-x_1} + \frac{S}{2S-x_2} + \cdots + \frac{S}{2S-x_n}.$$

In this problem (and many like it) one gets a nice hint from the fact that there is a simple expression for the sum of the denominators on the right-hand side.

Exercise 8.3 (Integral Analogs and Homogeneity in Σ)

(a) Show that for all nonnegative sequences $\{a_k : 1 \leq k \leq n\}$ one has

$$\left\{\sum_{k=1}^{n} a_k^{1/2}\right\}^2 \leq \left\{\sum_{k=1}^{n} a_k^{1/3}\right\}^3, \tag{8.29}$$

and be sure to notice the differences between this bound and the power mean inequality (8.10) with $s = 1/3$ and $t = 1/2$.

(b) By analogy with the bound (8.29), one might carelessly guess that for nonnegative f that one has an integral bound

$$\left\{\int_0^1 f^{1/2}(x)\,dx\right\}^2 \leq \left\{\int_0^1 f^{1/3}(x)\,dx\right\}^3. \tag{8.30}$$

Show by example that the bound (8.30) does not hold in general.

The likelihood of an integral analog can often be explained by a heuristic principle which Hardy, Littlewood, and Pólya (1952, p. 4) describes as "homogeneity in Σ." The principle suggests that we consider Σ in a bound such as (8.29) as a formal symbol. In this case we see that the left side is "homogeneous of order two in Σ" while the right side is "homogeneous of order three in Σ." The two sides are therefore incompatible, and one should not expect any integral analog. On the other hand, in Cauchy's inequality and Hölder's inequality, both sides are homogeneous of order one in Σ. It is therefore natural — even inevitable — that we should have integral analogs for these bounds.

Exercise 8.4 (Pólya's Minimax Characterization)

Suppose you must guess the value of an unknown number x in the interval $[a, b] \subset (0, \infty)$ and suppose you will be forced to pay a fine based on the *relative error* of your guess. How should you guess if you want to minimize the worst fine that you would have to pay?

If you guess is p, then the maximum fine you would have to pay is

$$F(p) = \max_{x \in [a,b]} \left\{\frac{|p - x|}{x}\right\}, \tag{8.31}$$

so your analytical challenge is to find the value p^* such that

$$F(p^*) = \min_p F(p) = \min_p \max_{x \in [a,b]} \left\{\frac{|p - x|}{x}\right\}. \tag{8.32}$$

One expects p^* to be some well-known mean, but which one is it?

The Ladder of Power Means 133

Exercise 8.5 (The Geometric Mean as a Minimum)
Prove that the geometric mean has the representation

$$\left\{\prod_{k=1}^{n} a_k\right\}^{1/n} = \min\left\{\frac{1}{n}\sum_{k=1}^{n} a_k x_k : (x_1, x_2, \ldots, x_n) \in D\right\}, \quad (8.33)$$

where D is the region of $\mathbb{R}^n$ defined by

$$D = \left\{(x_1, x_2, \ldots, x_n) : \prod_{k=1}^{n} x_k = 1,\ x_k \geq 0, k = 1, 2, \ldots, n\right\}.$$

For practice with this characterization of the geometric mean, use it to give another proof that the geometric mean is superadditive; that is, show that the formula (8.33) implies the bound (2.31) on page 34.

Exercise 8.6 (More on the Method of Halves)
The method of halves applies to more than just inequalities; it can also be used to prove some elegant identities. As an illustration, show that the familiar half-angle formula $\sin x = 2\sin(x/2)\cos(x/2)$ implies the infinite product identity

$$\frac{\sin x}{x} = \prod_{k=1}^{\infty} \cos(x/2^k), \quad (8.34)$$

and verify in turn that this implies the poignant formula

$$\frac{2}{\pi} = \frac{\sqrt{2}}{2} \cdot \frac{\sqrt{2+\sqrt{2}}}{2} \cdot \frac{\sqrt{2+\sqrt{2+\sqrt{2}}}}{2} \cdots.$$

Incidentally, the product formula (8.34) for $\sin(x)/x$ is known as Viète's identity, and it has been known since 1593.

Exercise 8.7 (Differentiation of an Inequality)
In general one cannot differentiate the two sides of an inequality and expect any meaningful consequences, but there are special situations where "differentiation of an inequality" does make sense. There are even times when such differentiations have lead to spectacular new results. The aspirations of this exercise are more modest, but they point the way to what is possible.

(a) Consider a function f that is differentiable at t_0 and that satisfies the bound $f(t_0) \leq f(t)$ for all $t \in [t_0, t_0 + \Delta)$ and some $\Delta > 0$. Show that one then has $0 \leq f'(t_0)$.

(b) Use the preceding observation to show that the power mean inequality implies that for all $x_k > 0$ and all nonnegative p_k with total $p_1 + p_2 + \cdots + p_n = 1$, one has

$$\left\{\sum_{k=1}^{n} p_k x_k\right\} \log\left\{\sum_{k=1}^{n} p_k x_k\right\} \leq \left\{\sum_{k=1}^{n} p_k x_k \log x_k\right\}. \tag{8.35}$$

Exercise 8.8 (A Niven–Zuckerman Lemma for pth Powers)
Consider a sequence of n-tuples of nonnegative real numbers

$$(a_{1k}, a_{2k}, \ldots, a_{nk}) \qquad k = 1, 2, \ldots.$$

Suppose there is a constant $\mu \geq 0$ for which one has

$$a_{1k} + a_{2k} + \cdots + a_{nk} \to n\mu \qquad \text{as } k \to \infty, \tag{i}$$

and suppose for some $1 < p < \infty$ such that one also has

$$a_{1k}^p + a_{2k}^p + \cdots + a_{nk}^p \to n\mu^p \qquad \text{as } k \to \infty. \tag{ii}$$

Show that these conditions imply that one then has the n-term limit

$$\lim_{k \to \infty} a_{jk} = \mu \qquad \text{for all } 1 \leq j \leq n.$$

This exercise provides an example of a *consistency principle* which in this case asserts that if the sum of the coordinates of a vector and the sum of the corresponding pth powers have limits that are consistent with the possibility that all of the coordinates converge to a common constant, then that must indeed be the case. The consistency principle has many variations and, like the optimality principle of Exercise 2.8, page 33, it provides useful heuristic guidance even when it does not formally apply.

Exercise 8.9 (Points Crowded in an Interval)
Given n points in the interval $[-1, 1]$, we know that some pairs must be close together, and there are many ways to quantify this crowding. An uncommon yet insightful way once exploited by Paul Erdős is to look at the sum of the reciprocal gaps.

(a) Suppose that $-1 \leq x_1 < x_2 < \cdots < x_n \leq 1$, and show that

$$\sum_{1 \leq j < k \leq n} \frac{1}{x_k - x_j} \geq \frac{1}{8} n^2 \log n.$$

(b) Show that for any permutation $\sigma : [n] \to [n]$ one has the bound

$$\max_{1 < k \leq n} \sum_{j=1}^{k-1} \frac{1}{|x_{\sigma(k)} - x_{\sigma(j)}|} \geq \frac{1}{8} n \log n.$$

9
Hölder's Inequality

Four results provide the central core of the classical theory of inequalities, and we have already seen three of these: the Cauchy–Schwarz inequality, the AM-GM inequality, and Jensen's inequality. The quartet is completed by a result which was first obtained by L.C. Rogers in 1888 and which was derived in another way a year later by Otto Hölder. Cast in its modern form, the inequality asserts that for all nonnegative a_k and b_k, $k = 1, 2, \ldots, n$, one has the bound

$$\sum_{k=1}^{n} a_k b_k \leq \left(\sum_{k=1}^{n} a_k^p \right)^{1/p} \left(\sum_{k=1}^{n} b_k^q \right)^{1/q}, \qquad (9.1)$$

provided that the powers $p > 1$ and $q > 1$ satisfy the relation

$$\frac{1}{p} + \frac{1}{q} = 1. \qquad (9.2)$$

Ironically, the articles by Rogers and Hölder leave the impression that these authors were mainly concerned with the extension and application of the AM-GM inequality. In particular, they did not seem to view their version of the bound (9.1) as singularly important, though Rogers did value it enough to provide two proofs. Instead, the opportunity fell to Frigyes Riesz to cast the inequality (9.1) in its modern form and to recognize its fundamental role. Thus, one can argue that the bound (9.1) might better be called Rogers's inequality, or perhaps even the Rogers–Hölder–Riesz inequality. Nevertheless, long ago, the moving hand of history began to write "Hölder's inequality," and now, for one to use another name would be impractical, though from time to time some acknowledgment of the historical record seems appropriate.

The first challenge problem is easy to anticipate: one must prove the inequality (9.1), and one must determine the circumstances where equality can hold. As usual, readers who already know a proof of Hölder's inequality are invited to discover a new one. Although, new proofs of Hölder's inequality appear less often than those for the Cauchy–Schwarz inequality or the AM-GM inequality, one can have confidence that they can be found.

Problem 9.1 (Hölder's Inequality)
First prove Riesz's version (9.1) of the inequality of Rogers (1888) and Hölder (1889), then prove that one has equality for a nonzero sequence $a_1, a_2, \ldots, a_n$ if and only if there exists a constant $\lambda \in \mathbb{R}$ such that

$$\lambda a_k^{1/p} = b_k^{1/q} \qquad \text{for all } 1 \leq k \leq n. \tag{9.3}$$

BUILDING ON THE PAST

Surely one's first thought is to try to adapt one of the many proofs of Cauchy's inequality; it may even be instructive to see how some of these come up short. For example, when $p \neq 2$, Schwarz's argument is a nonstarter since there is no quadratic polynomial in sight. Similarly, the absence of a quadratic form means that one is unlikely to find an effective analog of Lagrange's identity.

This brings us to our most robust proof of Cauchy's inequality, the one that starts with the so-called "humble bound,"

$$xy \leq \frac{1}{2}x^2 + \frac{1}{2}y^2 \qquad \text{for all } x, y \in \mathbb{R}. \tag{9.4}$$

This bound may now remind us that the general AM-GM inequality (2.9), page 23, implies that

$$x^\alpha y^\beta \leq \frac{\alpha}{\alpha+\beta} x^{\alpha+\beta} + \frac{\beta}{\alpha+\beta} y^{\alpha+\beta} \tag{9.5}$$

for all $x \geq 0$, $y \geq 0$, $\alpha > 0$, and $\beta > 0$. If we then set $u = x^\alpha$, $v = y^\beta$, $p = (\alpha+\beta)/\alpha$, and $q = (\alpha+\beta)/\beta$, then we find for all $p > 1$ that one has the handy inference

$$\frac{1}{p} + \frac{1}{q} = 1 \implies uv \leq \frac{1}{p}u^p + \frac{1}{q}v^q \qquad \text{for all } u, v \in \mathbb{R}^+. \tag{9.6}$$

This is the perfect analog of the "humble bound" (9.4). It is known as Young's inequality, and it puts us well on the way to a solution of our challenge problem.

Hölder's Inequality

Another Additive to Multiplicative Transition

The rest of the proof of Hölder's inequality follows a familiar pattern. If we make the substitutions $u \mapsto a_k$ and $v \mapsto b_k$ in the bound (9.6) and sum over $1 \leq k \leq n$, then we find

$$\sum_{k=1}^{n} a_k b_k \leq \frac{1}{p} \sum_{k=1}^{n} a_k^p + \frac{1}{q} \sum_{k=1}^{n} b_k^q, \qquad (9.7)$$

and to pass from this additive bound to a multiplicative bound we can apply the *normalization device* with which we have already scored two successes. We can assume without loss of generality that neither of our sequences is identically zero, so the normalized variables

$$\hat{a}_k = a_k \bigg/ \left(\sum_{k=1}^{n} a_k^p \right)^{1/p} \quad \text{and} \quad \hat{b}_k = b_k \bigg/ \left(\sum_{k=1}^{n} b_k^q \right)^{1/q},$$

are well defined. Now, if we simply substitute these values into the additive bound (9.7), we find that easy arithmetic guides us quickly to the completion of the direct half of the challenge problem.

Looking Back — Contemplating Conjugacy

In retrospect, Riesz's argument is straightforward, but the easy proof does not tell the whole story. In fact, Riesz's *formulation* carried much of the burden, and he was particularly wise to focus our attention on the pairs of powers p and q such that $1/p + 1/q = 1$. Such (p, q) pairs are now said to be *conjugate*, and many problems depend on the trade-offs we face when we choose one conjugate pair over another. This balance is already visible in the p-q generalization (9.6) of the "humble bound" (9.4), but soon we will see deeper examples.

Backtracking and the Case of Equality

To complete the challenge problem, we still need to determine the circumstances where one has equality. To begin, we first note that equality trivially holds if $b_k = 0$ for all $1 \leq k \leq n$, but in that case the identity (9.3) is satisfied $\lambda = 0$; thus, we may assume with loss of generality that both sequences are nonzero.

Next, we note that equality is attained in Hölder's inequality (9.1) if and only if equality holds in the additive bound (9.7) when it is applied to the normalized variables $\hat{a}_k$ and $\hat{b}_k$. By the termwise bound (9.6), we further see that equality holds in the additive bound (9.7) if and only if

$$\sum_{k=1}^{n} a_k b_k = \{\sum_{k=1}^{n} a_k^p\}^{1/p} \{\sum_{k=1}^{n} b_k^p\}^{1/q} \qquad a_k^p = b_k^q \{(\sum_{k=1}^{n} b_k^q)^{1/q} (\sum_{k=1}^{n} a_k^p)^{-1/p}\}$$

$$\Updownarrow \qquad\qquad\qquad\qquad \Updownarrow$$

$$\sum_{k=1}^{n} \hat{a}_k \hat{b}_k = 1 \qquad\qquad \hat{a}_k^p = \hat{b}_k^q \quad k = 1, 2, \ldots, n$$

$$\Updownarrow \qquad\qquad\qquad\qquad \Updownarrow$$

$$\sum_{k=1}^{n} \hat{a}_k \hat{b}_k = \frac{1}{p}\sum_{k=1}^{n} \hat{a}_k^p + \frac{1}{q}\sum_{k=1}^{n} \hat{b}_k^q \quad\Longleftrightarrow\quad \hat{a}_k \hat{b}_k = \hat{a}_k^p/p + \hat{b}_k^p/q \ k = 1, 2, \ldots, n$$

Fig. 9.1. The case for equality in Hölder's inequality is easily framed as a blackboard display, and such a semi-graphical presentation has several advantages over a monologue of "if and only if" assertions. In particular, it helps us to see the argument at a glance, and it encourages us to question each of the individual inferences.

we have

$$\hat{a}_k \hat{b}_k = \frac{1}{p}\hat{a}_k^p + \frac{1}{q}\hat{b}_k^q \qquad \text{for all } k = 1, 2, \ldots, n.$$

Next, by the condition for equality in the special AM-GM bound (9.5), we find that for each $1 \leq k \leq n$ we must have $\hat{a}_k^p = \hat{b}_k^q$. Finally, when we peel away the normalization indicated by the hats, we see that $\lambda a_k^p = b_k^q$ for all $1 \leq k \leq n$ where λ is given explicitly by

$$\lambda = \left(\sum_{k=1}^{n} b_k^q\right)^{1/q} \Big/ \left(\sum_{k=1}^{n} a_k^p\right)^{1/p}.$$

This is characterization that we anticipated, and the solution of the challenge problem is complete.

A Blackboard Tool for Better Checking

Backtracking arguments, such as the one just given, are notorious for harboring gaps, or even outright errors. It seems that after working through a direct argument, many of us are just too tempted to believe that nothing could go wrong when the argument is "reversed." Unfortunately, there are times when this is wishful thinking.

A semi-graphical "blackboard display" such as that of Figure 9.1 may be of help here. Many of us have found ourselves nodding passively to

Hölder's Inequality 139

a monologue of "if and only if" statements, but the visible inferences of a blackboard display tend to provoke more active involvement. Such a display shows the whole argument at a glance, yet each inference is easily isolated.

A CONVERSE FOR HÖLDER

In logic, everyone knows that the converse of the inference $A \Rightarrow B$ is the inference $B \Rightarrow A$, but in the theory of inequalities the notion of a converse is more ambiguous. Nevertheless, there is a result that deserves to be called the *converse Hölder inequality*, and it provides our next challenge problem.

Problem 9.2 (The Hölder Converse — The Door to Duality)
Show that if $1 < p < \infty$ and if C is a constant such that

$$\sum_{k=1}^{n} a_k x_k \leq C \left\{ \sum_{k=1}^{n} |x_k|^p \right\}^{1/p} \tag{9.8}$$

for all x_k, $1 \leq k \leq n$, then for $q = p/(p-1)$ one has the bound

$$\left\{ \sum_{k=1}^{n} |a_k|^q \right\}^{1/q} \leq C. \tag{9.9}$$

HOW TO UNTANGLE THE UNWANTED VARIABLES

This problem helps to explain the inevitability of Riesz's conjugate pairs (p, q), and, to some extent, the simple conclusion is surprising. Nonlinear constraints are notoriously awkward, and here we see that we have x-variables tangled up on both sides of the hypothesis (9.8). We need a trick if we want to eliminate them.

One idea that sometimes works when we have free variables on both sides of a relation is to conspire to make the two sides as similar as possible. This "principle of similar sides" is necessarily vague, but here it may suggest that for each $1 \leq k \leq n$ we should choose x_k such that $a_k x_k = |x_k|^p$; in other words, we set $x_k = \text{sign}(a_k)|a_k|^{p/(p-1)}$ where $\text{sign}(a_k)$ is 1 if $a_k \geq 0$ and it is -1 if $a_k < 0$. With this choice the condition (9.8) becomes

$$\sum_{k=1}^{n} |a_k|^{p/(p-1)} \leq C \left\{ \sum_{k=1}^{n} |a_k|^{p/(p-1)} \right\}^{1/p}. \tag{9.10}$$

We can assume without loss of generality that the sum on the right is nonzero, so it is safe to divide by that sum. The relation $1/p + 1/q = 1$ then confirms that we have indeed proved our target bound (9.9).

A Shorthand Designed for Hölder's Inequality

Hölder's inequality and the duality bound (9.9) can be recast in several forms, but to give the nicest of these it will be useful to introduce some shorthand. If $\mathbf{a} = (a_1, a_2, \ldots, a_n)$ is an n-tuple of real numbers, and $1 \le p < \infty$ we will write

$$\|\mathbf{a}\|_p = \left(\sum_{k=1}^{n} |a_k|^p \right)^{1/p}, \qquad (9.11)$$

while for $p = \infty$ we simply set $\|\mathbf{a}\|_\infty = \max_{1 \le k \le n} |a_k|$. With this notation, Hölder's inequality (9.1) for $1 \le p < \infty$ then takes on the simple form

$$\left| \sum_{k=1}^{n} a_k b_k \right| \le \|\mathbf{a}\|_p \|\mathbf{b}\|_q,$$

where for $1 < p < \infty$ the pair (p, q) are the usual conjugates which are determined by the relation

$$\frac{1}{p} + \frac{1}{q} = 1 \qquad \text{when } 1 < p < \infty,$$

but for $p = 1$ we just simply set $q = \infty$.

The quantity $\|a\|_p$ is called the p-norm, or the ℓ^p-norm, of the n-tuple, but, to justify this name, one needs to check that the function $\mathbf{a} \mapsto \|\mathbf{a}\|_p$ does indeed satisfy all of the properties required by the definition a norm; specifically, one needs to verify the three properties:

(i) $\|\mathbf{a}\|_p = 0$ if and only if $\mathbf{a} = \mathbf{0}$,

(ii) $\|\alpha \mathbf{a}\|_p = |\alpha| \, \|\mathbf{a}\|_p$ for all $\alpha \in \mathbb{R}$, and

(iii) $\|\mathbf{a} + \mathbf{b}\|_p \le \|\mathbf{a}\|_p + \|\mathbf{b}\|_p$ for all real n-tuples $\mathbf{a}$ and $\mathbf{b}$.

The first two properties are immediate from the definition (9.11), but the third property is more substantial. It is known as Minkowski's inequality, and, even though it is not difficult to prove, the result is a fundamental one which deserves to be framed as a challenge problem.

Hölder's Inequality

Problem 9.3 (Minkowski's Inequality)

Show that for each $\mathbf{a} = (a_1, a_2, \ldots, a_n)$ and $\mathbf{b} = (b_1, b_2, \ldots, b_n)$ one has

$$\|\mathbf{a} + \mathbf{b}\|_p \leq \|\mathbf{a}\|_p + \|\mathbf{b}\|_p, \tag{9.12}$$

or, in longhand, show that for all $p \geq 1$ one has the bound

$$\left(\sum_{k=1}^{n} |a_k + b_k|^p \right)^{1/p} \leq \left(\sum_{k=1}^{n} |a_k|^p \right)^{1/p} + \left(\sum_{k=1}^{n} |b_k|^p \right)^{1/p}. \tag{9.13}$$

Moreover, show that if $\|\mathbf{a}\|_p \neq 0$ and if $p > 1$, then one has equality in the bound (9.12) if and only if (1) there exist a constant $\lambda \in \mathbb{R}$ such that $|b_k| = \lambda |a_k|$ for all $k = 1, 2, \ldots, n$, and (2) a_k and b_k have the same sign for each $k = 1, 2, \ldots, n$.

RIESZ'S ARGUMENT FOR MINKOWSKI'S INEQUALITY

There are many ways to prove Minkowski's inequality, but the method used by F. Riesz is a compelling favorite — especially if one is asked to prove Minkowski's inequality immediately after a discussion of Hölder's inequality. One simply asks, "How can Hölder help?" Soon thereafter, algebra can be our guide.

Since we seek an upper bound which is the sum of two terms, it is reasonable to break our sum into two parts:

$$\sum_{k=1}^{n} |a_k + b_k|^p \leq \sum_{k=1}^{n} |a_k| |a_k + b_k|^{p-1} + \sum_{k=1}^{n} |b_k| |a_k + b_k|^{p-1}. \tag{9.14}$$

This decomposition already gives us Minkowski's inequality (9.13) for $p = 1$, so we may now assume $p > 1$. If we then apply Hölder's inequality separately to each of the bounding sums (9.14), we find for the first sum that

$$\sum_{k=1}^{n} |a_k| |a_k + b_k|^{p-1} \leq \left(\sum_{k=1}^{n} |a_k|^p \right)^{1/p} \left(\sum_{k=1}^{n} |a_k + b_k|^p \right)^{(p-1)/p}$$

while for the second we find

$$\sum_{k=1}^{n} |b_k| |a_k + b_k|^{p-1} \leq \left(\sum_{k=1}^{n} |b_k|^p \right)^{1/p} \left(\sum_{k=1}^{n} |a_k + b_k|^p \right)^{(p-1)/p}.$$

Thus, in our shorthand notation the factorization (9.14) gives us

$$\|\mathbf{a} + \mathbf{b}\|_p^p \leq \|\mathbf{a}\|_p \cdot \|\mathbf{a} + \mathbf{b}\|_p^{p-1} + \|\mathbf{b}\|_p \cdot \|\mathbf{a} + \mathbf{b}\|_p^{p-1}. \tag{9.15}$$

Since Minkowski's inequality (9.12) is trivial when $\|\mathbf{a}+\mathbf{b}\|_p = 0$, we can assume without loss of generality that $\|\mathbf{a}+\mathbf{b}\|_p \neq 0$. We then divide both sides of the bound (9.15) by $\|\mathbf{a}+\mathbf{b}\|_p^{p-1}$ to complete the proof.

A Hidden Benefit: The Case of Equality

One virtue of Riesz's method for proving Minkowski's inequality (9.12), is that his argument may be worked backwards to determine the case of equality. Conceptually the plan is simple, but some of the details can seem fussy.

To begin, we note that equality in Minkowski's bound (9.12) implies equality in our first step (9.14) and that $|a_k + b_k| = |a_k| + |b_k|$ for each $1 \leq k \leq n$. Thus, we may assume that a_k and b_k are the same sign for all $1 \leq k \leq n$, and in fact there is no loss of generality if we assume $a_k \geq 0$ and $b_k \geq 0$ for all $1 \leq k \leq n$.

Equality in Minkowski's bound (9.12) also implies that we have equality in both of our applications of Hölder's inequality, so, assuming that $\|\mathbf{a}+\mathbf{b}\|_p \neq 0$, we deduce that there exists $\lambda \geq 1$ such that

$$\lambda |a_k|^p = \{|a_k + b_k|^{p-1}\}^q = |a_k + b_k|^p$$

and there exists $\lambda' \geq 1$ such that

$$\lambda' |b_k|^p = \{|a_k + b_k|^{p-1}\}^q = |a_k + b_k|^p.$$

From these identities, we see that if we set $\lambda'' = \lambda/\lambda'$ then we have $\lambda''|a_k|^p = |b_k|^p$ for all $k = 1, 2, \ldots, n$.

This is precisely the characterization which we hoped to prove. Still, on principle, every backtrack argument deserves to be put to the test; one should prod the argument to see that it is truly airtight. This is perhaps best achieved with help from a semi-graphical display analogous to Figure 9.1.

Subadditivity and Quasilinearization

Minkowski's inequality tells us that the function $h : \mathbb{R}^n \to \mathbb{R}$ defined by $h(\mathbf{a}) = \|\mathbf{a}\|_p$ is *subadditive* in the sense that one has the bound

$$h(\mathbf{a}+\mathbf{b}) \leq h(\mathbf{a}) + h(\mathbf{b}) \qquad \text{for all } \mathbf{a}, \mathbf{b} \in \mathbb{R}^n.$$

Subadditive relations are typically much more obvious than Riesz's proof, and one may wonder if there is some way to see Minkowski's inequality at a glance. The next challenge problem confirms this suspicion and throws added precision into the bargain.

Hölder's Inequality   143

Problem 9.4 (Quasilinearization of the ℓ^p Norm)
Show that for all $1 \leq p \leq \infty$ one has the identity

$$\|\mathbf{a}\|_p = \max\left\{ \sum_{k=1}^n a_k x_k : \|\mathbf{x}\|_q = 1 \right\}, \tag{9.16}$$

where $\mathbf{a} = (a_1, a_2, \ldots, a_n)$ and where p and q are conjugate (so one has $q = p/(p-1)$ when $p > 1$, but $q = \infty$ when $p = 1$ and $q = 1$ when $p = \infty$). Finally, explain why this identity yields Minkowski's inequality without any further computation.

QUASILINEARIZATION IN CONTEXT

Before addressing the problem, it may be useful to add some context. If V is a vector space (such as $\mathbb{R}^n$) and if $L : V \times W \to \mathbb{R}$ is a function which is additive in its first variable, $L(\mathbf{a} + \mathbf{b}, \mathbf{w}) = L(\mathbf{b}, \mathbf{w}) + L(\mathbf{b}, \mathbf{w})$, then the function $h : V \to \mathbb{R}$, defined by

$$h(\mathbf{a}) = \max_{\mathbf{w} \in W} L(\mathbf{a}, \mathbf{w}), \tag{9.17}$$

will always be subadditive simply because two choices are always at least as good as one:

$$h(\mathbf{a} + \mathbf{b}) = \max_{\mathbf{w} \in W} L(\mathbf{a} + \mathbf{b}, \mathbf{w}) = \max_{\mathbf{w} \in W} \{L(\mathbf{a}, \mathbf{w}) + L(\mathbf{b}, \mathbf{w})\}$$
$$\leq \max_{\mathbf{w}_0 \in W} L(\mathbf{a}, \mathbf{w}_0) + \max_{\mathbf{w}_1 \in W} L(\mathbf{b}, \mathbf{w}_1) = h(\mathbf{a}) + h(\mathbf{b}).$$

The formula (9.17) is said to be a *quasilinear representation* of h, and many of the most fundamental quantities in the theory of inequalities have analogous representations.

CONFIRMATION OF THE IDENTITY

The existence of a quasilinear representation (9.16) for the function $h(\mathbf{a}) = \|\mathbf{a}\|_p$ is an easy consequence of Hölder's inequality and its converse. Nevertheless, the logic is slippery, and it is useful to be explicit. To begin, we consider the set

$$S = \left\{ \sum_{k=1}^n a_k x_k : \sum_{k=1}^n |x_k|^q \leq 1 \right\},$$

and we note that Hölder's inequality implies $s \leq \|\mathbf{a}\|_p$ for all $s \in S$. This gives us our first bound, $\max\{s \in S\} \leq \|\mathbf{a}\|_p$. Next, just by the definition of S and by scaling we have

$$\sum_{k=1}^{n} a_k y_k \leq \|\mathbf{y}\|_q \max\{s \in S\} \quad \text{for all } \mathbf{y} \in \mathbb{R}^n. \tag{9.18}$$

Thus, by the converse Hölder bound (9.9) for the conjugate pair (q, p) — as opposed to the pair (p, q) in the statement of the bound (9.9) — we have our second bound, $\|\mathbf{a}\|_p \leq \max\{s \in S\}$. The first and second bounds now combine to give us the quasilinear representation (9.16) for $h(\mathbf{a}) = \|\mathbf{a}\|_p$.

A Stability Result for Hölder's Inequality

In many areas of mathematics one finds both *characterization results* and *stability results*. A characterization result typically provides a concrete characterization of the solutions of some equation, while the associated stability result asserts that if the equation "almost holds" then the characterization "almost applies."

There are many examples of stability results in the theory of inequalities. We have already seen that the case of equality in the AM-GM bound has a corresponding stability result (Exercise 2.12, page 35), and it is natural to ask if Hölder's inequality might also be amenable to such a development.

To make this suggestion specific, we first note that the 1-trick and Hölder's inequality imply that for each $p > 1$ and for each sequence of nonnegative real numbers $a_1, a_2, \ldots, a_n$ one has the bound

$$\sum_{j=1}^{n} a_j \leq n^{(p-1)/p} \left(\sum_{j=1}^{n} a_j^p \right)^{1/p}.$$

If we then define the *difference defect* $\delta(\mathbf{a})$ by setting

$$\delta(\mathbf{a}) \stackrel{\text{def}}{=} \sum_{j=1}^{n} a_j^p - n^{1-p} \left(\sum_{j=1}^{n} a_j \right)^p, \tag{9.19}$$

then one has $\delta(\mathbf{a}) \geq 0$, but, more to the point, the criterion for equality in Hölder's bound now tells us that $\delta(\mathbf{a}) = 0$ if and only if there is a constant μ such that $a_j = \mu$ for all $j = 1, 2, \ldots, n$. That is, the condition $\delta(\mathbf{a}) = 0$ *characterizes* the vector $\mathbf{a} = (a_1, a_2, \ldots, a_n)$ as a constant vector.

This characterization leads in turn to a variety of stability results, and our next challenge problem focuses on one of the most pleasing of these. It also introduces an exceptionally general technique for exploiting estimates of sums of squares.

Problem 9.5 (A Stability Result for Hölder's Inequality)

Show that if $p \geq 2$ and if $a_j \geq 0$ for all $1 \leq j \leq n$, then there exists a constant $\lambda = \lambda(\mathbf{a}, p)$ such that

$$a_j \in [(\lambda - \delta^{\frac{1}{2}})^{2/p}, (\lambda + \delta^{\frac{1}{2}})^{2/p}] \quad \text{for all } j = 1, 2, \ldots, n. \tag{9.20}$$

In other words, show that if the difference defect $\delta = \delta(\mathbf{a})$ is small, then the sequence $a_1, a_2, \ldots, a_n$ is almost constant.

ORIENTATION

There are many ways to express the idea that a sequence is almost constant, and the specific formula (9.20) used here is just one of several possibilities. Nevertheless, this choice does give us a hint about how we might proceed.

The relation (9.20) may be written more sensibly as $(a_j^{p/2} - \lambda)^2 \leq \delta(\mathbf{a})$, and we can prove all of the individual bounds (9.20) in a single step if we can prove the stronger conjecture that there exists a constant λ for which we have the bound

$$\sum_{j=1}^{n}(a_j^{p/2} - \lambda)^2 \leq \delta(\mathbf{a}). \tag{9.21}$$

It is possible, of course, that the inequality (9.21) asks for too much, but it is such a nice conjecture that it deserves some attention.

WHY IS IT NICE?

First of all, if $p = 2$, then one finds by direct computation from the definition of $\delta(\mathbf{a})$ that the bound (9.21) is actually an identity, provided that one takes $\lambda = (a_1 + a_2 + \cdots + a_n)/n$. It is always a good sign when a conjecture is known to be true in some special case.

A more subtle charm of the conjecture (9.21) is that it asks us indirectly if a certain quadratic polynomial has a real root. Namely, if the inequality (9.21) holds for some real λ, then by continuity there must also exist a real λ that satisfies the *equation*

$$\sum_{j=1}^{n}(a_j^{p/2} - \lambda)^2 = \delta(\mathbf{a}) \stackrel{\text{def}}{=} \sum_{j=1}^{n} a_j^p - n^{1-p}\left(\sum_{j=1}^{n} a_j\right)^p.$$

After algebraic expansion and simplification, we therefore find that the conjecture (9.21) is true if and only if there is a real root of the equation

$$n\lambda^2 - 2\lambda \sum_{j=1}^n a_j^{p/2} + n^{1-p}\left(\sum_{j=1}^n a_j\right)^p = 0. \tag{9.22}$$

Since a quadratic equation $A\lambda^2 + 2B\lambda + C = 0$ has a real root if and only if $AC \leq B^2$, we see that the solution to the challenge problem will be complete if we can show

$$n^{2-p}\left(\sum_{j=1}^n a_j\right)^p \leq \left(\sum_{j=1}^n a_j^{p/2}\right)^2. \tag{9.23}$$

Fortunately, it is easy to see that this bound holds; in fact, it is a just another corollary of Hölder's inequality and the 1-trick. To be explicit, one just applies Hölder's inequality with $p' = p/2$ and $q' = p/(p-2)$ to the sum $a_1 \cdot 1 + a_2 \cdot 1 + \cdots + a_n \cdot 1$.

INTERPOLATION

The ℓ^1 norm and the ℓ^∞ norm represent the two natural extremes among the ℓ^p norms, and it is reasonable to guess that in favorable circumstances one should be able to combine an ℓ^1 inequality and an ℓ^∞ inequality to get an analogous inequality for the ℓ^p norm where $1 < p < \infty$.

Our final challenge problem provides an important example of this possibility. It also points the way to one of the most pervasive themes in the theory of inequalities — *interpolation*.

Problem 9.6 (An Illustration of ℓ^1-ℓ^∞ Interpolation)

Let c_{jk}, $1 \leq j \leq m$, $1 \leq k \leq n$, be an array of nonnegative real numbers such that

$$\sum_{j=1}^m \left|\sum_{k=1}^n c_{jk} x_k\right| \leq A \sum_{k=1}^n |x_k| \quad \text{and} \quad \max_{1 \leq j \leq m}\left|\sum_{k=1}^n c_{jk} x_k\right| \leq B \max_{1 \leq k \leq n} |x_k|$$

for all x_k, $1 \leq k \leq n$. If $1 < p < \infty$ and $q = p/(1-p)$ show that one also has the interpolation bound

$$\left(\sum_{j=1}^m \left|\sum_{k=1}^n c_{jk} x_k\right|^p\right)^{1/p} \leq A^{1/p} B^{1/q} \left(\sum_{k=1}^n |x_k|^p\right)^{1/p} \tag{9.24}$$

for all x_k, $1 \leq k \leq n$.

Hölder's Inequality

SEARCH FOR A SIMPLER FORMULATION

The feature of the inequality (9.24) which may seem troublesome is the presence of the pth roots; one quickly starts to hunger for a way to make them disappear. The root on the right side is not a problem since by scaling $\mathbf{x}$ we can assume without loss of generality that $\|\mathbf{x}\|_p \leq 1$, but what can we do about the pth root on the left side?

Luckily, we have a tool that is well suited to the task. The converse of Hölder's inequality (page 139) tells us that to prove the bound (9.24) it suffices to show that, for all real vectors $\mathbf{x}$ and $\mathbf{y}$ such that $\|\mathbf{x}\|_p \leq 1$ and $\|\mathbf{y}\|_q \leq 1$, one has

$$\sum_{j=1}^{m}\sum_{k=1}^{n} c_{jk} x_k y_j \leq A^{1/p} B^{1/q}. \tag{9.25}$$

Moreover, since we assume that $c_{jk} \geq 0$ for all j and k, it suffices to prove the bound just for $\|\mathbf{x}\|_p \leq 1$ and $\|\mathbf{y}\|_q \leq 1$ with $x_k \geq 0$ and $y_j \geq 0$ for all j and k.

The reformulation (9.25) offers signs of real progress; in particular, the pth roots are gone. We now face a problem of the kind we have met several times before; we simply need to estimate a sum subject to some nonlinear constraints.

FROM FORMULATION TO FINISH

In the past, the splitting trick has been a great help with such bounds, and here it is natural to take a clue from the relation $1/p + 1/q = 1$. By splitting and by Hölder's inequality we find

$$\sum_{j=1}^{m}\sum_{k=1}^{n} c_{jk} x_k y_j = \sum_{j=1}^{m}\sum_{k=1}^{n} (c_{jk} x_k^p)^{1/p} (c_{jk} y_j^q)^{1/q}$$

$$\leq \left(\sum_{j=1}^{m}\sum_{k=1}^{n} c_{jk} x_k^p \right)^{1/p} \left(\sum_{j=1}^{m}\sum_{k=1}^{n} c_{jk} y_j^q \right)^{1/q}, \tag{9.26}$$

and now we just need to estimate the last two factors.

The first factor is easy, since our first hypothesis and the assumption $\|\mathbf{x}\|_p \leq 1$ give us the bound

$$\sum_{j=1}^{m}\sum_{k=1}^{n} c_{jk} x_k^p \leq A \sum_{k=1}^{n} x_k^p \leq A. \tag{9.27}$$

Estimation of the second is not much harder since after one crude bound

148   Hölder's Inequality

our second hypothesis and the assumption $\|\mathbf{y}\|_q \leq 1$ give us

$$\sum_{j=1}^{m}\sum_{k=1}^{n} c_{jk} y_j^q \leq \sum_{j=1}^{m} y_j^q \left\{ \max_{1 \leq j \leq m} \sum_{k=1}^{n} c_{jk} \right\} \leq B \sum_{j=1}^{m} y_j^q \leq B. \qquad (9.28)$$

Finally, when we use the estimates (9.27) and (9.28) to estimate the product (9.26), we get our target bound (9.25), and thus we complete the solution of the first challenge problem.

EXERCISES

Exercise 9.1 (Doing the Sums for Hölder)

In Exercise 1.8 we saw that the effective use of Cauchy's inequality may depend on having an estimate for one of the bounding sums and, in this respect, Hölder's inequality is a natural heir. As a warm-up, check that for real a_j, $j = 1, 2, \ldots$, one has

$$\sum_{k=1}^{n} \frac{a_k}{\{k(k+1)\}^{1/5}} < \left(\sum_{k=1}^{n} |a_k|^{5/4} \right)^{4/5}, \qquad (a)$$

$$\sum_{k=1}^{n} \frac{a_k}{\sqrt{k}} < 6^{-1/4} \sqrt{\pi} \left(\sum_{k=1}^{n} |a_k|^{4/3} \right)^{3/4}, \quad \text{and} \qquad (b)$$

$$\sum_{k=0}^{\infty} a_k x^k \leq (1 - x^3)^{-1/3} \left(\sum_{k=0}^{\infty} |a_k|^{3/2} \right)^{2/3} \quad \text{for } 0 \leq x < 1. \qquad (c)$$

Exercise 9.2 (An Inclusion Radius Bound)

For a polynomial $P(z) = z^n + a_{n-1} z^{n-1} + \cdots + a_1 z + a_0$ with real or complex coefficients, the smallest value $r(P)$ such all roots of P are contained in the disk $\{z : |z| \leq r(P)\}$ is called the *inclusion radius* for P. Show that for any conjugate pair $p > 1$ and $q = p/(p-1) > 1$ one has the bound

$$r(P) < \left(1 + A_p^q \right)^{1/q} \qquad \text{where } A_p = \left(\sum_{n=0}^{n-1} |a_j|^p \right)^{1/p}. \qquad (9.29)$$

Hölder's Inequality

Exercise 9.3 (Cauchy Implies Hölder)

Prove that Cauchy's inequality implies Hölder's inequality. More specifically, show that Cauchy's inequality implies Hölder's inequality for $p \in \{8/1, 8/2, 8/3, \ldots, 8/6, 8/7\}$ by first showing

$$\left\{\sum_{j=1}^n a_j b_j c_j d_j e_j f_j g_j h_j\right\}^8 \leq \left\{\sum_{j=1}^n a_j^8\right\}\left\{\sum_{j=1}^n b_j^8\right\}\cdots\left\{\sum_{j=1}^n h_j^8\right\}.$$

By the same method, one can prove Hölder's inequality for all $p = 2^k/j$, $1 \leq j < 2^k$. One can then call on continuity to obtain Hölder's inequality for all $1 \leq p < \infty$.

This argument serves as a reminder that an ℓ^2-result may sometimes be applied iteratively to obtain an ℓ^p-result. The inequalities one finds this way are often proved more elegantly by other methods, but iteration is still a remarkably effective tools for the *discovery* of new bounds.

Exercise 9.4 (Interpolation Bound for Moment Sequences)

If $\phi : [0, \infty) \to [0, \infty)$ is an integrable function and $t \in (0, \infty)$, then the integral

$$\mu_t = \int_0^\infty x^t \phi(x)\, dx$$

is called the tth moment of ϕ. Show that if $t \in (t_0, t_1)$ then

$$\mu_t \leq \mu_{t_0}^{1-\alpha} \mu_{t_1}^\alpha \quad \text{where } t = (1-\alpha)t_0 + \alpha t_1 \text{ and } 0 < \alpha < 1.$$

In other words, the linearly interpolated moment is bounded by the geometric interpolation of two extreme moments.

Exercise 9.5 (Complex Hölder — and the Case of Equality)

Hölder's inequality for real numbers implies that for complex numbers $a_1, a_2, \ldots, a_n$ and $b_1, b_2, \ldots, b_n$ one has the bound

$$\left|\sum_{k=1}^n a_k b_k\right| \leq \left(\sum_{k=1}^n |a_k|^p\right)^{1/p} \left(\sum_{k=1}^n |b_k|^q\right)^{1/q} \tag{9.30}$$

when $p > 1$ and $q > 1$ satisfy $1/p + 1/q = 1$. What conditions on the complex numbers $a_1, a_2, \ldots, a_n$, and $b_1, b_2, \ldots, b_n$ are necessary and sufficient equality to hold in the bound (9.30)? Although this exercise is easy, it nevertheless offers one useful morsel of insight that should not be missed.

Exercise 9.6 (Jensen Implies Minkowski)

By Jensen's inequality, we know that for a convex ϕ and positive weights $w_1, w_2, \ldots, w_n$ one has

$$\phi\left(\frac{w_1 x_1 + w_2 x_2 + \cdots + w_n x_n}{w_1 + w_2 + \cdots + w_n}\right) \leq \frac{w_1 \phi(x_1) + w_2 \phi(x_2) + \cdots + w_n \phi(x_n)}{w_1 + w_2 + \cdots + w_n}. \quad (9.31)$$

Consider the *concave* function $\phi(x) = (1 + x^{1/p})^p$ on $[0, \infty]$, and show that by making the right choice of the weights w_k and the values x_k in Jensen's inequality (9.31) one obtains Minkowski's inequality.

Exercise 9.7 (Hölder's Inequality for Integrals)

Naturally there are integral versions of Hölder's inequality and, in keeping with the more modern custom, there is no cause for a name change when one switches from sums to integrals.

Let $w : D \to [0, \infty)$ be given, and reinforce your mastery of Hölder's inequality by checking that our earlier argument (page 137) also shows that, for all suitably integrable functions f and g from D to $\mathbb{R}$,

$$\int_D f(x) g(x) w(x) \, dx \leq \left(\int_D |f(x)|^p w(x) \, dx\right)^{1/p} \left(\int_D |g(x)|^q w(x) \, dx\right)^{1/q}$$

where, as usual, $1 < p < \infty$ and $p^{-1} + q^{-1} = 1$.

Exercise 9.8 (Legendre Transforms and Young's Inequality)

If $f : (a, b) \to \mathbb{R}$, then the function $g : \mathbb{R} \to \mathbb{R}$ defined by

$$g(y) = \sup_{x \in (a,b)} \{xy - f(x)\} \quad (9.32)$$

is called the *Legendre transform* of f. It is used widely in the theory of inequalities, and part of its charm is that it helps us relate products to sums. For example, the definition (9.32) gives us the immediate bound

$$xy \leq f(x) + g(y) \quad \text{for all } (x, y) \in (a, b) \times \mathbb{R}. \quad (9.33)$$

(a) Find the Legendre transform of $f(x) = x^p/p$ for $p > 1$ and compare the general bound (9.33) to Young's inequality (9.6).

(b) Find the Legendre transforms of $f(x) = e^x$ and $\phi(x) = x \log x - x$.

(c) Show that for any function f the Legendre transform g is convex.

Hölder's Inequality

Exercise 9.9 (Self-Generalizations of Hölder's Inequality)

Hölder's inequality is self-generalizing in the sense that it implies several apparently more general inequalities. This exercise address two of the most pleasing of these generalizations.

(a) Show that for positive p, q, bigger than r one has

$$\frac{1}{p} + \frac{1}{q} = \frac{1}{r} \quad \Rightarrow \quad \left\{\sum_{j=1}^{n}(a_j b_j)^r\right\}^{1/r} \leq \left\{\sum_{j=1}^{n} a_j^p\right\}^{1/p} \left\{\sum_{j=1}^{n} b_j^q\right\}^{1/q}.$$

(b) Given $p, q,$ and r are bigger than 1, show that if

$$\frac{1}{p} + \frac{1}{q} + \frac{1}{r} = 1,$$

then one has the triple produce inequality

$$\sum_{j=1}^{n} a_j b_j c_j \leq \left\{\sum_{j=1}^{n} a_j^p\right\}^{1/p} \left\{\sum_{j=1}^{n} b_j^q\right\}^{1/q} \left\{\sum_{j=1}^{n} c_j^r\right\}^{1/r}.$$

Exercise 9.10 (The Historical Hölder Inequality)

The inequality which Hölder actually proved in his 1889 article asserts that for $w_k \geq 0$, $y_k \geq 0$, and $p > 1$ one has

$$\sum_{k=1}^{n} w_k y_k \leq \left\{\sum_{k=1}^{n} w_k\right\}^{(p-1)/p} \left\{\sum_{k=1}^{n} w_k y_k^p\right\}^{1/p}. \quad (9.34)$$

Show, as Hölder did, that this inequality follows from the weighted version (9.31) of Jensen's inequality. Finally close the loop by showing that the historical version (9.34) of Hölder's inequality is equivalent to the modern version that was introduced by F. Riesz. That is, check that inequality (9.34) implies inequality (9.1), and vice versa.

Exercise 9.11 (Minkowski Implies Hölder)

The triangle inequality implies Cauchy's inequality, so it surely seems reasonable to guess that Minkowski's inequality might also imply Hölder's inequality. The guess is true, but the confirmation is a bit subtle. As a hint, consider what Minkowski's inequality (9.12) for ℓ^s says for the vectors $\theta(a_1^{p/s}, a_2^{p/s}, \ldots, a_n^{p/s})$ and $(1-\theta)(b_1^{q/s}, b_2^{q/s}, \ldots, b_n^{q/s})$ when s is very large.

Hölder's Inequality

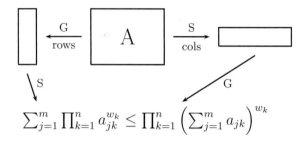

Fig. 9.2. Hölder's inequality for an array (9.35) is easier to keep in mind if one visualizes its meaning. In fact, it asserts a natural *commutativity relationship* between the summation operation S and the geometric mean operation G. As the figure suggests, if we let G act on rows and let S act on columns, then the inequality (9.35) tells us that by acting first with the geometric mean G we get a smaller number than if we act first with S.

Exercise 9.12 (Hölder's Inequality for an Array)

Any formula that generalizes Hölder's inequality to an array is likely to look complicated but, as Figure 9.2 suggests, it is still possible for such a formula to be *conceptually* simple.

Show that for nonnegative real numbers a_{jk}, $1 \leq j \leq m$, $1 \leq k \leq n$ and positive weights $w_1, \ldots, w_n$ that sum to 1, we have the bound

$$\sum_{j=1}^{m} \prod_{k=1}^{n} a_{jk}^{w_k} \leq \prod_{k=1}^{n} \left(\sum_{j=1}^{m} a_{jk} \right)^{w_k}. \tag{9.35}$$

Prove this inequality, and use it to prove the mixed mean inequality which asserts that for nonnegative x, y, z one has

$$\frac{x + (xy)^{\frac{1}{2}} + (xyz)^{\frac{1}{3}}}{3} \leq \left(x \cdot \frac{x+y}{2} \cdot \frac{x+y+z}{3} \right)^{1/3}. \tag{9.36}$$

Exercise 9.13 (Rogers's Inequality — the Proto-Hölder)

The inequality that L.C. Rogers proved in this 1888 article asserts that for $0 < r < s < t < \infty$ and for nonnegative a_k, b_k, $k = 1, 2, \ldots, n$, one has the bound

$$\left(\sum_{k=1}^{n} a_k b_k^s \right)^{t-r} \leq \left(\sum_{k=1}^{n} a_k b_k^r \right)^{t-s} \left(\sum_{k=1}^{n} a_k b_k^t \right)^{s-r},$$

which we may write more succinctly as

$$(S_s)^{t-r} \le (S_r)^{t-s}(S_t)^{s-r} \quad \text{where } S_p = \sum_{k=1}^n a_k b_k^p \text{ for } p > 0. \quad (9.37)$$

Rogers gave two proofs of his bound (9.37). In the first of these he called on the Cauchy–Binet formula [see (3.7), page 49], and the second he used the AM-GM inequality which he wrote in the form

$$x_1^{w_1} x_2^{w_2} \cdots x_n^{w_n} \le \left(\frac{w_1 x_1 + w_2 x_2 + \cdots + w_n x_n}{w_1 + w_2 + \cdots + w_n} \right)^{w_1 + w_2 + \cdots + w_n}$$

where the values $w_1, w_2, \ldots, w_n$ are assumed to be positive but which are otherwise arbitrary.

Now, follow in Rogers's footsteps and use the *very clever* substitutions $w_k = a_k b_k^s$ and $x_k = b_k^{t-s}$ to deduce the bound

$$\left(b_1^{a_1 b_1^s} b_2^{a_2 b_2^s} \cdots b_n^{a_n b_n^s} \right)^{t-s} \le (S_t/S_s)^{S_s}, \quad (9.38)$$

and use the substitutions $w_k = a_k b_k^s$ and $x_k = b_k^{r-s}$ to deduce the bound

$$\left(b_1^{a_1 b_1^s} b_2^{a_2 b_2^s} \cdots b_n^{a_n b_n^s} \right)^{r-s} \le (S_r/S_s)^{S_s}. \quad (9.39)$$

Finally, show how these two relations imply Rogers's inequality (9.37).

Exercise 9.14 (Interpolation for Positive Matrices)

Let $1 \le s_0, t_0, s_1, t_1 \le \infty$ be given and consider an $m \times n$ matrix T with nonnegative real entries c_{jk}, $1 \le j \le m$, $1 \le k \le n$. Show that if there exist constants M_0 and M_1 such that

$$\|T\mathbf{x}\|_{t_0} \le M_0 \|\mathbf{x}\|_{s_0} \quad \text{and} \quad \|T\mathbf{x}\|_{t_1} \le M_1 \|\mathbf{x}\|_{s_1} \quad (9.40)$$

for all $\mathbf{x} \in \mathbb{R}^m$, then for each $0 \le \theta \le 1$, one has the bound

$$\|T\mathbf{x}\|_t \le M_\theta \|\mathbf{x}\|_s \quad \text{for all } \mathbf{x} \in \mathbb{R}^m \quad (9.41)$$

where M_θ is defined by $M_\theta = M_1^\theta M_0^{1-\theta}$ and where s and t are given by

$$\frac{1}{s} = \frac{\theta}{s_1} + \frac{1-\theta}{s_0}, \quad \frac{1}{t} = \frac{\theta}{t_1} + \frac{1-\theta}{t_0}. \quad (9.42)$$

This problem takes some time to absorb, but the result is important, and it pays generous interest on all invested effort. Figure 9.3 should help one visualize the condition (9.42) and the constraints on the parameters $1 \le s_0, t_0, s_1, t_1 \le \infty$. One might also note that the bound (9.41) would follow trivially from the hypotheses (9.40) if $\theta = 0$ or $\theta = 1$. Moreover,

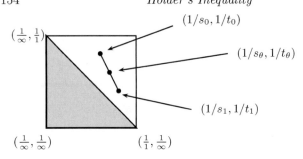

Fig. 9.3. The constraints $1 \le s_0, t_0, s_1, t_1 \le \infty$ mean that the reciprocals are contained in the unit square $S = [0,1] \times [0,1]$, and the exponent relation (9.42) tells us that $(1/s, 1/t)$ is on the line from $(1/s_1, 1/t_1)$ to $(1/s_0, 1/t_0)$. The parameter θ is then determined by the explicit interpolation formula $(1/s, 1/t) = \theta(1/s_1, 1/t_1) + (1-\theta)(1/s_0, 1/t_0)$.

the bound (9.41) automatically recaptures the inequality (9.24) from Challenge Problem 9.6; one only needs to set $t_1 = 1$, $s_1 = 1$, $M_1 = A$, $t_0 = \infty$, $s_0 = \infty$, $M_0 = B$, and $\theta = 1/p$.

Despite the apparent complexity of Exercise 9.14, one does not need to look far to find a plan for proving the interpolation formula (9.41). The strategy which worked for Problem 9.6 (page 146) seems likely to work here, even though it may put one's skill with the splitting trick to the test.

Finally, for anyone who may still be hesitant to take up the challenge of Exercise 9.14, there is one last appeal: first think about proving the more concrete inequality (9.43) given below. This inequality is typical of a large class of apparently tough problems which crumble quickly after one calls on the interpolation formula (9.41).

Exercise 9.15 (An ℓ^2 Interpolation Bound)

Let c_{jk}, $1 \le j \le m$, $1 \le k \le n$ be an array of nonnegative real numbers for which one has the implication

$$X_j = \sum_{k=1}^{n} c_{jk} x_k \quad \text{for all } j = 1, 2, \ldots, m \quad \Rightarrow \quad \sum_{j=1}^{n} |X_j|^2 \le \sum_{k=1}^{n} |x_k|^2.$$

Show that for all $1 \le p \le 2$ one then has the bound

$$\left(\sum_{j=1}^{m} |X_j|^q \right)^{1/q} \le M^{(2-p)/p} \left(\sum_{k=1}^{n} |x_k|^p \right)^{1/p} \tag{9.43}$$

where and $q = p/(p-1)$ and $M = \max |c_{jk}|$.

10
Hilbert's Inequality and Compensating Difficulties

Some of the most satisfying experiences in problem solving take place when one starts out on a natural path and then bumps into an unexpected difficulty. On occasion this deeper view of the problem forces us to look for an entirely new approach. Perhaps more often we only need to find a way to press harder on an appropriate variation of the original plan.

This chapter's introductory problem provides an instructive case; here we will discover *two difficulties*. Nevertheless, we manage to achieve our goal by pitting one difficulty against the other.

Problem 10.1 (Hilbert's Inequality)

Show that there is a constant C such that for every pair of sequences of real numbers $\{a_n\}$ and $\{b_n\}$ one has

$$\sum_{m=1}^{\infty} \sum_{n=1}^{\infty} \frac{a_m b_n}{m+n} < C \left(\sum_{m=1}^{\infty} a_m^2 \right)^{\frac{1}{2}} \left(\sum_{n=1}^{\infty} b_n^2 \right)^{\frac{1}{2}}. \qquad (10.1)$$

SOME HISTORICAL BACKGROUND

This famous inequality was discovered in the early 1900s by David Hilbert; specifically, Hilbert proved that the inequality (10.1) holds with $C = 2\pi$. Several years after Hilbert's discovery, Issai Schur provided a new proof which showed Hilbert's inequality actually holds with $C = \pi$. We will see shortly that no smaller value of C will suffice.

Despite the similarities between Hilbert's inequality and Cauchy's inequality, Hilbert's original proof did not call on Cauchy's inequality; he took an entirely different approach that exploited the evaluation of some cleverly chosen trigonometric integrals. Nevertheless, one can prove

156   Hilbert's Inequality and Compensating Difficulties

Hilbert's inequality through an appropriate application of Cauchy's inequality. The proof turns out to be both simple and instructive.

If S is any countable set and $\{\alpha_s\}$ and $\{\beta_s\}$ are collections of real numbers indexed by S, then Cauchy's inequality can be written as

$$\sum_{s \in S} \alpha_s \beta_s \leq \left(\sum_{s \in S} \alpha_s^2 \right)^{\frac{1}{2}} \left(\sum_{s \in S} \beta_s^2 \right)^{\frac{1}{2}}. \tag{10.2}$$

This modest reformulation of Cauchy's inequality sometimes helps us see the possibilities more clearly, and here, of course, one hopes that wise choices for S, $\{\alpha_s\}$, and $\{\beta_s\}$ will lead us from the bound (10.2) to the Hilbert's inequality (10.1).

An Obvious First Attempt

If we charge ahead without too much thought, we might simply take the index set to be $S = \{(m, n) : m \geq 1, n \geq 1\}$ and take α_s and β_s to be defined by the splitting

$$\alpha_s = \frac{a_m}{\sqrt{m+n}} \quad \text{and} \quad \beta_s = \frac{b_n}{\sqrt{m+n}} \quad \text{where } s = (m, n).$$

By design, the products $\alpha_s \beta_s$ recapture the terms one finds on the left-hand side of Hilbert's inequality, but the bound one obtains from Cauchy's inequality (10.2) turns out to be disappointing. Specifically, it gives us the double sum estimate

$$\left(\sum_{m=1}^{\infty} \sum_{n=1}^{\infty} \frac{a_m b_n}{m+n} \right)^2 \leq \sum_{m=1}^{\infty} \sum_{n=1}^{\infty} \frac{a_m^2}{m+n} \sum_{n=1}^{\infty} \sum_{m=1}^{\infty} \frac{b_n^2}{m+n} \tag{10.3}$$

but, unfortunately, both of the last two factors turn out to be infinite.

The first factor on the right side of the bound (10.3) diverges like a harmonic series when we sum on n, and the second factor diverges like a harmonic series when we sum on m. Thus, in itself, inequality (10.3) is virtually worthless. Nevertheless, if we look more deeply, we soon find that the complementary nature of these failings points the way to a wiser choice of $\{\alpha_s\}$ and $\{\beta_s\}$.

Exploiting Compensating Difficulties

The two sums on the right-hand side of the naive bound (10.3) diverge, but the good news is that they diverge for different reasons. In a sense, the first factor diverges because

$$\alpha_s = \frac{a_m}{\sqrt{m+n}}$$

Hilbert's Inequality and Compensating Difficulties

is too big as a function of n, whereas the second factor diverges because

$$\beta_s = \frac{b_n}{\sqrt{m+n}}$$

is too big as a function of m. All told, this suggests that we might improve on α_s and β_s if we multiply α_s by a decreasing function of n and multiply β_s by a decreasing function of m. Since we want to preserve the basic property that

$$\alpha_s \beta_s = \frac{a_m b_n}{m+n},$$

we may not need long to hit on the idea of introducing a parametric family of candidates such as

$$\alpha_s = \frac{a_m}{\sqrt{m+n}} \left(\frac{m}{n}\right)^\lambda \quad \text{and} \quad \beta_s = \frac{b_n}{\sqrt{m+n}} \left(\frac{n}{m}\right)^\lambda, \qquad (10.4)$$

where $s = (m, n)$ and where $\lambda > 0$ is a constant that can be chosen later. This new family of candidates turns out to lead us quickly to the proof of Hilbert's inequality.

EXECUTION OF THE PLAN

When we apply Cauchy's inequality (10.2) to the pair (10.4), we find

$$\left(\sum_{m=1}^\infty \sum_{n=1}^\infty \frac{a_m b_n}{m+n} \right)^2 \leq \sum_{m=1}^\infty \sum_{n=1}^\infty \frac{a_m^2}{m+n}\left(\frac{m}{n}\right)^{2\lambda} \sum_{n=1}^\infty \sum_{m=1}^\infty \frac{b_n^2}{m+n}\left(\frac{n}{m}\right)^{2\lambda},$$

so, when we consider the first factor on the right-hand side we see

$$\sum_{m=1}^\infty \sum_{n=1}^\infty \frac{a_m^2}{m+n}\left(\frac{m}{n}\right)^{2\lambda} = \sum_{m=1}^\infty a_m^2 \sum_{n=1}^\infty \frac{1}{m+n}\left(\frac{m}{n}\right)^{2\lambda}.$$

By the symmetry of the summands $a_m b_n/(m+n)$ in our target sum, we now see that the proof of Hilbert's inequality will be complete if we can show that for some choice of λ there is a constant $B_\lambda < \infty$ such that

$$\sum_{n=1}^\infty \frac{1}{m+n}\left(\frac{m}{n}\right)^{2\lambda} \leq B_\lambda \quad \text{for all } m \geq 1. \qquad (10.5)$$

Now we just need to estimate the sum (10.5), and we first recall that for any nonnegative decreasing function $f : [0, \infty) \to \mathbb{R}$, we have the integral bound

$$\sum_{n=1}^\infty f(n) \leq \int_0^\infty f(x)\, dx.$$

In the specific case of $f(x) = m^{2\lambda}x^{2\lambda}(m+x)^{-1}$, we therefore find

$$\sum_{n=1}^{\infty} \frac{1}{m+n}\left(\frac{m}{n}\right)^{2\lambda} \leq \int_0^{\infty} \frac{1}{m+x}\frac{m^{2\lambda}}{x^{2\lambda}}\,dx = \int_0^{\infty} \frac{1}{(1+y)}\frac{1}{y^{2\lambda}}\,dy, \quad (10.6)$$

where the last equality comes from the change of variables $x = my$. The integral on the right side of the inequality (10.6) is clearly convergent when λ satisfies $0 < \lambda < 1/2$ and, by our earlier observation (10.5), the existence of any such λ would suffice to complete the proof of Hilbert's inequality (10.1).

Seizing an Opportunity

Our problem has been solved as stated, but we would be derelict in our duties if we did not take a moment to find the value of the constant C that is provided by our proof. When we look over our argument, we actually find that we have proved that Hilbert's inequality (10.1) must hold for any $C = C_\lambda$ with

$$C_\lambda = \int_0^{\infty} \frac{1}{(1+y)}\frac{1}{y^{2\lambda}}\,dy \quad \text{for } 0 < \lambda < 1/2. \quad (10.7)$$

Naturally, we should find the value of λ that provides the smallest of these.

By a quick and lazy consultation of *Mathematica* or *Maple*, we discover that we are in luck. The integral for C_λ turns out to both simple and explicit:

$$\int_0^{\infty} \frac{1}{(1+y)}\frac{1}{y^{2\lambda}}\,dy = \frac{\pi}{\sin 2\pi\lambda} \quad \text{for } 0 < \lambda < 1/2. \quad (10.8)$$

Now, since $\sin 2\pi\lambda$ is maximized when $\lambda = 1/4$, we see that the smallest value attained by C_λ with $0 < \lambda < 1/2$ is equal to

$$C = C_{1/4} = \int_0^{\infty} \frac{1}{(1+y)}\frac{1}{\sqrt{y}}\,dy = \pi. \quad (10.9)$$

Quite remarkably, our direct assault on Hilbert's inequality has almost effortlessly provided the sharp constant $C = \pi$ that was discovered by Schur.

This is a fine achievement for Cauchy's inequality, but it should not be oversold. Many proofs of Hilbert's inequality are now available, and some of these are quite brief. Nevertheless, for the connoisseur of techniques for exploiting Cauchy's inequality, this proof of Hilbert's inequality is a sweet victory.

Finally, there is a small point that we should note in passing. The

Hilbert's Inequality and Compensating Difficulties

integral (10.8) is actually a textbook classic; both Bak and Newman (1997) and Cartan (1995) use it to illustrate the standard technique for integrating $R(x)/x^\alpha$ over $[0, \infty)$ where $R(x)$ is a rational function and $0 < \alpha < 1$. This integral also has a connection to a noteworthy gamma function identity that is described in Exercise 10.8.

Of Miracles and Converses

For a Cauchy–Schwarz argument to be precise enough to show that one can take $C = \pi$ in Hilbert's inequality may seem to require a miracle, but there is another way of looking at the relation between the two sides of Hilbert's inequality that makes it clear that no miracle was required. With the right point of view, one can see that both π and the special integrals (10.8) have an inevitable role. To develop this connection, we will take on the challenge of proving a converse to our first problem.

Problem 10.2 *Suppose that the constant C satisfies*

$$\sum_{m=1}^{\infty} \sum_{n=1}^{\infty} \frac{a_m b_n}{m+n} < C \left(\sum_{m=1}^{\infty} a_m^2 \right)^{\frac{1}{2}} \left(\sum_{n=1}^{\infty} b_n^2 \right)^{\frac{1}{2}} \qquad (10.10)$$

for all pairs of sequences of real numbers $\{a_n\}$ and $\{b_n\}$. Show that $C \geq \pi$.

If we plug any pair of sequences $\{a_n\}$ and $\{b_n\}$ into the inequality (10.10) we will get some lower bound on c, but we will not get too far with this process unless we find some systematic way to guide our choices. What we would really like is a parametric family of pairs $\{a_n(\epsilon)\}$ and $\{b_n(\epsilon)\}$ that provide us with a sequence of lower bounds on C that approach π as $\epsilon \to 0$. This surely sounds good, but how do we find appropriate candidates for $\{a_n(\epsilon)\}$ and $\{b_n(\epsilon)\}$?

Stress Testing an Inequality

Two basic ideas can help us narrow our search. First, we need to be able to calculate (or estimate) the sums that appear in the inequality (10.10). We cannot do many sums, so this definitely limits our search. The second idea is more subtle; we need to put the inequality *under stress*. This general notion has many possible interpretations, but here it at least suggests that we should look for sequences $\{a_n(\epsilon)\}$ and $\{b_n(\epsilon)\}$ such that all the quantities in the inequality (10.10) tend to infinity as $\epsilon \to 0$. This particular strategy for stressing the inequality (10.10) may not seem too compelling when one faces it for the first time, but experience with even a few examples is enough to convince most people that the principle contains more than a drop of wisdom.

Without a doubt, the most natural candidates for $\{a_n(\epsilon)\}$ and $\{b_n(\epsilon)\}$ are given by the identical twins

$$a_n(\epsilon) = b_n(\epsilon) = n^{-\frac{1}{2}-\epsilon}.$$

For this choice, one may easily work out the estimates that are needed to understand the right-hand side of Hilbert's inequality. Specifically, we see that as $\epsilon \to 0$ we have

$$\left(\sum_{m=1}^{\infty} a_m^2(\epsilon)\right)^{\frac{1}{2}} \left(\sum_{n=1}^{\infty} b_n^2(\epsilon)\right)^{\frac{1}{2}} = \sum_{n=1}^{\infty} \frac{1}{n^{1+2\epsilon}} \sim \int_1^{\infty} \frac{dx}{x^{1+2\epsilon}} = \frac{1}{2\epsilon}. \quad (10.11)$$

CLOSING THE LOOP

To complete the solution of Problem 10.2, we only need to show that the corresponding sum for the left-hand side of Hilbert's inequality (10.10) is asymptotic to $\pi/2\epsilon$ as $\epsilon \to 0$. This is indeed the case, and the computation is instructive. We lay out the result as a lemma.

Double Sum Lemma.

$$\sum_{m=1}^{\infty} \sum_{n=1}^{\infty} \frac{1}{n^{\frac{1}{2}+\epsilon}} \frac{1}{m^{\frac{1}{2}+\epsilon}} \frac{1}{m+n} \sim \frac{\pi}{2\epsilon} \quad \text{as } \epsilon \to 0.$$

For the proof, we first note that integral comparisons tell us that it suffices to show

$$I(\epsilon) = \int_1^{\infty} \int_1^{\infty} \frac{1}{x^{\frac{1}{2}+\epsilon}} \frac{1}{y^{\frac{1}{2}+\epsilon}} \frac{1}{x+y} \, dx dy \sim \frac{\pi}{2\epsilon} \quad \text{as } \epsilon \to 0,$$

and the change of variables $u = y/x$ also tells us that

$$I(\epsilon) = \int_1^{\infty} x^{-1-2\epsilon} \left[\int_{1/x}^{\infty} u^{-\frac{1}{2}-\epsilon} \frac{du}{1+u}\right] dx. \quad (10.12)$$

This integral would be easy to calculate if we could replace the lower limit $1/x$ of the inside integral by 0, and, to estimate how much damage such a change would cause, we first note that

$$0 < \int_0^{1/x} u^{-\frac{1}{2}-\epsilon} \frac{du}{1+u} < \int_0^{1/x} u^{-\frac{1}{2}-\epsilon} du = \frac{x^{-\frac{1}{2}+\epsilon}}{\frac{1}{2}-\epsilon}.$$

When we use this bound in equation (10.12) and write the result using

Hilbert's Inequality and Compensating Difficulties

big O notation of Landau (say, as defined on page 120), then we find

$$I(\epsilon) = \int_1^\infty x^{-1-2\epsilon} \left\{ \int_0^\infty u^{-\frac{1}{2}-\epsilon} \frac{du}{1+u} \right\} dx + O\left(\int_1^\infty x^{-\frac{3}{2}-\epsilon} dx \right)$$

$$= \frac{1}{2\epsilon} \int_0^\infty u^{-\frac{1}{2}-\epsilon} \frac{du}{1+u} + O(1).$$

Finally, for $\epsilon \to 0$, we see from our earlier experience with the integral (10.9) that we have

$$\int_0^\infty u^{-\frac{1}{2}-\epsilon} \frac{du}{1+u} \to \int_0^\infty u^{-\frac{1}{2}} \frac{du}{1+u} = \pi,$$

so the proof of the lemma is complete.

FINDING THE CIRCLE IN HILBERT'S INEQUALITY

Any time π appears in a problem that has no circle in sight, there is a certain sense of mystery. Sometimes this mystery remains without a satisfying resolution, but, in the case of Hilbert's inequality, a geometric explanation for the appearance of π was found in 1993 by Krysztof Oleszkiewicz. This discovery is a bit off of our central theme, but it does build on the calculations we have just completed, and it is too lovely to miss.

Quarter Circle Lemma. For all $m \geq 1$, we have the bound

$$\sum_{n=1}^\infty \frac{1}{m+n} \left(\frac{m}{n} \right)^{\frac{1}{2}} < \pi. \qquad (10.13)$$

For the proof, we first note that the shaded triangle of Figure 10.1 is similar to the triangle T determined by $(0,0)$, $(\sqrt{m}, \sqrt{n-1})$, and $(\sqrt{m}, \sqrt{n})$, and the area of T is simply $\frac{1}{2}\sqrt{m}(\sqrt{n} - \sqrt{n-1})$. Thus, one finds by scaling that the area A_n of the shaded triangle is given by

$$A_n = \left(\frac{\sqrt{m}}{\sqrt{n+m}} \right)^2 \frac{1}{2}\sqrt{m}(\sqrt{n} - \sqrt{n-1}). \qquad (10.14)$$

Since $1/\sqrt{x}$ is decreasing on $[0, \infty)$, we have

$$\sqrt{n} - \sqrt{n-1} = \frac{1}{2} \int_{n-1}^n \frac{dx}{\sqrt{x}} > \frac{1}{2\sqrt{n}}$$

so, in the end, we find

$$A_n > \frac{1}{4} \frac{m}{m+n} \frac{\sqrt{m}}{\sqrt{n}}. \qquad (10.15)$$

Finally, what makes this geometric bound most interesting is that all

Hilbert's Inequality and Compensating Difficulties

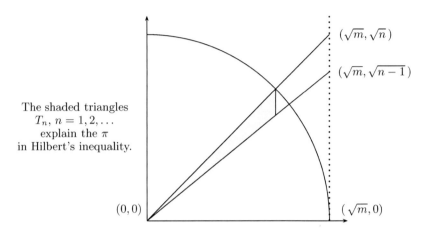

The shaded triangles T_n, $n = 1, 2, \ldots$ explain the π in Hilbert's inequality.

Fig. 10.1. The shaded triangle is similar to the triangle determined by the three points $(0,0)$, $(\sqrt{m}, \sqrt{n-1})$, and $(\sqrt{m}, \sqrt{n})$ so we can determine its area by geometry. Also, the triangles T_n have disjoint interiors so the sum of their areas cannot exceed $\pi/4$. These facts give us the proof of the Quarter Circle Lemma.

of the shaded triangles are contained in the quarter circle. They have disjoint interiors, so we find that the sum of their areas is bounded by $\pi m/4$, the area of the quarter circle with radius $\sqrt{m}$ that contains them.

Exercises

Exercise 10.1 (Guaranteed Positivity)
Show that for any real numbers $a_1, a_2, \ldots, a_n$ one has

$$\sum_{j,k=1}^{n} \frac{a_j a_k}{j+k} \geq 0 \tag{10.16}$$

and, more generally, show that for positive $\lambda_1, \lambda_2, \ldots, \lambda_n$ one has

$$\sum_{j,k=1}^{n} \frac{a_j a_k}{\lambda_j + \lambda_k} \geq 0. \tag{10.17}$$

Obviously the second inequality implies the first, so the bound (10.16) is mainly a hint which makes the link to Hilbert's inequality. As a better hint, one might consider the possibility of representing $1/\lambda_j$ as an integral.

Exercise 10.2 (Insertion of a Fudge Factor)

There are many ways to continue the theme of Exercise 10.1, and this exercise is one of the most useful. It provides a generic way to leverage an inequality such as Hilbert's.

Show that if the complex array $\{a_{jk} : 1 \leq j \leq m, 1 \leq k \leq n\}$ satisfies the bound

$$\left| \sum_{j,k} a_{jk} x_j y_k \right| \leq M \|x\|_2 \|y\|_2, \qquad (10.18)$$

then one also has the bound

$$\left| \sum_{j,k} a_{jk} h_{jk} x_j y_k \right| \leq \alpha \beta M \|x\|_2 \|y\|_2 \qquad (10.19)$$

provided that the factors h_{jk} have an integral representation of the form

$$h_{jk} = \int_D f_j(x) g_k(x)\, dx \qquad (10.20)$$

for which for all j and k one has the bounds

$$\int_D |f_j(x)|^2\, dx \leq \alpha^2 \quad \text{and} \quad \int_D |g_k(x)|^2\, dx \leq \beta^2. \qquad (10.21)$$

Exercise 10.3 (Max Version of Hilbert's Inequality)

Show that for every pair of sequences of real numbers $\{a_n\}$ and $\{b_n\}$ one has

$$\sum_{m=1}^{\infty} \sum_{n=1}^{\infty} \frac{a_m b_n}{\max(m,n)} < 4 \left(\sum_{m=1}^{\infty} a_m^2 \right)^{\frac{1}{2}} \left(\sum_{n=1}^{\infty} b_n^2 \right)^{\frac{1}{2}}, \qquad (10.22)$$

and show that 4 may not be replaced by a smaller constant.

Exercise 10.4 (Integral Version)

Prove the integral form of Hilbert's inequality. That is, show that for any $f, g : [0, \infty) \to \mathbb{R}$, one has

$$\int_0^\infty \int_0^\infty \frac{f(x)g(y)}{x+y}\, dx dy < \pi \left(\int_0^\infty |f(x)|^2\, dx \right)^{\frac{1}{2}} \left(\int_0^\infty |g(y)|^2\, dy \right)^{\frac{1}{2}}.$$

The discrete Hilbert inequality (10.1) can be used to prove a continuous version, but the strict inequality would be lost in the process. Typically, it is better to mimic the earlier argument rather than to apply the earlier result.

Exercise 10.5 (Homogeneous Kernel Version)

If the function $K : [0, \infty) \times [0, \infty) \to [0, \infty)$ has the homogeneity property $K(\lambda x, \lambda y) = \lambda^{-1} K(x, y)$ for all $\lambda > 0$, then for any pair of functions $f, g : [0, \infty) \to \mathbb{R}$, one has

$$\int_0^\infty \int_0^\infty K(x,y) f(x) g(y) \, dx dy$$
$$< C \left(\int_0^\infty |f(x)|^2 \, dx \right)^{\frac{1}{2}} \left(\int_0^\infty |g(y)|^2 \, dy \right)^{\frac{1}{2}},$$

where the constant C is given by common value of the integrals

$$\int_0^\infty K(1,y) \frac{1}{\sqrt{y}} \, dy = \int_0^\infty K(y,1) \frac{1}{\sqrt{y}} \, dy = \int_1^\infty \frac{K(1,y) + K(y,1)}{\sqrt{y}} \, dy.$$

Exercise 10.6 (The Method of "Parameterized Parameters")

For any positive weights w_k, $k = 1, 2, \ldots, n$, Cauchy's inequality can be restated as a bound on the square of a general sum,

$$(a_1 + a_2 + \cdots + a_n)^2 \leq \left\{ \sum_{k=1}^n \frac{1}{w_k} \right\} \left\{ \sum_{k=1}^n a_k^2 w_k \right\}, \qquad (10.23)$$

and given such a bound it is sometimes useful to note the values w_k, $k = 1, 2, \ldots, n$, can be regarded as *free parameters*. The natural question then becomes, "What can be done with this freedom?" Oddly enough, one may then benefit from introducing yet another real parameter t so that we can write each weight w_k as $w_k(t)$. This purely psychological step hopes to simplify our search for a wise choice of the w_k by refocusing our attention on desirable properties of the *functions* $w_k(t)$, $k = 1, 2, \ldots, n$.

Here we want to squeeze information out of the bound (10.23), and one concrete idea is to look for choices where (1) the first factor of the product (10.23) is bounded uniformly in t and where (2) one can calculate the minimum value over all t of the second factor. These may seem like tall orders, but they can be filled and the next three steps show how this plan leads to some marvelous inferences.

(a) Show that if one takes $w_k(t) = t + k^2/t$ for $k = 1, 2, \ldots, n$ then the first factor of the inequality (10.23) is bounded by $\pi/2$ for all $t \geq 0$ and all $n = 1, 2, \ldots$.

(b) Show that for this choice we also have the identity

$$\min_{t : t \geq 0} \left\{ \sum_{k=1}^n a_k^2 w_k(t) \right\} = 2 \left\{ \sum_{k=1}^n a_k^2 \right\}^{\frac{1}{2}} \left\{ \sum_{k=1}^n k^2 a_k^2 \right\}^{\frac{1}{2}}.$$

Hilbert's Inequality and Compensating Difficulties   165

(c) Combine the preceding observations to conclude that

$$\left\{\sum_{k=1}^{n} a_k\right\}^4 \leq \pi^2 \left\{\sum_{k=1}^{n} a_k^2\right\}\left\{\sum_{k=1}^{n} k^2 a_k^2\right\}. \quad (10.24)$$

This curious bound is known as Carlson's inequality, and it has been known since 1934. Despite several almost arbitrary steps on the path to the inequality (10.24), the value π^2 cannot be replaced by a smaller one, as one can prove by the stress testing method (page 159), though not without thought.

Exercise 10.7 (Hilbert's Inequality via the Toeplitz Method)
Show that the elementary integral

$$\frac{1}{2\pi} \int_0^{2\pi} (t - \pi) e^{int} dt = \frac{1}{in},$$

for $n \neq 0$, implies that for real a_k, b_k, $1 \leq k \leq N$ one has the integral representation

$$I = \frac{1}{2\pi} \int_0^{2\pi} (t - \pi) \sum_{k=1}^{N} a_k\, e^{ikt} \sum_{k=1}^{N} b_k\, e^{ikt} dt = \sum_{m=1}^{N} \sum_{n=1}^{N} \frac{a_m\, b_n}{m+n},$$

then show that this representation and Schwarz's inequality yield a quick and easy proof of Hilbert's inequality.

Exercise 10.8 (Functional Equation for the Gamma Function)
Recall that the gamma function is defined by the integral

$$\Gamma(\lambda) = \int_0^\infty x^{\lambda-1} e^{-x}\, dx,$$

and use an integral representation for $1/(1+y)$ to show that

$$\int_0^\infty \frac{1}{(1+y)} \frac{1}{y^{2\lambda}}\, dy = \Gamma(2\lambda)\Gamma(1-2\lambda) \quad \text{for } 0 < \lambda < 1/2. \quad (10.25)$$

As a consequence, one finds that the evaluation of the integral (10.8) yields the famous functional equation for the Gamma function,

$$\Gamma(2\lambda)\Gamma(1-2\lambda) = \frac{\pi}{\sin 2\pi\lambda}.$$

11
Hardy's Inequality and the Flop

The *flop* is a simple algebraic manipulation, but many who master it feel that they are forever changed. This is not to say that the flop is particularly miraculous; in fact, it is perfectly ordinary. What may distinguish the flop among mathematical techniques is that it works at two levels: it is *tactical* in that it is just a step in an argument, and it is *strategic* in that it suggests general plans which can have a variety of twists and turns.

To illustrate the flop, we call on a concrete challenge problem of independent interest. This time the immediate challenge is to prove an inequality of G.H. Hardy which he discovered while looking for a new proof of the famous inequality of Hilbert that anchored the preceding chapter. Hardy's inequality is now widely used in both pure and applied mathematics, and many would consider it to be equal in importance to Hilbert's inequality.

Problem 11.1 (Hardy's Inequality)

Show that every integrable function $f : (0, T) \to \mathbb{R}$ satisfies the inequality

$$\int_0^T \left\{ \frac{1}{x} \int_0^x f(u)\, du \right\}^2 dx \leq 4 \int_0^T f^2(x)\, dx \tag{11.1}$$

and show, moreover, that the constant 4 cannot be replaced with any smaller value.

To familiarize this inequality, one should note that it provides a concrete interpretation of the general idea that the average of a function typically behaves as well (or at least not much worse) than the function itself. Here we see that the square integral of the average is never more than four times the square integral of the original.

Hardy's Inequality and the Flop

To deepen our understanding of the bound (11.1), we might also see if we can confirm that the constant 4 is actually the best one can do. One natural idea is to try the stress testing method (page 159) which helped us before. Here the test function that seems to occur first to almost everyone is simply the power map $x \mapsto x^\alpha$. When we substitute this function into an inequality of the form

$$\int_0^T \left\{ \frac{1}{x} \int_0^x f(u)\,du \right\}^2 dx \le C \int_0^T f^2(x)\,dx, \qquad (11.2)$$

we see that it implies

$$\frac{1}{(\alpha+1)^2 (2\alpha+1)} \le \frac{C}{(2\alpha+1)} \qquad \text{for all } \alpha \text{ such that } 2\alpha + 1 > 0.$$

Now, by letting $\alpha \to -1/2$, we see that for the bound (11.2) to hold in general one must have $C \ge 4$. Thus, we have another pleasing victory for the stress testing technique. Knowing that a bound cannot be improved always adds some extra zest to the search for a proof.

Integration by Parts — and On Speculation

Any time we work with an integral we must keep in mind the many alternative forms that it can take after a change of variables or other transformation. Here we want to bound the integral of a product of two functions, so integration by parts naturally suggests itself, especially after the integral is rewritten as

$$I = \int_0^T \left\{ \int_0^x f(u)\,du \right\}^2 \frac{1}{x^2}\,dx = -\int_0^T \left\{ \int_0^x f(u)\,du \right\}^2 \left(\frac{1}{x}\right)' dx.$$

There is no way to know *a priori* if an integration by parts will provide us with a more convenient formulation of our problem, but there is also no harm in trying, so, for the moment, we simply compute

$$I = 2 \int_0^T \left\{ \int_0^x f(u)\,du \right\} f(x) \frac{1}{x}\,dx - \left|_0^T \left\{ \int_0^x f(u)\,du \right\}^2 \frac{1}{x} \right. . \qquad (11.3)$$

Now, to simplify the last expression, we first note that we may assume that f is square integrable, or else our target inequality (11.1) is trivially true. Also, we note that for any square integrable f, Schwarz's inequality and the 1-trick tell us that for any $x \ge 0$ we have

$$\left| \int_0^x f(u)\,du \right| \le x^{\frac{1}{2}} \left\{ \int_0^x f^2(u)\,du \right\}^{\frac{1}{2}} = o(x^{\frac{1}{2}}) \qquad \text{as } x \to 0,$$

so our integration by parts formula (11.3) may be simplified to

$$I = 2\int_0^T \left\{\int_0^x f(u)\,du\right\} f(x)\frac{1}{x}\,dx - \frac{1}{T}\left\{\int_0^T f(u)\,du\right\}^2.$$

This form of the integral I may not look any more convenient than the original representation, but it does suggest a bold action. The last term is nonpositive, so we can simply discard it from the identity to get

$$\int_0^T \left\{\frac{1}{x}\int_0^x f(u)\,du\right\}^2 dx \leq 2\int_0^T \left\{\frac{1}{x}\int_0^x f(u)\,du\right\} f(x)\,dx. \quad (11.4)$$

We now face a bottom line question: Is this new bound (11.4) strong enough to imply our target inequality (11.1)? The answer turns out to be both quick and instructive.

APPLICATION OF THE FLOP

If we introduce functions φ and ψ by setting

$$\varphi(x) = \frac{1}{x}\int_0^x f(u)\,du \quad \text{and} \quad \psi(x) = f(x), \quad (11.5)$$

then the new inequality (11.4) can be written crisply as

$$\int_0^T \varphi^2(x)\,dx \leq C\int_0^T \varphi(x)\psi(x)\,dx, \quad (11.6)$$

where $C = 2$. The critical feature of this inequality is that the function φ is raised to a higher power on the left side of the equation than on the right. This is far from a minor detail; it opens up the possibility of a maneuver which has featured in thousands of investigations.

The key observation is that by applying Schwarz's inequality to the right-hand side of the inequality (11.6), we find

$$\int_0^T \varphi^2(x)\,dx \leq C\left\{\int_0^T \varphi^2(x)\,dx\right\}^{\frac{1}{2}}\left\{\int_0^T \psi^2(x)\,dx\right\}^{\frac{1}{2}} \quad (11.7)$$

so, if $\varphi(x)$ is not identically zero, we can divide both sides of this inequality by

$$\left\{\int_0^T \varphi^2(x)\,dx\right\}^{\frac{1}{2}} \neq 0.$$

This division gives us

$$\left\{\int_0^T \varphi^2(x)\,dx\right\}^{\frac{1}{2}} \leq C\left\{\int_0^T \psi^2(x)\,dx\right\}^{\frac{1}{2}}, \quad (11.8)$$

Hardy's Inequality and the Flop

and, when we square this inequality and replace C, φ, and ψ with their defining values (11.5), we see that the "postflop" inequality (11.8) is exactly the same as the target inequality (11.1) which we hoped to prove.

A Discrete Analog

One can always ask if a given result for real or complex functions has an analog for finite or infinite sequences, and the answer is often routine. Nevertheless, there are also times when one meets unexpected difficulties that lead to new insight. We will face just such a situation in our second challenge problem.

Problem 11.2 (The Discrete Hardy Inequality)

Show that for any sequence of nonnegative real numbers $a_1, a_2, \ldots, a_N$ one has the inequality

$$\sum_{n=1}^{N} \left\{ \frac{1}{n}(a_1 + a_2 + \cdots + a_n) \right\}^2 \leq 4 \sum_{n=1}^{N} a_n^2. \qquad (11.9)$$

Surely the most natural way to approach this problem is to mimic the method we used for the first challenge problem. Moreover, our earlier experience also provides mileposts that can help us measure our progress. In particular, it is reasonable to guess that to prove the inequality (11.9) by an application of a flop, then we might do well to look for a "preflop" inequality of the form

$$\sum_{n=1}^{N} \left\{ \frac{1}{n}(a_1+a_2+\cdots+a_n) \right\}^2 \leq 2 \sum_{n=1}^{N} \left\{ \frac{1}{n}(a_1+a_2+\cdots+a_n) \right\} a_n, \qquad (11.10)$$

which is the natural analog of our earlier preflop bound (11.4).

Following the Natural Plan

Summation by parts is the natural analog of integration by parts, although it is a bit less mechanical. Here, for example, we must decide how to represent $1/n^2$ as a difference; after all, we can either write

$$\frac{1}{n^2} = s_n - s_{n+1} \quad \text{where } s_n = \sum_{k=n}^{\infty} \frac{1}{k^2}$$

or, alternatively, we can look at the initial sum and write

$$\frac{1}{n^2} = \tilde{s}_n - \tilde{s}_{n-1} \quad \text{where } \tilde{s}_n = \sum_{k=1}^{n} \frac{1}{k^2}.$$

The only universal basis for a sound choice is experimentation, so, for the moment, we simply take the first option.

Now, if we let T_N denote the sum on the left-hand side of the target inequality (11.9), then we have

$$T_N = \sum_{n=1}^{N} (s_n - s_{n+1})(a_1 + a_2 + \cdots + a_n)^2,$$

so, by distributing the sums and shifting the indices, we have

$$T_N = \sum_{n=1}^{N} s_n (a_1 + a_2 + \cdots + a_n)^2 - \sum_{n=2}^{N+1} s_n (a_1 + a_2 + \cdots + a_{n-1})^2.$$

When we bring the sums back together, we see that T_N equals

$$s_1 a_1^2 - s_{N+1}(a_1 + a_2 + \cdots + a_n)^2 + \sum_{n=2}^{N} s_n \{2(a_1 + a_2 + \cdots + a_{n-1})a_n + a_n^2\}$$

and, since $s_{N+1}(a_1 + a_2 + \cdots + a_n)^2 \geq 0$, we at last find

$$\sum_{n=1}^{N} \left\{ \frac{1}{n}(a_1 + a_2 + \cdots + a_n) \right\}^2 \leq 2 \sum_{n=1}^{N} \{s_n(a_1 + a_2 + \cdots + a_n)\} a_n. \quad (11.11)$$

This bound looks much like out target preflop inequality (11.10), but there is a small problem: on the right side we have s_n where we hoped to have $1/n$. Since $s_n = 1/n + O(1/n^2)$, we seem to have made progress, but the prize (11.10) is not in our hands.

SO NEAR ... YET

One natural way to try to bring our plan to its logical conclusion is simply to replace the sum s_n in the inequality (11.11) by an honest upper bound. The most systematic way to estimate s_n is by integral comparison, but there is also an instructive telescoping argument that gives an equivalent result. The key observation is that for $n \geq 2$ we have

$$s_n = \sum_{k=n}^{\infty} \frac{1}{k^2} \leq \sum_{k=n}^{\infty} \frac{1}{k(k-1)}$$

$$= \sum_{k=n}^{\infty} \left\{ \frac{1}{k-1} - \frac{1}{k} \right\} = \frac{1}{n-1} \leq \frac{2}{n},$$

and, since $s_1 = 1 + s_2 \leq 1 + 1/(2-1) = 2$, we see that

$$\sum_{k=n}^{\infty} \frac{1}{k^2} \leq \frac{2}{n} \quad \text{for all } n \geq 1. \quad (11.12)$$

Hardy's Inequality and the Flop

Now, when we use this bound in our summation by parts inequality (11.11), we find

$$\sum_{n=1}^{N}\left\{\frac{1}{n}(a_1+a_2+\cdots+a_n)\right\}^2 \le 4\sum_{n=1}^{N}\left\{\frac{1}{n}(a_1+a_2+\cdots+a_n)\right\}a_n, \quad (11.13)$$

and this is *almost* the inequality (11.10) that we wanted to prove. The only difference is that the constant 2 in the preflop inequality (11.10) has been replaced by a 4. Unfortunately, this difference is enough to keep us from our ultimate goal. When we apply the flop to the inequality (11.13), we fail to get the constant that is required in our challenge problem; we get an 8 where a 4 is needed.

Taking the Flop as Our Guide

Once again, the obvious plan has come up short, and we must look for some way to improve our argument. Certainly we can sharpen our estimate for s_n, but, before worrying about small analytic details, we should look at the structure of our plan. We used summation by parts because we hoped to replicate a successful argument that used integration by parts, but the most fundamental component of our argument simply calls on us to prove the preflop inequality

$$\sum_{n=1}^{N}\left\{\frac{1}{n}(a_1+a_2+\cdots+a_n)\right\}^2 \le 2\sum_{n=1}^{N}\left\{\frac{1}{n}(a_1+a_2+\cdots+a_n)\right\}a_n. \quad (11.14)$$

There is no law that says that we must prove this inequality by starting with the left-hand side and using summation by parts. If we stay flexible, perhaps we can find a fresh approach.

Flexible and Hopeful

To begin our fresh approach, we may as well work toward a clearer view of our problem; certainly some of the clutter may be removed by setting $A_n = (a_1 + a_2 + \cdots + a_n)/n$. Also, if we consider the term-by-term differences Δ_n between the summands in the preflop inequality (11.14), then we have the simple identity $\Delta_n = A_n^2 - 2A_n a_n$. The proof of the preflop inequality (11.14) therefore comes down to showing that the sum of the increments Δ_n over $1 \le n \le N$ is bounded by zero.

We now have a concrete goal — but not much else. Still, we may recall that one of the few ways we have to simplify sums is by telescoping. Thus, even though no telescoping sums are presently in sight, we might want to explore the algebra of the difference Δ_n while keeping the possibility of telescoping in mind. If we now try to write Δ_n just in

172   Hardy's Inequality and the Flop

terms of A_n and A_{n-1}, then we have
$$\begin{aligned}\Delta_n &= A_n^2 - 2A_n a_n \\ &= A_n^2 - 2A_n\bigl(nA_n - (n-1)A_{n-1}\bigr) \\ &= (1-2n)A_n^2 + 2(n-1)A_n A_{n-1},\end{aligned}$$
but unfortunately the product $A_n A_{n-1}$ emerges as a new trouble spot. Nevertheless, we can eliminate this product if we recall the "humble bound" and note that if we replace $A_n A_{n-1}$ by $(A_n^2 + A_{n-1}^2)/2$ we have
$$\begin{aligned}\Delta_n &\leq (1-2n)A_n^2 + (n-1)\bigl(A_n^2 + A_{n-1}^2\bigr) \\ &= (n-1)A_{n-1}^2 - nA_n^2.\end{aligned}$$
After a few dark moments, we now find that we are the beneficiaries of some good luck: the last inequality is one that telescopes beautifully. When we sum over n, we find
$$\sum_{n=1}^{N} \Delta_n \leq \sum_{n=1}^{N} \bigl\{(n-1)A_{n-1}^2 - nA_n^2\bigr\} = -NA_N^2,$$
and, by the negativity of the last term, the proof of the preflop inequality (11.14) is complete. Finally, we know already that the flop will take us from the inequality (11.14) to the inequality (11.9) of our challenge problem, so the solution of the problem is also complete.

A Brief Look Back

Familiarity with the flop gives one access to a rich class of strategies for proving inequalities for integrals and for sums. In our second challenge problem, we made some headway through imitation of the strategy that worked in the continuous case, but definitive progress only came when we focused squarely on the flop and when we worked toward a direct proof of the preflop inequality
$$\sum_{n=1}^{N}\left\{\frac{1}{n}(a_1 + a_2 + \cdots + a_n)\right\}^2 \leq 2\sum_{n=1}^{N}\left\{\frac{1}{n}(a_1 + a_2 + \cdots + a_n)\right\}a_n.$$
The new focus was a fortunate one, and we found that the preflop inequality could be obtained by a pleasing telescoping argument that used little more than the bound $xy \leq (x^2 + y^2)/2$.

In the first two examples the flop was achieved with help from Cauchy's inequality or Schwarz inequality, but the basic idea is obviously quite general. In the next problem (and in several of the exercises) we will see that Hölder's inequality is perhaps the flop's more natural partner.

Hardy's Inequality and the Flop

CARLESON'S INEQUALITY — WITH CARLEMAN'S AS A COROLLARY

Our next challenge problem presents itself with no flop in sight; there is not even a product to be seen. Nevertheless, one soon discovers that the product — and the flop — are not far away.

Problem 11.3 (Carleson's Convexity Inequality)
Show that if $\varphi : [0,\infty) \to \mathbb{R}$ is convex and $\varphi(0) = 0$, then for all $-1 < \alpha < \infty$ one has the integral bound

$$I = \int_0^\infty x^\alpha \exp\left(-\frac{\varphi(x)}{x}\right) dx \leq e^{\alpha+1} \int_0^\infty x^\alpha \exp\left(-\varphi'(x)\right) dx \quad (11.15)$$

where, as usual, $e = 2.71828\ldots$ is the natural base.

The shape of the inequality (11.15) is uncharacteristic of any we have met before, so one may be at a loss for a reasonable plan. To be sure, convexity always gives us something useful; in particular, convexity provides an estimate of the shift difference $\varphi(y+t) - \varphi(y)$. Unfortunately this estimate does not seem to help us much here.

The way Carleson cut the Gordian knot was to consider instead the *scale shift difference* $\varphi(py) - \varphi(y)$ where $p > 1$ is a parameter that we can optimize later. This is a clever idea, yet conceived, it easily becomes a part of our permanent toolkit.

A FLOP OF A DIFFERENT FLAVOR

Carleson set up his estimation of the integral I by first making the change of variables $x \mapsto py$ and then using the convexity estimate,

$$\varphi(py) \geq \varphi(y) + (p-1)y\varphi'(y), \quad (11.16)$$

which is illustrated in Figure 11.1. The exponential of this sum gives us a product, so Hölder's inequality and the flop are almost ready to act.

Still, some care is needed to avoid integrals which may be divergent, so we first restrict our attention to a finite interval $[0, A]$ to note that

$$I_A = \int_0^A x^\alpha \exp\left(-\frac{\varphi(x)}{x}\right) dx = p^{\alpha+1} \int_0^{A/p} y^\alpha \exp\left(-\frac{\varphi(py)}{py}\right) dy$$

$$\leq p^{\alpha+1} \int_0^A y^\alpha \exp\left(\frac{-\varphi(y) - (p-1)y\varphi'(y)}{py}\right) dy,$$

where in the second step we used the convexity bound (11.16) and extended the range of integration from $[0, A/p]$ to $[0, A]$. If we introduce

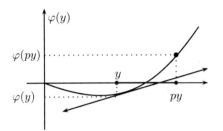

Fig. 11.1. The convexity bound $\varphi(py) \geq \varphi(y) + (p-1)y\varphi'(y)$ for $p > 1$ tells us how φ changes under a scale shift. It also cooperates wonderfully with changes of variables, Hölder's inequality, and the flop.

the conjugate $q = p/(p-1)$ and apply Hölder's inequality to the natural splitting suggested by $1/p + 1/q = 1$, we then find

$$p^{-\alpha-1} I_A \leq \int_0^A \left\{ y^{\alpha/p} \exp\left(-\frac{\varphi(y)}{py}\right) \right\} \left\{ y^{\alpha/q} \exp\left(-\frac{(p-1)}{p}\varphi'(y)\right) \right\} dy$$

$$\leq I_A^{1/p} \left\{ \int_0^A y^\alpha \exp\left(-\varphi'(y)\right) dy \right\}^{1/q}.$$

Since $I_A < \infty$, we may divide by $I_A^{1/p}$ to complete the flop. Upon taking the qth power of the resulting inequality, we find

$$I_A = \int_0^A y^\alpha \exp\left(-\frac{\varphi(y)}{y}\right) dy \leq p^{(\alpha+1)p/(p-1)} \int_0^A y^\alpha \exp\left(-\varphi'(y)\right) dy,$$

and this is actually more than we need.

To obtain the stated form (11.15) of Carleson's inequality, we first let $A \to \infty$ and then let $p \to 1$. The familiar relation $\log(1+\epsilon) = \epsilon + O(\epsilon^2)$ implies that $p^{p/(p-1)} \to e$ as $p \to 1$, so the solution of the challenge problem is complete.

An Informative Choice of φ

Part of the charm of Carleson's inequality is that it provides a sly generalization of the famous Carleman's inequality, which we have met twice before (pages 27 and 128). In fact, one only needs to make a wise choice of φ.

Given the hint of this possibility and a little time for experimentation, one is quite likely to hit on the candidate suggested by Figure 11.2. For

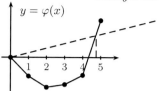

A simple, but useful, observation: the slope $\varphi(x)/x$ of the chord increases with x.

Fig. 11.2. If $y = \varphi(x)$ is the curve given by the linear interpolation of the points $(n, s(n))$ where $s(n) = \log(1/a_1) + \log(1/a_2) + \cdots + \log(1/a_n)$, then on the interval $(n-1, n)$ we have $\varphi'(x) = \log(1/a_n)$. If we assume that $a_n \geq a_{n+1}$ then $\varphi'(x)$ is non-decreasing and $\varphi(x)$ is convex. Also, since $\varphi(0) = 0$, the chord slope $\varphi(x)/x$ is monotone increasing.

the function φ defined there, we have identity

$$\int_{n-1}^{n} \exp(-\varphi'(x))\,dx = a_k \tag{11.17}$$

and, since $\varphi(x)/x$ is nondecreasing, we also have the bound

$$\left(\prod_{k=1}^{n} a_k\right)^{1/n} = \exp\left(\frac{-\varphi(n)}{n}\right) \leq \int_{n-1}^{n} \exp\left(\frac{-\varphi(x)}{x}\right) dx. \tag{11.18}$$

When we sum the relations (11.17) and (11.18), we then find by invoking Carleson's inequality (11.15) with $\alpha = 0$ that

$$\sum_{n=1}^{\infty} \left(\prod_{k=1}^{n} a_k\right)^{1/n} \leq \int_{0}^{\infty} \exp\left(\frac{-\varphi(x)}{x}\right) dx$$

$$\leq e \int_{0}^{\infty} \exp(-\varphi'(x))\,dx = e \sum_{n=1}^{\infty} a_n.$$

Thus we recover Carleman's inequality under the added assumption that $a_1 \geq a_2 \geq a_3 \cdots$. Moreover, this assumption incurs no loss of generality, as one easily confirms in Exercise 11.7.

Exercises

Exercise 11.1 (The L^p Flop and a General Principle)

Suppose that $1 < \alpha < \beta$ and suppose that the bounded nonnegative functions φ and ψ satisfy the inequality

$$\int_{0}^{T} \varphi^{\beta}(x)\,dx \leq C \int_{0}^{T} \varphi^{\alpha}(x)\psi(x)\,dx. \tag{11.19}$$

Show that one can "clear φ to the left" in the sense that one has

$$\int_{0}^{T} \varphi^{\beta}(x)\,dx \leq C^{\beta/(\beta-\alpha)} \int_{0}^{T} \psi^{\beta/(\beta-\alpha)}(x)\,dx. \tag{11.20}$$

176   Hardy's Inequality and the Flop

The bound (11.20) is just one example of a general (but vague) principle: If we have a factor on both sides of an equation and if it appears to a smaller power on the "right" than on the "left," then we can *clear the factor to the left* to obtain a new — and potentially useful — bound.

Exercise 11.2 (Rudimentary Example of a General Principle)
The principle of Exercise 11.1 can be illustrated with the simplest of tools. For example, show for nonnegative x and y that
$$2x^3 \leq y^3 + y^2 x + y x^2 \quad \text{implies } x^3 \leq 2y^3.$$

Exercise 11.3 (An Exam-Time Discovery of F. Riesz)
Show that there is a constant A (not depending on u and v) such that for each pair of functions u and v on $[-\pi, \pi]$ for which one has
$$\int_{-\pi}^{\pi} v^4(\theta)\, d\theta \leq \int_{-\pi}^{\pi} u^4(\theta)\, d\theta + 6 \int_{-\pi}^{\pi} u^2(\theta) v^2(\theta)\, d\theta, \qquad (11.21)$$
one also has the bound
$$\int_{-\pi}^{\pi} v^4(\theta)\, d\theta \leq A \int_{-\pi}^{\pi} u^4(\theta)\, d\theta. \qquad (11.22)$$

According to J.E. Littlewood (1988, p. 194), F. Riesz was trying to set an examination problem when he observed almost by accident that the bound (11.21) holds for the real u and imaginary v parts of $f(e^{i\theta})$ when $f(z)$ is a continuous function that is analytic in the unit disk. This observation and the inference (11.22) subsequently put Riesz on the trail of some of his most important discoveries.

Exercise 11.4 (The L^p Norm of the Average)
Show that if $f : [0, \infty) \to \mathbb{R}^+$ is integrable and $p > 1$, then one has
$$\int_0^\infty \left\{ \frac{1}{x} \int_0^x f(u)\, du \right\}^p dx \leq \left(\frac{p}{p-1} \right)^p \int_0^\infty f^p(x)\, dx. \qquad (11.23)$$

Exercise 11.5 (Hardy and the Qualitative Version of Hilbert)
Use the discrete version (11.9) of Hardy's inequality to prove that
$$S = \sum_{n=1}^\infty a_n^2 < \infty \quad \text{implies that} \quad \sum_{n=1}^\infty \sum_{n=1}^\infty \frac{a_n a_m}{m+n} \quad \text{converges.}$$

This was the qualitative version of Hilbert's inequality that Hardy had in mind when he first considered the Problems 11.1 and 11.2.

Hardy's Inequality and the Flop

Exercise 11.6 (Optimality? — It Depends on Context)
Many inequalities which cannot be improved in general will nevertheless permit improvements under special circumstances. An elegant illustration of this possibility was given in a 1991 *American Mathematical Monthly* problem posed by Walther Janous. Readers were challenged to prove that for all $0 < x < 1$ and all $N \geq 1$, one has the bound

$$\sum_{j=1}^{N} \left(\frac{1 + x + x^2 + \cdots + x^{j-1}}{j} \right)^2 \leq (4 \log 2)(1 + x^2 + x^4 + \cdots + x^{2N-2}).$$

(a) Prove that a direct application of Hardy's inequality provides a similar bound where $4 \log 2$ is replaced by 4. Since $\log 2 = 0.693\ldots$, we then see that Janous's bound beats Hardy's in this particular instance.

(b) Prove Janous's inequality and show that one cannot replace $4 \log 2$ with a constant $C < 4 \log 2$.

Exercise 11.7 (Confirmation of the Obvious)
Show that if $a_1 \geq a_2 \geq a_3 \cdots$ and if $b_1, b_2, b_3, \ldots$ is any rearrangement of the sequence $a_1, a_2, a_3, \ldots$, then for each $N = 1, 2, \ldots$ one has

$$\sum_{n=1}^{N} \left(\prod_{k=1}^{n} b_k \right)^{1/n} \leq \sum_{n=1}^{N} \left(\prod_{k=1}^{n} a_k \right)^{1/n}. \tag{11.24}$$

Thus, in the proof of Carleman's inequality, one can assume without lose of generality that $a_1 \geq a_2 \geq a_3 \cdots$ since a rearrangement does not change the right side.

Exercise 11.8 (Kronecker's Lemma)
Prove that for any sequence $a_1, a_2, \ldots$ of real or complex numbers one has the inference

$$\sum_{n=1}^{\infty} \frac{a_n}{n} \text{ converges } \Rightarrow \lim_{n \to \infty} (a_1 + a_2 + \cdots + a_n)/n = 0. \tag{11.25}$$

Like Hardy's inequality, this result tells us how to convert one type of information about averages to another type of information. This implication is particularly useful in probability theory where it is used to draw a connection between the convergence of certain random sums and the famous law of large numbers.

12
Symmetric Sums

The kth elementary symmetric function of the n variables $x_1, x_2, \ldots, x_n$ is the polynomial defined the formula

$$e_k(x_1, x_2, \ldots, x_n) = \sum_{1 \leq i_1 < i_2 < \cdots < i_k \leq n} x_{i_1} x_{i_2} \cdots x_{i_k}.$$

The first three of these polynomials are simply

$$e_0(x_1, x_2, \ldots, x_n) = 1, \quad e_1(x_1, x_2, \ldots, x_n) = x_1 + x_2 + \ldots + x_n,$$

and $\quad e_2(x_1, x_2, \ldots, x_n) = \sum_{1 \leq j < k \leq n} x_j x_k,$

while the nth elementary symmetric function is simply the full product

$$e_n(x_1, x_2, \ldots, x_n) = x_1 x_2 \cdots x_n.$$

These functions are used in virtually every part of the mathematical sciences, yet they draw much of their importance from the connection they provide between the *coefficients* of a polynomial and functions of its *roots*. To be explicit, if the polynomial $P(t)$ is written as the product $P(t) = (t - x_1)(t - x_2) \cdots (t - x_n)$, then it also has the representation

$$P(t) = t^n - e_1(\mathbf{x}) t^{n-1} + \cdots + (-1)^k e_k(\mathbf{x}) t^{n-k} + \cdots + (-1)^n e_n(\mathbf{x}), \quad (12.1)$$

where for brevity we have written $e_k(\mathbf{x})$ in place of $e_k(x_1, x_2, \ldots, x_n)$.

THE CLASSICAL INEQUALITIES OF NEWTON AND MACLAURIN

The elementary polynomials have many connections with the theory of inequalities. Two of the most famous of these date back to the great Isaac Newton (1642–1727) and the Scottish prodigy Colin Maclaurin

Symmetric Sums   179

(1696–1746). Their namesake inequalities are best expressed in terms of the averages

$$E_k(\mathbf{x}) = E_k(x_1, x_2, \ldots, x_n) = \frac{e_k(x_1, x_2, \ldots, x_n)}{\binom{n}{k}},$$

which bring us to our first challenge problem.

Problem 12.1 (Inequalities of Newton and Maclaurin)
Show that for all $\mathbf{x} \in \mathbb{R}^n$ one has Newton's inequalities

$$E_{k-1}(\mathbf{x}) \cdot E_{k+1}(\mathbf{x}) \leq E_k^2(\mathbf{x}) \quad \text{for } 0 < k < n \quad (12.2)$$

and check that they imply Maclaurin's inequalities which assert that

$$E_n^{1/n}(\mathbf{x}) \leq E_{n-1}^{1/(n-1)}(\mathbf{x}) \leq \cdots \leq E_2(\mathbf{x})^{1/2} \leq E_1(\mathbf{x}) \quad (12.3)$$

for all $\mathbf{x} = (x_1, x_2, \ldots, x_n)$ such that $x_k \geq 0$ for all $1 \leq k \leq n$.

ORIENTATION AND THE AM-GM CONNECTION

If we take $n = 3$ and set $\mathbf{x} = (x, y, z)$, then Maclaurin's inequalities simply say

$$(xyz)^{1/3} \leq \left(\frac{xy + xz + yz}{3}\right)^{1/2} \leq \frac{x + y + z}{3},$$

which is a sly refinement of the AM-GM inequality. In the general case, Maclaurin's inequalities insert a whole line of ever increasing expressions between the geometric mean $(x_1 x_2 \cdots x_n)^{1/n}$ and the arithmetic mean $(x_1 + x_2 + \cdots + x_n)/n$.

FROM NEWTON TO MACLAURIN BY GEOMETRY

For a vector $\mathbf{x} \in \mathbb{R}^n$ with only nonnegative coordinates, the values $\{E_k(\mathbf{x}) : 0 \leq k \leq n\}$ are also nonnegative, so we can take logarithms of Newton's inequalities to deduce that

$$\frac{\log E_{k-1}(\mathbf{x}) + \log E_{k+1}(\mathbf{x})}{2} \leq \log E_k(\mathbf{x}) \quad (12.4)$$

for all $1 \leq k < n$. In particular, we see for $\mathbf{x} \in [0, \infty)^n$ that Newton's inequalities are equivalent to the assertion that the piecewise linear curve determined by the point set $\{(k, \log E_k(\mathbf{x})) : 0 \leq k \leq n\}$ is concave.

If L_k denotes the line determined by the points $(0,0) = (0, \log E_1(\mathbf{x}))$ and $(k, \log E_k(\mathbf{x}))$, then as Figure 12.1 suggests, the slope of L_{k+1} is never larger than the slope of L_k for any $k = 1, 2, \ldots, n-1$. Since the

180   *Symmetric Sums*

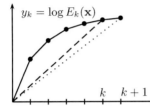

The inequalities of Maclaurin call on the observation that the successive chords from the origin have nonincreasing slopes.

Fig. 12.1. If $E_k(\mathbf{x}) \geq 0$ for all $1 \leq k \leq n$, then Newton's inequalities are equivalent to the assertion that the piecewise linear curve determined by the points (k, y_k), $1 \leq k \leq n$, is concave. Maclaurin's inequalities capitalize on just one part of this geometry.

slope of L_k is $\log E_k(\mathbf{x})/k$, we find $\log E_k(\mathbf{x})/k \leq \log E_{k+1}(\mathbf{x})/(k+1)$, and this is precisely the kth of Maclaurin's inequalities.

The real challenge is to prove Newton's inequalities. As one might expect for a result that is both ancient and fundamental, there are many possible approaches. Most of these depend on calculus in one way or another, but Newton never published a proof of his namesake inequalities, so we do not know if his argument relied on his "method of fluxions."

POLYNOMIALS AND THEIR DERIVATIVES

Even if Newton took a different path, it does make sense to ask what the derivative $P'(t)$ might tell us about the about the special polynomials $E_k(x_1, x_2, \ldots, x_n)$, $1 \leq k \leq n$. If we write the identity (12.1) in the form

$$P(t) = (t - x_1)(t - x_2) \cdots (t - x_n)$$
$$= \sum_{k=0}^{n} (-1)^k \binom{n}{k} E_k(x_1, x_2, \ldots, x_n) t^{n-k}, \qquad (12.5)$$

then its derivative is almost a perfect clone. More precisely, we have

$$Q(t) = \frac{1}{n} P'(t) = \sum_{k=0}^{n-1} (-1)^k \binom{n}{k} \frac{n-k}{n} E_k(x_1, x_2, \ldots, x_n) t^{n-k-1}$$
$$= \sum_{k=0}^{n-1} (-1)^k \binom{n-1}{k} E_k(x_1, x_2, \ldots, x_n) t^{n-k-1},$$

where in the second line we used the familiar identity

$$\binom{n}{k} \frac{n-k}{n} = \frac{n!}{k!(n-k)!} \frac{n-k}{n} = \frac{(n-1)!}{k!(n-k-1)!} = \binom{n-1}{k}.$$

Symmetric Sums

If the values x_k, $k = 1, 2, \ldots, n$ are elements of the interval $[a, b]$, then the polynomial $P(t)$ has n real roots in $[a, b]$, and Rolle's theorem tells us that the derivative $P'(x)$ must have $n - 1$ real roots in $[a, b]$. If we denote these roots by $\{y_1, y_2, \ldots, y_{n-1}\}$, then we also have the identity

$$Q(t) = \frac{1}{n} P'(t) = (t - y_1)(t - y_2) \cdots (t - y_{n-1})$$

$$= \sum_{k=0}^{n-1} (-1)^k \binom{n-1}{k} E_k(y_1, y_2, \ldots, y_{n-1}) t^{n-k-1}.$$

If we now equate the coefficients in our two formulas for $Q(t)$, we find that for all $0 \leq k \leq n - 1$ we have the truly remarkable identity

$$E_k(x_1, x_2, \ldots, x_n) = E_k(y_1, y_2, \ldots, y_{n-1}). \qquad (12.6)$$

WHY IS IT SO REMARKABLE?

The left-hand side of the identity (12.6) is a function of the n vector $\mathbf{x} = (x_1, x_2, \ldots, x_n)$ while the right side is a function of the $n - 1$ vector $\mathbf{y} = (y_1, y_2, \ldots, y_{n-1})$. Thus, if we can prove a relation such as

$$0 \leq F(E_0(\mathbf{y}), E_1(\mathbf{y}), \ldots, E_{n-1}(\mathbf{y})) \qquad \text{for all } \mathbf{y} \in [a, b]^{n-1},$$

then it follows that we also have the relation

$$0 \leq F(E_0(\mathbf{x}), E_1(\mathbf{x}), \ldots, E_{n-1}(\mathbf{x})) \qquad \text{for all } \mathbf{x} \in [a, b]^n.$$

That is, any inequality — or identity — which provides a relation between the $n - 1$ quantities $E_0(\mathbf{y}), E_1(\mathbf{y}), \ldots, E_{n-1}(\mathbf{y})$ and which is valid for all values of $\mathbf{y} \in [a, b]^{n-1}$ extends automatically to a corresponding relation for the $n-1$ quantities $E_0(\mathbf{x}), E_1(\mathbf{x}), \ldots, E_{n-1}(\mathbf{x})$ which is valid for all values of $\mathbf{x} \in [a, b]^n$.

This presents a rare but valuable situation where to prove a relation for functions of n variables it suffices to prove an analogous relation for functions of just $n - 1$ variables. This observation can be used in an *ad hoc* way to produce many special identities which otherwise would be completely baffling, and it can also be used systematically to provide seamless induction proofs for results such as Newton's inequalities.

INDUCTION ON THE NUMBER OF VARIABLES

Consider now the induction hypothesis H_n which asserts that

$$E_{j-1}(x_1, x_2, \ldots, x_n) E_{j+1}(x_1, x_2, \ldots, x_n) \leq E_j^2(x_1, x_2, \ldots, x_n) \qquad (12.7)$$

for all $x \in \mathbb{R}^n$ and all $1 < j < n$. For $n = 1$ this assertion is empty, so our induction argument begins with H_2, in which case we just need to prove one inequality,

$$E_0(x_1,x_2)E_2(x_1,x_2) \leq E_1^2(x_1,x_2) \quad \text{or} \quad x_1 x_2 \leq \left(\frac{x_1+x_2}{2}\right)^2. \qquad (12.8)$$

As we have seen a dozen times before, this holds for all real x_1 and x_2 because of the trivial bound $(x_1-x_2)^2 > 0$.

Logically, we could now address the general induction step, but we first need a clear understanding of the underlying pattern. Thus, we consider the hypothesis H_3 which consists of the two assertions:

$$E_0(x_1,x_2,x_3)E_2(x_1,x_2,x_3) \leq E_1^2(x_1,x_2,x_3), \qquad (12.9)$$

$$E_1(x_1,x_2,x_3)E_3(x_1,x_2,x_3) \leq E_2^2(x_1,x_2,x_3). \qquad (12.10)$$

Now the "remarkable identity" (12.6) springs into action. The assertion (12.9) says for three variables what the inequality (12.8) says for two, therefore (12.6) tells us that our first inequality (12.9) is true. We have obtained half of the hypothesis H_3 virtually for free.

To complete the proof of H_3, we now only need prove the second bound (12.10). To make the task clear we first rewrite the bound (12.10) in longhand as

$$\left\{\frac{x_1+x_2+x_3}{3}\right\}\{x_1 x_2 x_3\} \leq \left\{\frac{x_1 x_2 + x_1 x_3 + x_2 x_3}{3}\right\}^2. \qquad (12.11)$$

This bound is trivial if $x_1 x_2 x_3 = 0$, so there is no loss of generality if we assume $x_1 x_2 x_3 \neq 0$. We can then divide our bound by $(x_1 x_2 x_3)^2$ to get

$$\frac{1}{3}\left\{\frac{1}{x_1 x_2} + \frac{1}{x_1 x_3} + \frac{1}{x_2 x_3}\right\} \leq \frac{1}{9}\left\{\frac{1}{x_1} + \frac{1}{x_2} + \frac{1}{x_3}\right\}^2$$

which may be expanded and simplified to

$$\frac{1}{x_1 x_2} + \frac{1}{x_1 x_3} + \frac{1}{x_2 x_3} \leq \frac{1}{x_1^2} + \frac{1}{x_2^2} + \frac{1}{x_3^2}.$$

At this stage of our *Master Class*, this inequality is almost obvious. For a thematic proof, one can apply Cauchy's inequality to the pair of vectors $(1/x_1, 1/x_3, 1/x_2)$ and $(1/x_2, 1/x_1, 1/x_3)$, or, a bit more generally, one can sum the three AM-GM bounds

$$\frac{1}{x_j x_k} \leq \frac{1}{2}\left\{\frac{1}{x_j^2} + \frac{1}{x_k^2}\right\} \qquad 1 \leq j < k \leq 3.$$

Thus, the proof of H_3 is complete, and, moreover, we have found a pattern that should guide us through the general induction step.

A Pattern Confirmed

The general hypothesis H_n consists of $n-1$ inequalities which may be viewed in two groups. First, for $\mathbf{x} = (x_1, x_2, \ldots, x_n)$ we have the $n-2$ inequalities which involve only $E_j(\mathbf{x})$ with $0 \leq j < n$,

$$E_{k-1}(\mathbf{x})E_{k+1}(\mathbf{x}) \leq E_k^2(\mathbf{x}) \qquad \text{for } 1 \leq k < n-1, \tag{12.12}$$

then we have one final inequality which involves $E_n(\mathbf{x})$,

$$E_{n-2}(\mathbf{x})E_n(\mathbf{x}) \leq E_{n-1}^2(\mathbf{x}). \tag{12.13}$$

In parallel with the analysis of H_3, we now see that all of the inequalities in the first group (12.12) follow from the induction hypothesis H_n and the identity (12.6). All of the inequalities of H_n have come to us for free, except for one.

If we write the bound (12.13) in longhand and use $\hat{x}_j$ as a symbol to suggest that x_j is omitted, then we see that it remains for us to prove that we have the relation

$$\frac{2}{n(n-1)} \Bigg\{ \sum_{1 \leq j < k \leq n} x_1 \cdots \hat{x}_j \cdots \hat{x}_k \cdots x_n \Bigg\} x_1 x_2 \cdots x_n$$
$$\leq \Bigg\{ \frac{1}{n} \sum_{j=1}^{n} x_1 x_2 \cdots \hat{x}_j \cdots x_n \Bigg\}^2. \tag{12.14}$$

In parallel with our earlier experience, we note that there is no loss of generality in assuming $x_1 x_2 \cdots x_n \neq 0$. After division by $(x_1 x_2 \cdots x_n)^2$ and some simplification, we see that the bound (12.14) is equivalent to

$$\frac{1}{\binom{n}{2}} \sum_{1 \leq j < k \leq n} \frac{1}{x_j x_k} \leq \Bigg\{ \frac{1}{n} \sum_{j=1}^{n} \frac{1}{x_j} \Bigg\}^2. \tag{12.15}$$

We could now stick with the pattern that worked for H_3, but there is a more graceful way to finish which is almost staring us in the face. If we adopt the language of symmetric functions, the target bound (12.15) may be written more systematically as

$$E_0(1/x_1, 1/x_2, \ldots, 1/x_n)E_2(1/x_1, 1/x_2, \ldots, 1/x_n)$$
$$\leq E_1^2(1/x_1, 1/x_2, \ldots, 1/x_n),$$

and one now sees that this inequality is covered by the first bound of the group (12.12). Thus, the proof of Newton's inequalities is complete.

Equality in the Bounds of Newton or Maclaurin

From Figure 12.1, we see that we have equality in the kth Maclaurin bound $y_{k+1}/(k+1) \leq y_k/k$ if and only if the dotted and the dashed lines have the same slope. By the concavity of the piecewise linear curve through the points $\{(j, y_j) : 0 \leq j \leq n\}$, this is possible if and only if the three points $(k-1, y_{k-1})$, (k, y_k), and $(k+1, y_{k+1})$ all lie on a straight line. This is equivalent to the assertion $y_k = (y_{k-1} + y_{k+1})/2$, so, by geometry, we find that equality holds in the kth Maclaurin bound if and only if it holds in the kth Newton bound.

It takes only a moment to check that equality holds in each of Newton's bounds when $x_1 = x_2 = \cdots = x_n$, and there are several ways to prove that this is the only circumstance where equality is possible. For us, perhaps the easiest way to prove this assertion is by making some small changes to our induction argument. In fact, the diligent reader will surely want to confirm that our induction argument can be repeated almost word for word while including induction hypothesis (12.7) the condition for strict inequality.

Passage to Muirhead

David Hilbert once said, "The art of doing mathematics consists in finding that special case which contains all the germs of generality." The next challenge problem is surely more modest than the examples that Hilbert had in mind, but in this chapter and the next we will see that it amply illustrates Hilbert's point.

Problem 12.2 (A Symmetric Appetizer)
Show that for nonnegative x, y, and z one has the bound

$$x^2y^3 + x^2z^3 + y^2x^3 + y^2z^3 + z^2x^3 + z^2y^3$$
$$\leq xy^4 + xz^4 + yx^4 + yz^4 + zx^4 + zy^4, \tag{12.16}$$

and take inspiration from your discoveries to generalize this result as widely as you can.

Making Connections

We have already met several problems where the AM-GM inequality helped us to understand the relationship between two homogeneous polynomials, and if we hope to use a similar idea here we need to show that each summand on the left can be written as a weighted geometric mean of the summands on the right. After some experimentation, one

Symmetric Sums

is sure to observe that for any nonnegative a and b we have the product representation $a^2b^3 = (ab^4)^{\frac{2}{3}}(a^4b)^{\frac{1}{3}}$. The weighted AM-GM inequality (2.9) then gives us the bound

$$a^2b^3 = (ab^4)^{\frac{2}{3}}(a^4b)^{\frac{1}{3}} \leq \frac{2}{3}ab^4 + \frac{1}{3}a^4b, \qquad (12.17)$$

and now we just need to see how this may be applied.

If we replace (a, b) in turn by the ordered pairs (x, y) and (y, x), then the sum of the resulting bounds gives us $x^2y^3 + y^2x^3 \leq xy^4 + x^4y$ and, in exactly the same way, we can get two analogous inequalities by summing the bound (12.17) for the two pairs (x, z) and (z, x), and the two pairs (y, z) and (z, y). Finally, the sum of the resulting three bounds then gives us our target inequality (12.16).

PASSAGE TO AN APPROPRIATE GENERALIZATION

This argument can be applied almost without modification to any symmetric sum of two-term products $x^a y^b$, but one may feel some uncertainty about sums that contain triple products such as $x^a y^b z^c$. Such sums may have many terms, and complexity can get the best of us unless we develop a systematic approach.

Fortunately, geometry points the way. From Figure 12.2 one sees at a glance that $(2, 3) = \frac{2}{3}(1, 4) + \frac{1}{3}(4, 1)$, and, by exponentiation, we see that this recaptures us our decomposition $a^2b^3 = (ab^4)^{\frac{2}{3}}(a^4b)^{\frac{1}{3}}$. Geometry makes quick work of such two-term decompositions, but the real benefit of the geometric point of view is that it suggests useful representation for products of three or more variables. The key is to find the right analog of Figure 12.2.

In abstract terms, the solution of the first challenge problem pivoted on the observation that $(2, 3)$ is in the convex hull of $(1, 4)$ and its permutation $(4, 1)$. Now, more generally, given any pair of n-vectors $\alpha = (\alpha_1, \alpha_2, \ldots, \alpha_n)$ and $\beta = (\beta_1, \beta_2, \ldots, \beta_n)$, we can consider an analogous situation where α is contained in the convex hull $H(\beta)$ of the set of points $(\beta_{\tau(1)}, \beta_{\tau(2)}, \ldots, \beta_{\tau(n)})$ which are determined by letting τ run over the set $\mathcal{S}_n$ of all $n!$ permutations of $\{1, 2, \ldots, n\}$.

This suggestion points us to a far reaching generalization of our first challenge problem. The result is due to another Scot, Robert Franklin Muirhead (1860–1941). It has been known since 1903, and, at first, it may look complicated. Nevertheless, with experience one finds that it has both simplicity and a timeless grace.

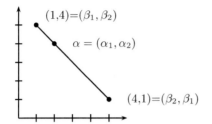

Fig. 12.2. If the point (α_1, α_2) is in the convex hull of (β_1, β_2) and (β_2, β_1) then $x^{\alpha_1} y^{\alpha_2}$ is bounded by a linear combination of $x^{\beta_1} y^{\beta_2}$ and $x^{\beta_2} y^{\beta_1}$. This leads to some engaging inequalities when applied to symmetric sums of products, and there are exceptionally revealing generalizations of these bounds.

Problem 12.3 (Muirhead's inequality)

Given that $\alpha \in H(\beta)$ where $\alpha = (\alpha_1, \alpha_2, \ldots, \alpha_n)$ and $\beta = (\beta_1, \beta_2, \ldots, \beta_n)$, show that for all positive $x_1, x_2, \ldots, x_n$ one has the bound

$$\sum_{\sigma \in \mathcal{S}_n} x_{\sigma(1)}^{\alpha_1} x_{\sigma(2)}^{\alpha_2} \cdots x_{\sigma(n)}^{\alpha_n} \leq \sum_{\sigma \in \mathcal{S}_n} x_{\sigma(1)}^{\beta_1} x_{\sigma(2)}^{\beta_2} \cdots x_{\sigma(n)}^{\beta_n} \qquad (12.18)$$

A QUICK ORIENTATION

To familiarize this notation, one might first check that Muirhead's inequality does indeed contain the bound given by our second challenge problem (page 184). In that case, $\mathcal{S}_3$ is the set of six permutations of the set $\{1, 2, 3\}$, and we have $(x_1, x_2, x_3) = (x, y, z)$. We also have

$$(\alpha_1, \alpha_2, \alpha_3) = (2, 3, 0) \quad \text{and} \quad (\beta_1, \beta_2, \beta_3) = (1, 4, 0),$$

and since $(2, 3, 0) = \tfrac{2}{3}(1, 4, 0) + \tfrac{1}{3}(4, 1, 0)$ we find that $\alpha \in H(\beta)$. Finally, one has the α-sum

$$\sum_{\sigma \in \mathcal{S}_3} x_{\sigma(1)}^{\alpha_1} x_{\sigma(2)}^{\alpha_2} x_{\sigma(3)}^{\alpha_3} = x^2 y^3 + x^2 z^3 + y^2 x^3 + y^2 z^3 + z^2 x^3 + z^2 y^3,$$

while the β-sum is given by

$$\sum_{\sigma \in \mathcal{S}_3} x_{\sigma(1)}^{\beta_1} x_{\sigma(2)}^{\beta_2} x_{\sigma(3)}^{\beta_3} = xy^4 + xz^4 + yx^4 + yz^4 + zx^4 + zy^4,$$

so Muirhead's inequality (12.18) does indeed give us a generalization of our first challenge bound (12.16).

Finally, before we address the proof, we should note that there is no constraint on the sign of the coordinates of α and β in Muirhead's inequality. Thus, for example, if we take $\alpha = (1/2, 1/2, 0)$ and take

Symmetric Sums

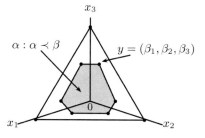

The set $H(\beta)$ in $\mathbb{R}^3$ is a subset of the 2-dimensional hyperplane in $\mathbb{R}^3$ spanned by the six points obtained by permuting the coordinates of $\beta = (\beta_1, \beta_2, \beta_3)$.

Fig. 12.3. The geometry of the condition $\alpha \in H(\beta)$ is trivial in dimension 2, and this figure shows how it may be visualized in dimension 3. In higher dimensions, geometric intuition is still suggestive, but algebra serves as our unfailing guide.

$\beta = (-1, 2, 0)$, then Muirhead's inequality tells us that for positive x, y, and z one has

$$2\left(\sqrt{xy} + \sqrt{xz} + \sqrt{yz}\right) \leq \frac{x^2}{y} + \frac{x^2}{z} + \frac{y^2}{x} + \frac{y^2}{z} + \frac{z^2}{x} + \frac{z^2}{y}. \qquad (12.19)$$

This instructive bound can be proved in many ways; for example, both Cauchy's inequality and the AM-GM bound provide easy derivations. Nevertheless, it is Muirhead's inequality which makes the bound most immediate and which embeds the bound in the richest context.

PROOF OF MUIRHEAD'S INEQUALITY

We were led to conjecture Muirhead's inequality by the solution of our first challenge problem, so we naturally hope to prove it by leaning on our earlier argument. First, just to make the hypothesis $\alpha \in H(\beta)$ concrete, we note that it is equivalent to the assertion that

$$(\alpha_1, \alpha_2, \ldots, \alpha_n) = \sum_{\tau \in \mathcal{S}_n} p_\tau (\beta_{\tau(1)}, \beta_{\tau(2)}, \ldots, \beta_{\tau(n)})$$

$$\text{where} \quad p_\tau \geq 0 \quad \text{and} \quad \sum_{\tau \in \mathcal{S}_n} p_\tau = 1.$$

Now, if we use the jth coordinate of this identity to express $x_{\sigma(j)}^{\alpha_j}$ as a product, then we can take the product over all j to obtain the identity

$$x_{\sigma(1)}^{\alpha_1} x_{\sigma(2)}^{\alpha_2} \cdots x_{\sigma(n)}^{\alpha_n} = \prod_{\tau \in \mathcal{S}_n} \left(x_{\sigma(1)}^{\beta_{\tau(1)}} x_{\sigma(2)}^{\beta_{\tau(2)}} \cdots x_{\sigma(n)}^{\beta_{\tau(n)}} \right)^{p_\tau}.$$

From this point the AM-GM inequality and arithmetic do the rest of the work. In particular, we have

$$\sum_{\sigma \in \mathcal{S}_n} x_{\sigma(1)}^{\alpha_1} x_{\sigma(2)}^{\alpha_2} \cdots x_{\sigma(n)}^{\alpha_n} \leq \sum_{\sigma \in \mathcal{S}_n} \sum_{\tau \in \mathcal{S}_n} p_\tau x_{\sigma(1)}^{\beta_{\tau(1)}} x_{\sigma(2)}^{\beta_{\tau(2)}} \cdots x_{\sigma(n)}^{\beta_{\tau(n)}}$$

$$= \sum_{\tau \in \mathcal{S}_n} p_\tau \sum_{\sigma \in \mathcal{S}_n} x_{\sigma(1)}^{\beta_{\tau(1)}} x_{\sigma(2)}^{\beta_{\tau(2)}} \cdots x_{\sigma(n)}^{\beta_{\tau(n)}}$$

$$= \sum_{\tau \in \mathcal{S}_n} p_\tau \sum_{\sigma \in \mathcal{S}_n} x_{\sigma(1)}^{\beta_1} x_{\sigma(2)}^{\beta_2} \cdots x_{\sigma(n)}^{\beta_n}$$

$$= \sum_{\sigma \in \mathcal{S}_n} x_{\sigma(1)}^{\beta_1} x_{\sigma(2)}^{\beta_2} \cdots x_{\sigma(n)}^{\beta_n},$$

and, as one surely hoped, the two ends of this chain give us Muirhead's inequality (12.18).

LOOKING BACK: BENEFITS OF SYMMETRY

There is nothing difficult in the individual steps of the calculations that give us Muirhead's inequality (12.18), but the sudden disappearance of the p_τ may seem like exceptionally good luck. To be sure, we are not strangers to the benefits that sometimes flow from changing the order of summation, but, as this example points out, those benefits can be particularly striking when symmetric sums are involved.

In many cases, dramatic simplifications arise simply from the observation that "the permutation of a permutation is a permutation." Sometimes we need to check that a one-to-one correspondence works as we hope it should, but even this step just takes patience. The miracle is already in the mix.

With experience, one finds that Muirhead's inequality (12.18) is a remarkably effective tool for understanding the relations between symmetric sums. Nevertheless, applications of Muirhead's inequality come at a price: somehow one must check the hypothesis $\alpha \in H(\beta)$. In many useful cases this can be done by inspection, but before Muirhead's inequality can come into its own, one needs a systematic way to test Muirhead's condition $\alpha \in H(\beta)$. Remarkably enough, there is an equivalent condition that lends itself to almost automatic checking. It is known as *majorization*, and it provides the central theme of our next chapter.

EXERCISES

Exercise 12.1 (On Polynomials with Positive Roots)
Show that if the real polynomial $P(x) = x^n + a_1 x^{n-1} + \cdots + a_{n-1} x + a_n$ has only positive roots, then one has the bound $na_n \leq a_1 a_{n-1}$.

Exercise 12.2 (Three Muirhead Short Stories)

(a) Show that for nonnegative $a, b,$ and c one has

$$8abc \leq (a+b)(b+c)(c+a). \tag{12.20}$$

(b) Show that for real a_j, $1 \leq j \leq n$, one has

$$2 \sum_{1 \leq j < k \leq n} a_j a_k \leq (n-1) \sum_{j=1}^{n} a_j^2. \tag{12.21}$$

(c) Show that for nonnegative a_j, $1 \leq j \leq n$, one has

$$(a_1 a_2 \cdots a_n)^{1/n} \leq \frac{2}{n(n-1)} \sum_{1 \leq j < k \leq n} \sqrt{a_j a_k}. \tag{12.22}$$

Exercise 12.3 (The Homogenization Trick)

Show that if the positive quantities x, y, and z satisfy the relation $xyz = 1$ then one has the inequality

$$x^2 + y^2 + z^2 \leq x^3 + y^3 + z^3. \tag{12.23}$$

The salient feature of this bound is that the left side is homogeneous of order 2 but the right side is homogeneous of order 3. Somehow the constraint $xyz = 1$ must make up for this incompatibility.

It may be unclear how to exploit the constraint $xyz = 1$, but one trick which works remarkably often is to use the side condition to construct a homogeneous problem which generalizes the problem at hand. One then solves the homogeneous problem with the help of Muirhead's inequality or related tools.

Exercise 12.4 (Power Sum Inequalities)

Show that for positive numbers x_k, $1 \leq k \leq n$, the power sums defined by $S_m(\mathbf{x}) = x_1^m + x_2^m + \cdots + x_n^m$ satisfy the bounds

$$S_m^2(\mathbf{x}) \leq S_{m-1}(\mathbf{x}) S_{m+1}(\mathbf{x}) \quad \text{for all } m = 1, 2, \ldots. \tag{12.24}$$

These may remind us of Newton's inequalities, but they are more elementary. They also tell us that the sequence $\{\log S_m(\mathbf{x})\}$ is convex, while Newton's inequalities tell us that $\{\log E_m(\mathbf{x})\}$ is concave.

Exercise 12.5 (Symmetric Problems & Symmetric Solutions)

Consider a real symmetric polynomial $p(x, y)$ such that $p(x, y) \to \infty$ as $|x| \to \infty$ and $|y| \to \infty$. It is reasonable to suspect that p attains its

minimum at a "symmetric point." That is, one might conjecture that there is a $t \in \mathbb{R}$ such that
$$p(t,t) = \min_{x,y} p(x,y).$$
This conjecture was proved for polynomials of degree three or less by Victor Yacovlevich Bunyakovsky in 1854, some five years before the publication of his famous *Mémoire* on integral inequalities. Bunyakovsky also provided a counterexample which shows that the conjecture is false for a polynomial with degree four. Can you find such an example?

Exercise 12.6 (Symmetry — Destroyed by Design)
Participants in the 1999 Canadian Olympiad were asked to show that if x, y, and z are nonnegative real numbers for which $x+y+z=1$, then one has the bound
$$f(x,y,z) = x^2 y + y^2 z + z^2 x \leq \frac{4}{27}.$$
As a hint, first check by calculus that $f(x,y,z)$ is maximized on the set $x+y=1$ by taking $x=2/3$ and $y=1/3$, so the crucial step is to show that without loss of generality one can assume that $z=0$.

Exercise 12.7 (Creative Bunching)
A problem in the popular text *Probability* by Jim Pitman requires one to show in essence that if x, y, and z are nonnegative real numbers for which $x+y+z=1$, then
$$\frac{1}{4} \leq x^3 + y^3 + z^3 + 6xyz.$$
Can you check this bound? Can you check it in more than one way?

Exercise 12.8 (Weierstrass's Polynomial Product Inequality)
Show that if the complex numbers $a_1, a_2, \ldots, a_n$ and $b_1, b_2, \ldots, b_n$ satisfy $|a_j| \leq 1$ and $|b_j| \leq 1$ for all $1 \leq j \leq n$ then
$$|a_1 a_2 \cdots a_n - b_1 b_2 \cdots b_n| \leq \sum_{j=1}^{n} |a_j - b_j|. \qquad (12.25)$$

13
Majorization and Schur Convexity

Majorization and Schur convexity are two of the most productive concepts in the theory of inequalities. They unify our understanding of many familiar bounds, and they point us to great collections of results which are only dimly sensed without their help. Although majorization and Schur convexity take a few paragraphs to explain, one finds with experience that both notions are stunningly simple. Still, they are not as well known as they should be, and they can become one's secret weapon.

Two Bare-Bones Definitions

Given an n-tuple $\gamma = (\gamma_1, \gamma_2, \ldots, \gamma_n)$, we let $\gamma_{[j]}$, $1 \leq j \leq n$, denote the jth largest of the n coordinates, so $\gamma_{[1]} = \max\{\gamma_j : 1 \leq j \leq n\}$, and in general one has $\gamma_{[1]} \geq \gamma_{[2]} \geq \cdots \geq \gamma_{[n]}$. Now, for any pair of real n-tuples $\alpha = (\alpha_1, \alpha_2, \ldots, \alpha_n)$ and $\beta = (\beta_1, \beta_2, \ldots, \beta_n)$, we say that α is *majorized* by β and we write $\alpha \prec \beta$ provided that α and β satisfy the following system of $n-1$ inequalities:

$$\alpha_{[1]} \leq \beta_{[1]},$$
$$\alpha_{[1]} + \alpha_{[2]} \leq \beta_{[1]} + \beta_{[2]},$$
$$\vdots \leq \vdots$$
$$\alpha_{[1]} + \alpha_{[2]} + \cdots + \alpha_{[n-1]} \leq \beta_{[1]} + \beta_{[2]} + \cdots + \beta_{[n-1]},$$

together with one final equality:

$$\alpha_{[1]} + \alpha_{[2]} + \cdots + \alpha_{[n]} = \beta_{[1]} + \beta_{[2]} + \cdots + \beta_{[n]}.$$

Thus, for example, we have the majorizations

$$(1,1,1,1) \prec (2,1,1,0) \prec (3,1,0,0) \prec (4,0,0,0) \qquad (13.1)$$

and, since the definition of the relation $\alpha \prec \beta$ depends only on the corresponding ordered values, $\{\alpha_{[j]}\}$ and $\{\beta_{[j]}\}$, we could just as well write the chain (13.1) as

$$(1,1,1,1) \prec (0,1,1,2) \prec (1,3,0,0) \prec (0,0,4,0).$$

To give a more generic example, one should also note that for any $(\alpha_1, \alpha_2, \ldots, \alpha_n)$ we have the two relations

$$(\bar{\alpha}, \bar{\alpha}, \ldots, \bar{\alpha}) \prec (\alpha_1, \alpha_2, \ldots, \alpha_n) \prec (\alpha_1 + \alpha_2 + \cdots + \alpha_n, 0, \ldots, 0)$$

where, as usual, we have set $\bar{\alpha} = (\alpha_1 + \alpha_2 + \ldots + \alpha_n)/n$. Moreover, it is immediate from the definition of majorization that relation $\prec$ is transitive: $\alpha \prec \beta$ and $\beta \prec \gamma$ imply that $\alpha \prec \gamma$. Consequently, the 4-chain (13.1) actually entails six valid relations.

Now, if $\mathcal{A} \subset \mathbb{R}^d$ and $f : \mathcal{A} \to \mathbb{R}$, we say that f is *Schur convex* on $\mathcal{A}$ provided that we have

$$f(\alpha) \leq f(\beta) \qquad \text{for all } \alpha, \beta \in \mathcal{A} \text{ for which } \alpha \prec \beta. \tag{13.2}$$

Such a function might more aptly be called Schur monotone rather than Schur convex, but the term Schur convex is now firmly rooted in tradition. By the same custom, if the first inequality of the relation (13.2) is reversed, we say that f is *Schur concave* on $\mathcal{A}$.

THE TYPICAL PATTERN AND A PRACTICAL CHALLENGE

If we were to follow our usual pattern, we would now call on some concrete problem to illustrate how majorization and Schur convexity are used in practice. For example, we might consider the assertion that for positive a, b, and c, one has the reciprocal bound

$$\frac{1}{a} + \frac{1}{b} + \frac{1}{c} \leq \frac{1}{x} + \frac{1}{y} + \frac{1}{z} \tag{13.3}$$

where $x = b + c - a$, $y = a + c - b$, $z = a + b - c$, and where we assume that x, y, and z are strictly positive.

This slightly modified version of the *American Mathematical Monthly* problem E2284 of Walker (1971) is a little tricky if approached from first principles, yet we will find shortly that it is an immediate consequence of the Schur convexity of the map $(t_1, t_2, t_3) \mapsto 1/t_1 + 1/t_2 + 1/t_3$ and the majorization $(a, b, c) \prec (x, y, z)$.

Nevertheless, before we can apply majorization and Schur convexity to problems like E2284, we need to develop some machinery. In particular, we need a practical way to check that a function is Schur convex. The method we consider was introduced by Issai Schur in 1923, but even now it accounts for a hefty majority of all such verifications.

Problem 13.1 (Schur's Criterion)

Given that the function $f : (a,b)^n \to \mathbb{R}$ is continuously differentiable and symmetric, show that it is Schur convex on $(a,b)^n$ if and only if for all $1 \leq j < k \leq n$ and all $\mathbf{x} \in (a,b)^n$ one has

$$0 \leq (x_j - x_k)\left(\frac{\partial f(\mathbf{x})}{\partial x_j} - \frac{\partial f(\mathbf{x})}{\partial x_k}\right). \qquad (13.4)$$

AN ORIENTING EXAMPLE

Schur's condition may be unfamiliar, but there is no mystery to its application. For example, if we consider the function

$$f(t_1, t_2, t_3) = 1/t_1 + 1/t_2 + 1/t_3$$

which featured in our discussion of Walker's inequality (13.3), then one easily computes

$$(t_j - t_k)\left(\frac{\partial f(\mathbf{t})}{\partial t_j} - \frac{\partial f(\mathbf{t})}{\partial t_k}\right) = (t_j - t_k)(1/t_k^2 - 1/t_j^2).$$

This quantity is nonnegative since (t_j, t_k) and $(1/t_j^2, 1/t_k^2)$ are oppositely ordered, and, accordingly, the function f is Schur convex.

INTERPRETATION OF A DERIVATIVE CONDITION

Since the condition (13.4) contains only first order derivatives, it may refer to the monotonicity of something, the question is *what*? The answer may not be immediate, but the partial sums in the defining conditions of majorization do provide a hint.

Given an n-tuple $\mathbf{w} = (w_1, w_2, \ldots, w_n)$, it will be convenient to write $\widetilde{w}_j = w_1 + w_2 + \cdots + w_j$ and to set $\widetilde{\mathbf{w}} = (\widetilde{w}_1, \widetilde{w}_2, \ldots, \widetilde{w}_n)$. In this notation we see that the majorization $\mathbf{x} \prec \mathbf{y}$ holds if and only if we have $\widetilde{x}_j \leq \widetilde{y}_j$ for all $1 \leq j < n$. One benefit of this "tilde transformation" is that is makes majorization look more like ordinary coordinate-by-coordinate comparison.

Now, since we have assumed that f is symmetric, we know that f is Schur convex on $(a,b)^n$ if and only if it is Schur convex on the set $\mathcal{B} = (a,b)^n \cap \mathcal{D}$ where $\mathcal{D} = \{(x_1, x_2, \ldots, x_n) : x_1 \geq x_2 \geq \cdots \geq x_n\}$. Also, if we introduce the set $\widetilde{\mathcal{B}} = \{\widetilde{\mathbf{x}} : \mathbf{x} \in \mathcal{B}\}$, then we can define a new function $\widetilde{f} : \widetilde{\mathcal{B}} \to \mathbb{R}$ by setting $\widetilde{f}(\widetilde{\mathbf{x}}) = f(\mathbf{x})$ for all $\widetilde{\mathbf{x}} \in \widetilde{\mathcal{B}}$. The point of the new function $\widetilde{f}$ is that it should translate the behavior of f into the simpler language of the "tilde coordinates."

The key observation is that $f(\mathbf{x}) \leq f(\mathbf{y})$ for all $\mathbf{x}, \mathbf{y} \in \mathcal{B}$ with $\mathbf{x} \prec \mathbf{y}$

if and only if we have $\widetilde{f}(\widetilde{\mathbf{x}}) \leq \widetilde{f}(\widetilde{\mathbf{y}})$ for all $\widetilde{\mathbf{x}}, \widetilde{\mathbf{y}} \in \widetilde{\mathcal{B}}$ such that

$$\widetilde{x}_n = \widetilde{y}_n \quad \text{and} \quad \widetilde{x}_j \leq \widetilde{y}_j \quad \text{for all } 1 \leq j < n.$$

That is, f is Schur convex on $\mathcal{B}$ if and only if the function $\widetilde{f}$ on $\widetilde{\mathcal{B}}$ is a nondecreasing function of its first $n-1$ coordinates.

Since we assume that f is continuously differentiable, we therefore find that f is Schur convex if and only if for each $\widetilde{\mathbf{x}}$ in the interior of $\widetilde{\mathcal{B}}$ we have

$$0 \leq \frac{\partial \widetilde{f}(\widetilde{\mathbf{x}})}{\partial \widetilde{x}_j} \quad \text{for all } 1 \leq j < n.$$

Further, because $\widetilde{f}(\widetilde{\mathbf{x}}) = f(\widetilde{x}_1, \widetilde{x}_2 - \widetilde{x}_1, \ldots, \widetilde{x}_n - \widetilde{x}_{n-1})$, the chain rule gives us

$$0 \leq \frac{\partial \widetilde{f}(\widetilde{\mathbf{x}})}{\partial \widetilde{x}_j} = \frac{\partial f(\mathbf{x})}{\partial x_j} - \frac{\partial f(\mathbf{x})}{\partial x_{j+1}} \quad \text{for all } 1 \leq j < n, \qquad (13.5)$$

so, if we take $1 \leq j < k \leq n$ and sum the bound (13.5) over the indices $j, j+1, \ldots, k-1$, then we find

$$0 \leq \frac{\partial f(\mathbf{x})}{\partial x_j} - \frac{\partial f(\mathbf{x})}{\partial x_k} \quad \text{for all } \mathbf{x} \in \mathcal{B}.$$

By the symmetry of f on $(a,b)^n$, this condition is equivalent to

$$0 \leq (x_j - x_k) \left(\frac{\partial f(\mathbf{x})}{\partial x_j} - \frac{\partial f(\mathbf{x})}{\partial x_k} \right) \quad \text{for all } \mathbf{x} \in (a,b)^n,$$

and the solution of the first challenge problem is complete.

A LEADING CASE: AM-GM VIA SCHUR CONCAVITY

To see how Schur's criterion works in a simple example, consider the function $f(x_1, x_2, \ldots, x_n) = x_1 x_2 \cdots x_n$ where $0 < x_j < \infty$ for $1 \leq j \leq n$. Here we see that Schur's differential (13.4) is just

$$(x_j - x_k)(f_{x_j} - f_{x_k}) = -(x_j - x_k)^2 (x_1 \cdots x_{j-1} x_{j+1} \cdots x_{k-1} x_{k+1} \cdots x_n),$$

and this is always nonpositive. Therefore, f is Schur *concave*.

We noted earlier that $\bar{\mathbf{x}} \prec \mathbf{x}$ where $\bar{\mathbf{x}}$ is the vector $(\bar{x}, \bar{x}, \ldots, \bar{x})$ and where $\bar{x}$ is the simple average $(x_1 + x_2 + \cdots + x_n)/n$, so the Schur concavity of f then gives us $f(\mathbf{x}) \leq f(\bar{\mathbf{x}})$. In longhand, this says $x_1 x_2 \cdots x_n \leq \bar{x}^n$, and this is the AM-GM inequality in its most classic form.

In this example, one does not use the full force of Schur convexity. In essence, we have used Jensen's inequality in disguise, but there is still a message here: almost every invocation of Jensen's inequality can be

Majorization and Schur Convexity   195

replaced by a call to Schur convexity. Surprisingly often, this simple translation brings useful dividends.

A SECOND TOOL: VECTORS AND THEIR AVERAGES

This proof of the AM-GM inequality could hardly have been more automatic, but we were perhaps a bit lucky to have known in advance that $\bar{\mathbf{x}} \prec \mathbf{x}$. Any application of Schur convexity (or Schur concavity) must begin with a majorization relation, but we cannot always count on having the required relation in our inventory. Moreover, there are times when the definition of majorization is not so easy to check.

For example, to complete our proof of Walker's inequality (13.3), we need to show that $(a, b, c) \prec (x, y, z)$, but since we do not have any information on the relative sizes of these coordinates, the direct verification of the definition is awkward. The next challenge problem provides a useful tool for dealing with this common situation.

Problem 13.2 (Muirhead Implies Majorization)

Show that Muirhead's condition implies that α is majorized by β; that is, show that one has the implication

$$\alpha \in H(\beta) \quad \Longrightarrow \quad \alpha \prec \beta. \tag{13.6}$$

FROM MUIRHEAD'S CONDITION TO A SPECIAL REPRESENTATION

Here we should first recall that the notation $\alpha \in H(\beta)$ simply means that there are nonnegative weights p_τ which sum to 1 for which we have

$$(\alpha_1, \alpha_2, \ldots, \alpha_n) = \sum_{\tau \in \mathcal{S}_n} p_\tau (\beta_{\tau(1)}, \beta_{\tau(2)}, \cdots \beta_{\tau(n)})$$

or, in other words, α is a weighted average of $(\beta_{\tau(1)}, \beta_{\tau(2)}, \cdots \beta_{\tau(n)})$ as τ runs over the set $\mathcal{S}_n$ of permutations of $\{1, 2, \ldots, n\}$. If we take just the jth component of this sum, then we find the identity

$$\alpha_j = \sum_{\tau \in \mathcal{S}_n} p_\tau \beta_{\tau(j)} = \sum_{k=1}^{n} \left\{ \sum_{\tau : \tau(j) = k} p_\tau \right\} \beta_k = \sum_{k=1}^{n} d_{jk} \beta_k, \tag{13.7}$$

where for brevity we have set

$$d_{jk} = \sum_{\tau : \tau(j) = k} p_\tau \tag{13.8}$$

and where the sum (13.8) runs over all permutations $\tau \in \mathcal{S}_n$ for which

$\tau(j) = k$. We obviously have $d_{jk} \geq 0$, and we also have the identities

$$\sum_{j=1}^{n} d_{jk} = 1 \quad \text{and} \quad \sum_{k=1}^{n} d_{jk} = 1 \qquad (13.9)$$

since each of these sums equals the sum of p_τ over all $\mathcal{S}_n$.

A matrix $D = \{d_{jk}\}$ of nonnegative real numbers which satisfies the conditions (13.9) is said to be *doubly stochastic* because each of its rows and each of its columns can be viewed as a probability distribution on the set $\{1, 2, \ldots, n\}$. Doubly stochastic matrices will be found to provide a fundamental link between majorization and Muirhead's condition.

If we regard α and β as column vectors, then in matrix notation the relation (13.7) says that

$$\alpha \in H(\beta) \quad \Longrightarrow \quad \alpha = D\beta \qquad (13.10)$$

where D is the doubly stochastic matrix defined by the sums (13.8). Now, to complete the solution of the first challenge problem we just need to show that the representation $\alpha = D\beta$ implies $\alpha \prec \beta$.

FROM THE REPRESENTATION $\alpha = D\beta$ TO THE MAJORIZATION $\alpha \prec \beta$

Since the relations $\alpha \in H(\beta)$ and $\alpha \prec \beta$ are unaffected by permutations of the coordinates of α and β, there is no loss of generality if we assume that $\alpha_1 \geq \alpha_2 \geq \cdots \geq \alpha_n$ and $\beta_1 \geq \beta_2 \geq \cdots \geq \beta_n$. If we then sum the representation (13.7) over the initial segment $1 \leq j \leq k$, then we find the identity

$$\sum_{j=1}^{k} \alpha_j = \sum_{j=1}^{k} \sum_{t=1}^{n} d_{jt} \beta_t = \sum_{t=1}^{n} c_t \beta_t \quad \text{where } c_t \stackrel{\text{def}}{=} \sum_{j=1}^{k} d_{jt}. \qquad (13.11)$$

Since c_t is the sum of the first k elements of the tth column of D, the fact that D is doubly stochastic then gives us

$$0 \leq c_t \leq 1 \quad \text{for all } 1 \leq t \leq n \quad \text{and} \quad c_1 + c_2 + \cdots + c_n = k. \qquad (13.12)$$

These constraints strongly suggest that the differences

$$\Delta_k \stackrel{\text{def}}{=} \sum_{j=1}^{k} \alpha_j - \sum_{j=1}^{k} \beta_j = \sum_{t=1}^{n} c_t \beta_t - \sum_{j=1}^{k} \beta_j$$

are nonpositive for each $1 \leq k \leq n$, but an honest proof can be elusive. One must somehow exploit the identity (13.12), and a simple (yet clever) way is to write

$$\Delta_k = \sum_{j=1}^{n} c_j \beta_j - \sum_{j=1}^{k} \beta_j + \beta_k \left(k - \sum_{j=1}^{n} c_j \right)$$

$$= \sum_{j=1}^{k} (\beta_k - \beta_j)(1 + c_j) + \sum_{j=k+1}^{n} c_j (\beta_j - \beta_k).$$

It is now evident that $\Delta_k \leq 0$ since for all $1 \leq j \leq k$ we have $\beta_j \geq \beta_k$ while for all $k < j \leq n$ we have $\beta_j \leq \beta_k$. It is trivial that $\Delta_n = 0$, so the relations $\Delta_k \leq 0$ for $1 \leq k < n$ complete our check of the definition. We therefore find that $\alpha \prec \beta$, and the solution of the second challenge problem is complete.

FINAL CONSIDERATION OF THE WALKER EXAMPLE

In Walker's *Monthly* problem (page 192) we have the three identities $x = b + c - a$, $y = a + c - b$, $z = a + b - c$, so to confirm the relation $(a, b, c) \in H[(x, y, z)]$, one only needs to notice that

$$\begin{bmatrix} a \\ b \\ c \end{bmatrix} = \frac{1}{2} \begin{bmatrix} y \\ z \\ x \end{bmatrix} + \frac{1}{2} \begin{bmatrix} z \\ x \\ y \end{bmatrix}. \qquad (13.13)$$

This tells us that $\alpha \prec \beta$, so the proof of Walker's inequality (13.3) is finally complete.

Our solution of the second challenge problem also tells us that the relation (13.13) implies that (a, b, c) is the image of (x, y, z) under some doubly stochastic transformation D, and it is sometimes useful to make such a representation explicit. Here, for example, we only need to express the identity (13.13) with permutation matrices and then collect terms:

$$\begin{bmatrix} a \\ b \\ c \end{bmatrix} = \frac{1}{2} \begin{pmatrix} 0 & 1 & 0 \\ 0 & 0 & 1 \\ 1 & 0 & 0 \end{pmatrix} \begin{bmatrix} x \\ y \\ z \end{bmatrix} + \frac{1}{2} \begin{pmatrix} 0 & 0 & 1 \\ 1 & 0 & 0 \\ 0 & 1 & 0 \end{pmatrix} \begin{bmatrix} x \\ y \\ z \end{bmatrix} = \begin{pmatrix} 0 & \frac{1}{2} & \frac{1}{2} \\ \frac{1}{2} & 0 & \frac{1}{2} \\ \frac{1}{2} & \frac{1}{2} & 0 \end{pmatrix} \begin{bmatrix} x \\ y \\ z \end{bmatrix}.$$

A CONVERSE AND AN INTERMEDIATE CHALLENGE

We now face an obvious question: Is is also true that $\alpha \prec \beta$ implies that $\alpha \in H(\beta)$? In due course, we will find that the answer is affirmative, but full justification of this fact will take several steps. Our next challenge problem addresses the most subtle of these. The result is due to the joint efforts of Hardy, Littlewood, and Pólya, and its solution requires a sustained effort. While working through it, one finds that majorization acquires new layers of meaning.

Problem 13.3 (The HLP Representation: $\alpha \prec \beta \Rightarrow \alpha = D\beta$)
Show that $\alpha \prec \beta$ implies that there exists a doubly stochastic matrix D such that $\alpha = D\beta$.

Hardy, Littlewood, and Pólya came to this result because of their interests in mathematical inequalities, but, ironically, the concept of majorization was originally introduced by economists who were interested in inequalities of a different sort — the inequalities of income which one finds in our society. Today, the role of majorization in mathematics far outstrips its role in economics, but consideration of income distribution can still add to our intuition.

INCOME INEQUALITY AND ROBIN HOOD TRANSFORMATIONS

Given a nation A we can gain some understanding of the distribution of income in that nation by setting α_1 equal to the percentage of total income which is received by the top 10% of income earners, setting α_2 equal to the percentage earned by the next 10%, and so on down to α_{10} which we set equal to the percentage of national income which is earned by the bottom 10% of earners. If β is defined similarly for nation B, then the relation $\alpha \prec \beta$ has an economic interpretation; it asserts that income is more unevenly distributed in nation B than in nation A. In other words, the relation $\prec$ provides a measure of *income inequality*.

One benefit of this interpretation is that it suggests how one might try to prove that $\alpha \prec \beta$ implies that $\alpha = D\beta$ for some doubly stochastic transformation D. To make the income distribution of nation B more like the income of nation A, one can simply draw on the philosophy of Robin Hood: one steals from the rich and gives to the poor. The technical task is to prove that this thievery can be done in scientifically correct proportions.

THE SIMPLEST CASE: $n = 2$

To see how such a *Robin Hood transformation* would work in the simplest case, we just take $\alpha = (\alpha_1, \alpha_2) = (\rho + \sigma, \rho - \sigma)$ and take $\beta = (\beta_1, \beta_2) = (\rho + \tau, \rho - \tau)$. There is no loss of generality in assuming $\alpha_1 \geq \alpha_2$, $\beta_1 \geq \beta_2$, and $\alpha_1 + \alpha_2 = \beta_1 + \beta_2$; moreover, no loss in assuming that α and β have the indicated forms. The immediate benefit of this choice is that we have $\alpha \prec \beta$ if and only if $\sigma \leq \tau$.

To find a doubly stochastic matrix D that takes β to α is now just a question of solving a linear system for the components of D. The system is overdetermined, but it does have a solution which one can confirm simply by checking the identity

$$D\beta = \begin{pmatrix} \frac{\tau+\sigma}{2\tau} & \frac{\tau-\sigma}{2\tau} \\ \frac{\tau-\sigma}{2\tau} & \frac{\tau+\sigma}{2\tau} \end{pmatrix} \begin{pmatrix} \rho+\tau \\ \rho-\tau \end{pmatrix} = \begin{pmatrix} \rho+\sigma \\ \rho-\sigma \end{pmatrix} = \alpha. \tag{13.14}$$

Thus, the case $n = 2$ is almost trivial. Nevertheless, it is rich enough to suggest an interesting approach to the general case. Perhaps one can show that an $n \times n$ doubly stochastic matrix D is the product of a finite number transformations each one of which changes only two coordinates.

An Inductive Construction

If we take $\alpha_1 \geq \alpha_2 \geq \cdots \geq \alpha_n$ and $\beta_1 \geq \beta_2 \geq \cdots \geq \beta_n$ where $\alpha \prec \beta$, then we can consider a proof by induction on the number N of coordinates j such that $\alpha_j \neq \beta_j$. Naturally we can assume that $N \geq 1$, or else we can simply take D to be the identity matrix.

Now, given $N \geq 1$, the definition of majorization implies that there must exist a pair of integers $1 \leq j < k \leq n$ for which we have the bounds

$$\beta_j > \alpha_j, \quad \beta_k < \alpha_k, \quad \text{and} \quad \beta_s = \alpha_s \quad \text{for all } j < s < k. \tag{13.15}$$

Figure 13.1 gives a useful representation of this situation; the essence of which is that the interval $[\alpha_k, \alpha_j]$ is properly contained in the interval $[\beta_k, \beta_j]$. The intervening values $\alpha_s = \beta_s$ for $j < s < k$ are omitted from the figure to minimize clutter, but the figure records several further values that are important in our construction. In particular, it marks out $\rho = (\beta_j + \beta_k)/2$ and $\tau \geq 0$ which we choose so that $\beta_j = \rho + \tau$ and $\beta_k = \rho - \tau$, and it indicates the value σ which is defined to be the maximum of $|\alpha_k - \rho|$ and $|\alpha_j - \rho|$.

We now take T to be the $n \times n$ doubly stochastic transformation which takes $\beta = (\beta_1, \beta_2, \ldots, \beta_n)$ to $\beta' = (\beta'_1, \beta'_2, \ldots, \beta'_n)$ where

$$\beta'_k = \beta_k + \sigma, \quad \beta'_j = \beta_j - \sigma, \quad \text{and} \quad \beta'_t = \beta_t \quad \text{for all } t \neq j, t \neq k.$$

The matrix representation for T is easily obtained from the matrix given by our 2×2 example. One just places the coefficients of the 2×2 matrix at the four coordinates of T which are determined by the j, k rows and the j, k columns. The rest of the diagonal is then filled with $n-2$ ones and then the remaining places are filled with $n^2 - n - 2$ zeros, so one

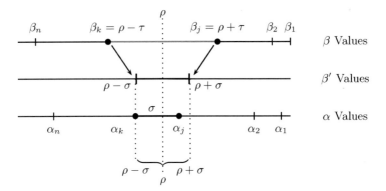

Fig. 13.1. The value ρ is the midpoint of $\beta_k = \rho - \tau$ and $\beta_j = \rho + \tau$ as well as the midpoint of $\alpha_k = \rho - \sigma$ and $\alpha_j = \rho + \sigma$. We have $0 < \sigma \leq \tau$, and the figure shows the case when $|\alpha_k - \rho|$ is larger than $|\alpha_j - \rho|$.

comes at last to a matrix with the shape

$$\begin{pmatrix} 1 & & & & & & & & \\ & \ddots & & & & & & & \\ & & 1 & & & & & & \\ & & & \frac{\tau+\sigma}{2\tau} & \cdots & \frac{\tau-\sigma}{2\tau} & & & \\ & & & \vdots & & \vdots & & & \\ & & & \frac{\tau-\sigma}{2\tau} & \cdots & \frac{\tau+\sigma}{2\tau} & & & \\ & & & & & & 1 & & \\ & & & & & & & \ddots & \\ & & & & & & & & 1 \end{pmatrix}. \qquad (13.16)$$

The Induction Step

We are almost ready to appeal to the induction step, but we still need to check that $\alpha \prec \beta' = T\beta$. If we use $s_t(\gamma) = \gamma_1 + \gamma_2 + \cdots + \gamma_t$ to simplify the writing of partial sums, then we have three basic observations:

$$s_t(\alpha) \leq s_t(\beta) = s_t(\beta') \qquad 1 \leq t < j \qquad \text{(a)}$$
$$s_t(\alpha) \leq s_t(\beta') \qquad j \leq t < k \qquad \text{(b)}$$
$$s_t(\alpha) \leq s_t(\beta) = s_t(\beta') \qquad k \leq t \leq n. \qquad \text{(c)}$$

Observations (a) and (c) are immediate, and to justify (b) we only need to note that $\alpha_j \leq \beta'_j$ and to recall that $\alpha_t = \beta'_t = \beta$ for $j < t < k$.

These bounds confirm that $\alpha \prec \beta'$ and, by the design of T, we know that the n-tuples α and β' agree in all but at most $N - 1$ coordinates. Hence, by induction, there is a doubly stochastic matrix D' such that $\alpha = D'\beta'$. Since $\beta' = T\beta$, we therefore have $\alpha = D'(T\beta) = (D'T)\beta$, and, since the product of two doubly stochastic matrices is doubly stochastic, we see that the matrix $D = D'T$ provides us with the solution to our challenge problem.

JENSEN'S INEQUALITY: REVISITED AND REFINED

The Hardy, Littlewood, Pólya representation $\alpha = D\beta$ is a statement about averages. Part of its message is that for each j the value α_j is an average of $\beta_1, \beta_2, \ldots, \beta_n$, but the identity $\alpha = D\beta$ actually tells us a bit more. Specifically, we also know that each column of D must sum to one, though for the moment it may not be clear how one might use this additional information.

We do know from our experience with Jensen's inequality that averages and convex functions can be combined to provide a large number of useful inequalities, and it is natural to ask if the representation $\alpha = D\beta$ might provide something even grander. Issai Schur confirmed this suggestion with a simple calculation which has become a classic part of the lore of majorization and which provides the final challenge problem of the chapter.

Problem 13.4 (Schur's Majorization Inequality)

Show that if $\phi : (a, b) \to \mathbb{R}$ is a convex function, then the function $f : (a, b)^n \to \mathbb{R}$ defined by the sum

$$f(x_1, x_2, \ldots, x_n) = \sum_{k=1}^{n} \phi(x_k) \qquad (13.17)$$

is Schur convex. Thus, for $\alpha, \beta \in (a, b)^n$ with $\alpha \prec \beta$ one has

$$\sum_{k=1}^{n} \phi(\alpha_k) \leq \sum_{k=1}^{n} \phi(\beta_k). \qquad (13.18)$$

ORIENTATION

To familiarize the bound (13.18), one should first note that if we take $\alpha = (\bar{x}, \bar{x}, \ldots, \bar{x})$ and $\beta = (x_1, x_2, \ldots, x_n)$, then it reduces to Jensen's inequality. Also, since the function $t \mapsto 1/t$ is convex on the set $(0, \infty)$, we see that Schur's majorization bound (13.18) also implies Walker's inequality (13.3), since we know now that the representation (13.13) implies $(a, b, c) \prec (x, y, z)$.

One should further note that if we assume that ϕ is differentiable, then the Schur convexity of f follows almost immediately from the differential criterion (13.4). In particular, by the convexity of ϕ the derivative ϕ' is nondecreasing, so the pairs (x_j, x_k) and $(\phi'(x_j), \phi'(x_k))$ are similarly ordered. Consequently, Schur's differential

$$(x_j - x_k)\bigl(f_{x_j}(\mathbf{x}) - f_{x_j}(\mathbf{x})\bigr) = (x_j - x_k)(\phi'(x_j) - \phi'(x_k))$$

is nonnegative, and f is Schur convex. Part of our challenge problem is thus to prove the Schur convexity of f without recourse to differentiability.

A DIRECT APPROACH

To prove the bound (13.18) by a direct appeal to convexity of ϕ, one needs to find an appropriate average, and the HLP representation (page 198) is a natural place to look. We are given $\alpha \prec \beta$, so the HLP representation tells us that there is a doubly stochastic matrix $D = \{d_{jk}\}$ such that $\alpha = D\beta$, or, in longhand, for each $j = 1, 2\ldots, n$ we have the representation

$$a_j = \sum_{k=1}^{n} d_{jk}\beta_k \quad \text{where } d_{j1} + d_{j2}\cdots + d_{jn} = 1.$$

Now, if we apply Jensen's inequality to these averages, we have

$$\phi(a_j) \leq \sum_{k=1}^{n} d_{jk}\phi(b_k),$$

and, except for the abstract quality of the d_{jk} factors, this bound is better than the one we seek. In particular, if we sum over j and change the order of summation, we find

$$\sum_{j=1}^{n} \phi(a_j) \leq \sum_{j=1}^{n}\sum_{k=1}^{n} d_{jk}\phi(b_k) = \sum_{k=1}^{n}\left\{\phi(b_k)\sum_{j=1}^{n} d_{jk}\right\} = \sum_{k=1}^{n} \phi(b_k),$$

just as we hoped to show.

No one would deny that Schur's majorization inequality (13.18) is a very easy result, but one should not be deceived by its simplicity. It strips away the secret of many otherwise mysterious bounds.

A Day-to-Day Example

The final challenge addresses a typical example of the flood of problems that one can solve — or invent — with help from the tools developed in this chapter.

Problem 13.5 *Given $x, y, z \in (0, 1)$ such that*

$$\max(x, y, z) \leq (x + y + z)/2 < 1, \qquad (13.19)$$

show that one has the bound

$$\left(\frac{1+x}{1-x}\right)\left(\frac{1+y}{1-y}\right)\left(\frac{1+z}{1-z}\right) \leq \left\{\frac{1 + \frac{1}{2}(x+y+z)}{1 - \frac{1}{2}(x+y+z)}\right\}^2. \qquad (13.20)$$

If this problem were met in another context, it might be quite puzzling. It is not obvious that the two sides are comparable, and the hypothesis (13.19) is unlike anything we have seen before. Still, with majorization in mind, one may not need long to hit on a fruitful plan.

In particular, one might think of exploiting the hypothesis (13.19) by noting that it gives us $(x, y, z) \prec (s, s, 0)$ where $s = (x+y+z)/2$. After this observation, it becomes clear that the bound (13.20) would follow from Schur's majorization inequality (13.18) if we could show that

$$\phi(t) = \log\left(\frac{1+t}{1-t}\right)$$

is a convex function on $(0, 1)$. This is easily confirmed by direct calculation of the second derivative,

$$\phi''(t) = \frac{4t}{(t^2 - 1)^2} > 0,$$

but it is also obvious from the Taylor expansion

$$\phi(t) = 2\left(t + \frac{t^3}{3} + \frac{t^5}{5} + \cdots\right).$$

Illustrative Exercises and a Vestige of Theory

Most of the chapter's exercises are designed to illustrate the applications of majorization and Schur convexity, but last the two exercises serve a different purpose. They are given to complete the picture of majorization theory that is illustrated by Figure 13.2. We have proved all of the implications pictured there except for the one which we have labelled as *Birkhoff's Theorem*.

This famous theorem asserts that every doubly stochastic matrix is a

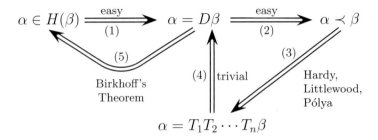

Fig. 13.2. Sometimes the definition of $\alpha \prec \beta$ is easy to check, but perhaps more often one relies on either the condition $\alpha = D\beta$ or the condition $\alpha \in H(\beta)$ to prove majorization.

convex combination of permutation matrices, and it closes the loop on the double implication $\alpha \prec \beta \Leftrightarrow \alpha \in H(\beta)$ asserting the equivalence of majorization and Muirhead's condition. Most day-to-day applications of majorization do not require Birkhoff's half of this equivalence, but Birkhoff's theorem has applications throughout pure and applied mathematics. It is sometimes called the fundamental theorem of doubly stochastic matrices.

EXERCISES

Exercise 13.1 (Two Doubly Stochastic Giveaways)
Show that for positive x, y, z one has the product bound
$$xyz \leq \left(x/2 + y/3 + z/6\right)\left(x/3 + 2y/3\right)\left(x/6 + 5y/6\right),$$
and the awe inspiring reciprocal bound
$$\left(\frac{2}{x+y}\right)^5 + \left(\frac{6}{3x+y+2z}\right)^5 + \left(\frac{6}{3x+3y+z}\right)^5 \leq \frac{1}{x^5} + \frac{1}{y^5} + \frac{1}{x^5}.$$

Exercise 13.2 (Finding the Majorization)
Given $1 \leq k \leq n$ and real numbers $x_j > 0$, $1 \leq j \leq n$, such that $\max(x_1, x_2, \ldots, x_n) \leq (x_1 + x_2 + \cdots + x_n)/k$, show that one has
$$\sum_{j=1}^{n} \frac{1}{1+x_j} \leq (n-k) + \frac{k^2}{k + x_1 + x_2 + \cdots + x_n}. \qquad (13.21)$$

Majorization and Schur Convexity   205

Exercise 13.3 (A Refinement of the 1-Trick)
Given integers $0 < m < n$ and real numbers $x_1, x_2, \ldots, x_n$ such that

$$\sum_{k=1}^{m} x_k = \frac{m}{n} \sum_{k=1}^{n} x_k + \delta \qquad (13.22)$$

where $\delta \geq 0$, show that the sum of squares has the lower bound

$$\sum_{k=1}^{n} x_k^2 \geq \frac{1}{n} \left(\sum_{k=1}^{n} x_k \right)^2 + \frac{\delta^2 n}{m(n-m)}. \qquad (13.23)$$

This refinement of the familiar 1-trick lower bound was crucial to the discovery and proof of the *Szemerédi's Regularity Lemma*, which is one of the cornerstones of modern combinatorial theory.

Exercise 13.4 (Symmetric Polynomials and Schur Concavity)
After observing that the kth elementary symmetric function

$$e_k(\mathbf{x}) = e_k(x_1, x_2, \ldots, x_n) = \sum_{1 \leq i_1 < i_2 < \cdots < i_k \leq n} x_{i_1} x_{i_2} \cdots x_{i_k}$$

satisfies the elegant "cancellation identity"

$$\frac{\partial e_k(\mathbf{x})}{\partial x_s} = e_{k-1}(x_1, x_2, \ldots, x_{s-1}, x_{s+1}, \ldots, x_n), \qquad (13.24)$$

show that $e_k(\mathbf{x})$ is Schur concave for $\mathbf{x} \in [0, \infty)^n$.

Exercise 13.5 (Schur Concavity and Measures of Dispersion)
Many methods have been proposed to measure dispersion. Statisticians, for example, often use the *sample variance*

$$s(\mathbf{x}) = \frac{1}{n-1} \sum_{j=1}^{n} (x_j - \bar{x})^2 \qquad \text{where } \bar{x} = (x_1 + x_2 + \cdots + x_n)/n$$

for $\mathbf{x} \in \mathbb{R}^n$, $n \geq 2$, while information theorists rely on the *entropy*

$$h(\mathbf{p}) = -\sum_{k=1}^{n} p_k \log p_k$$

to measure of dispersion of the probability distribution $(p_1, p_2, \ldots, p_n)$ where $p_k \geq 0$ and $p_1 + p_2 + \cdots + p_n = 1$. Show that both the sample variance $s(\mathbf{x})$ and the entropy $h(\mathbf{p})$ are Schur convex.

Exercise 13.6 (Another Inversion Preserving Form)

If $p_k \geq 0$, $p_1 + p_2 + \cdots + p_n = 1$, and $0 < \alpha$ show that

$$\frac{(n^2+1)^\alpha}{n^{\alpha-1}} \leq \sum_{k=1}^{n} \left(p_k + \frac{1}{p_k}\right)^\alpha. \tag{13.25}$$

Incidentally, way back in Exercise 1.6 we used Cauchy's inequality to deal with the case $\alpha = 1$. Remarkably often majorization helps one to put a consequence of Cauchy's inequality into a broader context.

Exercise 13.7 (A Birthday Problem)

Given n random people, what is the probability that two or more of them have the same birthday? Under the natural (but approximate!) model where the birthdays are viewed as an independent and uniformly distributed in the set $\{1, 2, \ldots, 365\}$, show that this probability is at least $1/2$ if $n \geq 23$. For the more novel bit, show this probability does not go down if one drops the assumption that the birthdays are uniformly distributed.

Exercise 13.8 (SDRs and the Marriage Problem)

If $S_1, S_2, \ldots, S_n$ is a collection of subsets of the set S, we say that the set $R = \{x_1, x_2, \ldots, x_n\} \subset S$ is a *system of distinct representatives* (or an *SDR*) provided that the elements of R are all distinct and $x_k \in S_k$ for each $1 \leq k \leq n$. Prove that a necessary and sufficient condition for the existence of an SDR is that one has the inequality

$$|A| \leq \left|\bigcup_{j \in A} S_j\right| \quad \text{for all } A \subset \{1, 2, \ldots, n\}, \tag{13.26}$$

where $|C|$ is used as shorthand for the cardinality of a set C.

The quaint term "marriage problem" comes from a 1949 article by Hermann Weyl who essentially put the issue as follows: given a set of girls and boys, it is possible for each girl to marry a boy she knows if and only if each subset of k girls knows at least k boys.

The marriage lemma is one of the most widely applied results in all of combinatorial theory, and it has many applications to the theory of inequalities. In particular, it is of great help with the final exercise which develops Birkhoff's Theorem.

Majorization and Schur Convexity

Exercise 13.9 (Birkhoff's Theorem)

Given a permutation $\sigma \in \mathcal{S}_n$, the *permutation matrix* associated with σ is the $n \times n$ matrix $P_\sigma = (P_\sigma(j,k) : 1 \leq j, k \leq n)$ with entries

$$P_\sigma(j,k) = \begin{cases} 1 & \text{if } \sigma(j) = k \\ 0 & \text{otherwise.} \end{cases}$$

Show that if D is an $n \times n$ doubly stochastic matrix, then there exist nonnegative weights $\{w_\sigma : \sigma \in \mathcal{S}_n\}$, such that

$$\sum_{\sigma \in \mathcal{S}_n} w_\sigma = 1 \quad \text{and} \quad \sum_{\sigma \in \mathcal{S}_n} w_\sigma P_\sigma = D. \tag{13.27}$$

In other words, every doubly stochastic matrix is an average of permutation matrices.

14
Cancellation and Aggregation

Cancellation is not often discussed as a self-standing topic, yet it is the source of some of the most important phenomena in mathematics. Given any sum of real or complex numbers, we can always obtain a bound by taking the absolute values of the summands, but such a step typically destroys the more refined elements of our problem. If we hope to take advantage of cancellation, we must consider summands in groups.

We begin with a classical result of Niels Henrik Abel (1802–1829) who is equally famous for his proof of the impossibility of solving the general quintic equation by radicals and for his brief tragic life. Abel's inequality is simple and well known, but it is also tremendously productive. Many applications of cancellation call on its guidance, either directly or indirectly.

Problem 14.1 (Abel's Inequality)

Let $z_1, z_2, \ldots, z_n$ denote a sequence of complex numbers with partial sums $S_k = z_1 + z_2 + \cdots + z_k$, $1 \leq k \leq n$. For each sequence of real numbers such that $a_1 \geq a_2 \geq \cdots \geq a_n \geq 0$ one has

$$|a_1 z_1 + a_2 z_2 + \cdots + a_n z_n| \leq a_1 \max_{1 \leq k \leq n} |S_k|. \qquad (14.1)$$

MAKING PARTIAL SUMS MORE VISIBLE

Part of the wisdom of Abel's inequality is that it shifts our focus onto the *maximal sequence* $M_n = \max_{1 \leq k \leq n} |S_k|$, $n = 1, 2, \ldots$, even when our primary concern might be for the sums $a_1 z_1 + a_2 z_2 + \cdots + a_n z_n$. Shortly we will find that there are subtle techniques for dealing with maximal sequences, but first we should attend to Abel's inequality and some of its consequences.

The challenge is to bound the modulus of $a_1 z_1 + a_2 z_2 + \cdots + a_n z_n$ with help from $\max_{1 \leq k \leq n} |S_k|$, so a natural first step is to use summation by parts to bring the partial sums $S_k = z_1 + z_2 + \cdots + z_k$ into view. Thus,

Cancellation and Aggregation 209

we first note that

$$a_1 z_1 + a_2 z_2 + \cdots + a_n z_n = a_1 S_1 + a_2(S_2 - S_1) + \cdots + a_n(S_n - S_{n-1})$$
$$= S_1(a_1 - a_2) + S_2(a_2 - a_3) + \cdots + S_{n-1}(a_{n-1} - a_n) + S_n a_n.$$

This identity (which is often called Abel's formula) now leaves little left for us to do. It shows that $|a_1 z_1 + a_2 z_2 + \cdots + a_n z_n|$ is bounded by

$$|S_1|(a_1 - a_2) + |S_2|(a_2 - a_3) + \cdots + |S_{n-1}|(a_{n-1} - a_n) + |S_n|a_n$$
$$\leq \max_{1 \leq k \leq n} |S_k| \{(a_1 - a_2) + (a_2 - a_3) + \cdots + (a_{n-1} - a_n) + a_n\}$$
$$= a_1 \max_{1 \leq k \leq n} |S_k|,$$

and the (very easy!) proof of Abel's inequality is complete.

APPLICATIONS OF ABEL'S INEQUALITY

Abel's inequality may be close to trivial, but its consequences can be surprisingly elegant. Certainly it is the tool of choice when one asks about the convergence of sums such as

$$Q = \sum_{k=1}^{\infty} \frac{(-1)^k}{\sqrt{k}} \quad \text{or} \quad R = \sum_{k=1}^{\infty} \frac{\cos(k\pi/6)}{\log(k+1)}.$$

For example, in the first case Abel's inequality gives the succinct bound

$$\left| \sum_{k=M}^{N} \frac{(-1)^k}{\sqrt{k}} \right| \leq \frac{1}{\sqrt{M}} \quad \text{for all } 1 \leq M \leq N < \infty. \tag{14.2}$$

This is more than one needs to show that the partial sums of Q form a Cauchy sequence, so the sum Q does indeed converge.

The second sum R may look harder, but it is almost as easy. Since the sequence $\{\cos(k\pi/6) : k = 1, 2, \ldots, \}$ is periodic with period 12, it is easy to check by brute force that

$$\max_{M,N} \left| \sum_{k=M}^{N} \cos(k\pi/6) \right| = 2 + \sqrt{3} = 3.732\ldots, \tag{14.3}$$

so Abel's inequality gives us another simple bound

$$\left| \sum_{k=M}^{N} \frac{\cos(k\pi/6)}{\log(k+1)} \right| \leq \frac{2 + \sqrt{3}}{\log(M+1)} \quad \text{for all } 1 \leq M \leq N < \infty. \tag{14.4}$$

This bound suffices to show the convergence of R and, moreover, one can check by numerical calculation that it has very little slack. For example, the constant $2 + \sqrt{3}$ cannot be replaced by a smaller one. Without foreknowledge of Abel's inequality, one probably would not guess that the partial sums of R would have such simple, sharp bounds.

THE ORIGINS OF CANCELLATION

Cancellation has widely diverse origins, but bounds for partial sums of complex exponentials may provide the single most common source. Such bounds lie behind the two introductory examples (14.2) and (14.3), and, although these are particularly easy, they still point toward an important theme.

Linear sums are the simplest exponential sums. Nevertheless, they can lead to subtle inferences, such as the bound (14.7) for the quadratic exponential sum which forms the core of our second challenge problem. To express the linear bound most simply, we use the common shorthand

$$\mathbf{e}(t) \stackrel{\text{def}}{=} \exp(2\pi i t) \quad \text{and} \quad ||t|| = \min\{|t - k| : k \in \mathbb{Z}\}; \qquad (14.5)$$

so, here, $||t||$ denotes the distance from $t \in \mathbb{R}$ to the nearest integer. This use of the "double bar" notation is traditional in this context, and it should not lead to any confusion with the notation for a vector norm.

Problem 14.2 (Linear and Quadratic Exponential Sums)

First, as a useful warm-up, show that for all $t \in \mathbb{R}$ and all integers M and N one has the bounds

$$\left| \sum_{k=M+1}^{M+N} \mathbf{e}(kt) \right| \leq \min\left\{ N, \frac{1}{|\sin \pi t|} \right\} \leq \min\left\{ N, \frac{1}{2||t||} \right\}, \qquad (14.6)$$

then, for a more engaging challenge, show that for $b, c \in \mathbb{R}$ and all integers $0 \leq M < N$ one also has a uniform bound for the quadratic exponential sums,

$$\left| \sum_{k=1}^{M} \mathbf{e}\left((k^2 + bk + c)/N\right) \right| \leq \sqrt{2N(1 + \log N)}. \qquad (14.7)$$

LINEAR EXPONENTIAL SUMS AND THEIR ESTIMATES

For a quick orientation, one should note that the bound (14.6) generalizes those which were used in the discussion of Abel's inequality. For example, since $|\operatorname{Re} w| \leq |w|$ we can set $t = 1/12$ in the bound (14.6) to obtain an estimate for the cosine sum

$$\left| \sum_{k=M+1}^{M+N} \cos(k\pi/6) \right| \leq \frac{1}{\sin(\pi/12)} = \frac{2\sqrt{2}}{\sqrt{3} - 1} = 3.8637\ldots.$$

Cancellation and Aggregation

This is remarkably close to the best possible bound (14.3), and the phenomenon it suggests is typical. If one must give a uniform estimate for a whole ensemble of linear sums, the estimate (14.6) is hard to beat, though, of course, it can be quite inefficient for many of the individual sums.

To prove the bound (14.6), one naturally begins with the formula for geometric summation,

$$\sum_{k=M+1}^{M+N} \mathbf{e}(kt) = \mathbf{e}((M+1)t)\left\{\frac{\mathbf{e}(Nt)-1}{\mathbf{e}(t)-1}\right\}$$

and, to bring the sine function into view, one has the factorization

$$\mathbf{e}((M+1)t)\frac{\mathbf{e}(Nt/2)}{\mathbf{e}(t/2)}\left\{\frac{\big(\mathbf{e}(Nt/2) - \mathbf{e}(-Nt/2)\big)/2i}{\big(\mathbf{e}(t/2) - \mathbf{e}(-t/2)\big)/2i}\right\}.$$

If we identify the bracketed fraction and take the absolute value, we find

$$\left|\sum_{k=M+1}^{M+N} \mathbf{e}(kt)\right| = \left|\frac{\sin(\pi Nt)}{\sin(\pi t)}\right| \leq \frac{1}{|\sin \pi t|}.$$

Finally, to get the second part of the bound (14.6), one only needs to notice that the graph of $t \mapsto \sin \pi t$ makes it obvious that $2||t|| \leq |\sin \pi t|$.

AN EXPLORATION OF QUADRATIC EXPONENTIAL SUMS

The geometric sum formula provided a ready-made plan for estimation of the linear sums, but the quadratic exponential sum (14.7) is further from our experience. Some experimentation seems appropriate before we try to settle on a plan.

If we consider a generic quadratic polynomial $P(k) = \alpha k^2 + \beta k + \gamma$ with $\alpha, \beta, \gamma \in \mathbb{R}$ and $k \in \mathbb{Z}$, we need to estimate the sum

$$S_M(P) \stackrel{\text{def}}{=} \sum_{k=1}^{M} \mathbf{e}(P(k)), \tag{14.8}$$

or, more precisely, we need to estimate the modulus $|S_M(P)|$ or its square $|S_M(P)|^2$. If we try brute force, we will need an n-term analog of the familiar formula $|c_1 + c_2|^2 = |c_1|^2 + |c_2|^2 + 2\text{Re}\{c_1\bar{c}_2\}$, and this calls for us to compute

$$\left|\sum_{n=1}^{M} c_n\right|^2 = \sum_{n=1}^{M} |c_n|^2 + \sum_{1 \leq m < n \leq M} \{c_m \bar{c}_n + \bar{c}_m c_n\}$$

$$= \sum_{n=1}^{M} |c_n|^2 + \sum_{1 \leq m < n \leq M} 2\operatorname{Re}\{c_n \bar{c}_m\}$$

$$= \sum_{n=1}^{M} |c_n|^2 + 2\operatorname{Re} \sum_{h=1}^{M-1} \sum_{m=1}^{M-h} c_{m+h} \bar{c}_m. \qquad (14.9)$$

If we specialize the formula (14.9) by setting $c_n = \mathbf{e}(P(n))$, then we come to the identity

$$|S_M(P)|^2 = M + 2\operatorname{Re} \sum_{h=1}^{M-1} \sum_{m=1}^{M-h} \mathbf{e}\left((P(m+h) - P(m))\right). \qquad (14.10)$$

This formula may seem complicated, but if one looks past the clutter, it suggests an interesting opportunity. The inside sum contains the exponentials of *differences* of a quadratic polynomial, and, since such differences are simply linear polynomials, we can estimate the inside sum with help from the basic bound (14.6).

The difference $P(m+h) - P(m) = 2\alpha m h + \alpha h^2 + \beta h$ brings us to the factorization $\mathbf{e}(P(m+h) - P(m)) = \mathbf{e}(\alpha h^2 + \beta h)\mathbf{e}(2\alpha m h)$, so for the inside sum of the identity (14.10) we have the bound

$$\left|\sum_{m=1}^{M-h} \mathbf{e}\left((P(m+h) - P(m))\right)\right| \leq \frac{1}{|\sin(\pi h \alpha)|}. \qquad (14.11)$$

Thus, for any real quadratic $P(k) = \alpha k^2 + \beta k + \gamma$ we have the estimate

$$|S_M(P)|^2 \leq M + 2 \sum_{h=1}^{M-1} \frac{1}{|\sin(\pi h \alpha)|} \leq N + \sum_{h=1}^{N-1} \frac{1}{||h\alpha||}, \qquad (14.12)$$

where $||\alpha h||$ is the distance from $\alpha h \in \mathbb{R}$ to the nearest integer.

After setting $\alpha = 1/N$, $\beta = b/N$, and $\gamma = c/N$ in the estimate (14.12), we find a bound for our target sum

$$\left|\sum_{k=1}^{M} \mathbf{e}\left((k^2 + bk + c)/N\right)\right|^2 \leq N + \sum_{h=1}^{N-1} \frac{1}{||h/N||}$$

$$\leq N + 2N \sum_{1 \leq h \leq N/2} \frac{1}{h}, \qquad (14.13)$$

Cancellation and Aggregation

where in the second step we used the fact that the fraction h/N is closest to 0 for $1 \le h \le N/2$ while for $N/2 < h < N$ it is closest to 1.

The logarithmic factor in the challenge bound (14.7) is no longer so mysterious; it is just the result of using the logarithmic bound for the harmonic series. Since $1 + 1/2 + \cdots + 1/m \le 1 + \log m$, we find that our estimate (14.13) not larger than $N + 2N(1 + \log(N/2))$ which is bounded by $2N(1 + \log N)$ since $(3 - 2\log 2) \le 2$. After taking square roots, the solution of the second challenge problem is complete.

THE ROLE OF AUTOCORRELATIONS

The proof of the quadratic bound (14.7) relied on the general relation

$$\left| \sum_{n=1}^{N} c_n \right|^2 \le \sum_{n=1}^{N} |c_n|^2 + 2 \sum_{h=1}^{N-1} \left| \sum_{m=1}^{N-h} c_{m+h} \bar{c}_m \right| \qquad (14.14)$$

which one obtains from the identity (14.9). This bound suggests that we focus on the *autocorrelation sums* which may be defined by setting

$$\rho_N(h) = \sum_{m=1}^{N-h} c_{m+h} \bar{c}_m \qquad \text{for all } 1 \le h < N. \qquad (14.15)$$

If these are small on average, then the sum $|c_1 + c_2 + \cdots + c_N|$ should also be relatively small.

Our proof of the quadratic bound (14.7) exploited this principle with help from the sharp estimate (14.11) for $|\rho_N(h)|$, but such quantitative bounds are often lacking. More commonly we only have *qualitative* information with which we hope to answer qualitative questions. For example, if we assume that $|c_k| \le 1$ for all $k = 1, 2, \ldots$ and assume that

$$\lim_{N \to \infty} \frac{\rho_N(h)}{N} = 0 \qquad \text{for all } h = 1, 2, \ldots, \qquad (14.16)$$

does it follow that $|c_1 + c_2 + \cdots + c_N|/N \to 0$ as $N \to \infty$? The answer to this question is *yes*, but the bound (14.14) cannot help us here.

LIMITATIONS AND A CHALLENGE

Although the bound (14.14) is natural and general, it has serious limitations. In particular, it requires one to sum $|\rho_N(h)|$ over the full range $1 \le h < N$, and consequently its effectiveness is greatly eroded if the available estimates for $|\rho_N(h)|$ grow too quickly with h. For example, in a case where one has $hN^{1/2} \le |\rho_N(h)| \le 2hN^{1/2}$ the limit conditions (14.16) are all satisfied, yet the bound provided by (14.14) is useless since it is larger than N^2.

Such limitations suggest that it could be quite useful to have an analog of the bound (14.14) where one only uses the autocorrelations $\rho_N(h)$ for $1 \leq h \leq H$ where H is a fixed integer. In 1931, J.G. van der Corput provided the world with just such an analog, and it forms the basis for our next challenge problem. We actually consider a streamlined version of van der Corput's which underscores the role of $\rho_N(h)$, the autocorrelation sum defined by formula (14.15).

Problem 14.3 (A Qualitative van der Corput Inequality)
Show that for each complex sequence $c_1, c_2, \ldots, c_N$ and for each integer $1 \leq H < N$ one has the inequality

$$\left| \sum_{n=1}^{N} c_n \right|^2 \leq \frac{4N}{H+1} \left\{ \sum_{n=1}^{N} |c_n|^2 + \sum_{h=1}^{H} |\rho_N(h)| \right\}. \qquad (14.17)$$

A QUESTION ANSWERED

Before we address the proof of the bound (14.17), we should check that it does indeed answer the question which was posed on page 213. If we assume that for each $h = 1, 2, \ldots$, one has $\rho_N(h)/N \to 0$ as $N \to \infty$ and if we assume that $|c_k| \leq 1$ for all k, then the bound (14.17) gives us

$$\limsup_{N \to \infty} \frac{1}{N^2} \left| \sum_{n=1}^{N} c_n \right|^2 \leq \frac{4}{H+1}. \qquad (14.18)$$

Here H is arbitrary, so we do find that $|c_1 + c_2 + \cdots + c_N|/N \to 0$ as $N \to \infty$, just as we hoped we would.

The cost — and the benefit — of van der Corput's inequality are tied to the parameter H. It makes the bound (14.17) more complicated than its naive precursor (14.14), but this is the price one pays for added flexibility and precision.

EXPLORATION AND PROOF

The challenge bound (14.17) does not come with any overt hints for its proof, and, until a concrete idea presents itself, almost all one can do is explore the algebra of similar expressions. In particular, one might try to understand more deeply the relationships between a sequence and shifts of itself.

To discuss such shifts without having to worry about boundary effects, it is often useful to take the finite sequence $c_1, c_2, \ldots, c_N$ and extend it to one which is doubly infinite by setting $c_k = 0$ for all $k \leq 0$ and all $k > N$. If we then consider the sequence along with its shifts, some natural relationships start to become evident. For example, if one considers the original sequence and the first two shifts, we get the picture

$$
\begin{array}{cccccccccc}
\cdots & c_{-2} & c_{-1} & c_0 & c_1 & c_2 & c_3 & \cdots & c_N & c_{N+1} & c_{N+2} & c_{N+3} & \cdots \\
\cdots & c_{-2} & c_{-1} & c_0 & c_1 & c_2 & c_3 & \cdots & c_N & c_{N+1} & c_{N+2} & c_{N+3} & \cdots \\
\cdots & c_{-2} & c_{-1} & c_0 & c_1 & c_2 & c_3 & \cdots & c_N & c_{N+1} & c_{N+2} & c_{N+3} & \cdots
\end{array}
$$

and when we sum along the "down-left" diagonals we see that the extended sequence satisfies the identity

$$3\sum_{n=1}^{N} c_n = \sum_{n=1}^{N+2} \sum_{h=0}^{2} c_{n-h}.$$

In the exactly same way, one can sum along the diagonals of an array with $H+1$ rows to show that the extended sequence satisfies

$$(H+1)\sum_{n=1}^{N} c_n = \sum_{n=1}^{N+H} \sum_{h=0}^{H} c_{n-h}. \tag{14.19}$$

This identity is not deep, but does achieve two aims: it represents a generic sum in terms of its shifts and it introduces a free parameter H.

AN APPLICATION OF CAUCHY'S INEQUALITY

If we take absolute values and square the sum (14.19), we find

$$(H+1)^2 \left|\sum_{n=1}^{N} c_n\right|^2 = \left|\sum_{n=1}^{N+H} \sum_{h=0}^{H} c_{n-h}\right|^2 \leq \left\{\sum_{n=1}^{N+H} \left|\sum_{h=0}^{H} c_{n-h}\right|\right\}^2,$$

and this invites us to apply Cauchy's inequality (and the 1-trick) to find

$$(H+1)^2 \left|\sum_{n=1}^{N} c_n\right|^2 \leq (N+H) \sum_{n=1}^{N+H} \left|\sum_{h=0}^{H} c_{n-h}\right|^2. \tag{14.20}$$

This estimate brings us close to our challenge bound (14.17); we just need to bring out the role of the autocorrelation sums. When we expand the absolute values and attend to the algebra, we find

$$\sum_{n=1}^{N+H} \left| \sum_{h=0}^{H} c_{n-h} \right|^2$$

$$= \sum_{n=1}^{N+H} \left\{ \sum_{j=0}^{H} c_{n-j} \sum_{k=0}^{H} \bar{c}_{n-k} \right\}$$

$$= \sum_{n=1}^{N+H} \left\{ \sum_{s=0}^{H} |c_{n-s}|^2 + 2\operatorname{Re} \sum_{s=0}^{H-1} \sum_{t=s+1}^{H} c_{n-s} \bar{c}_{n-t} \right\}$$

$$= (H+1) \sum_{n=1}^{N} |c_n|^2 + 2\operatorname{Re} \left\{ \sum_{s=0}^{H-1} \sum_{t=s+1}^{H} \sum_{n=1}^{N+H} c_{n-s} \bar{c}_{n-t} \right\}$$

$$\leq (H+1) \sum_{n=1}^{N} |c_n|^2 + 2 \sum_{s=0}^{H-1} \sum_{t=s+1}^{H} \left| \sum_{n=1}^{N+H} c_{n-s} \bar{c}_{n-t} \right|$$

$$= (H+1) \sum_{n=1}^{N} |c_n|^2 + 2 \sum_{h=1}^{H} (H+1-h) \left| \sum_{n=1}^{N} c_n \bar{c}_{n+h} \right|.$$

This estimate, the Cauchy bound (14.20), and the trivial observation that $|z| = |\bar{z}|$, now combine to give us

$$\left| \sum_{n=1}^{N} c_n \right|^2 \leq \frac{N+H}{H+1} \sum_{n=1}^{N} |c_n|^2 + \frac{2(N+H)}{H+1} \sum_{h=1}^{H} \left(1 - \frac{h}{H+1}\right) \left| \sum_{n=1}^{N-h} c_{n+h} \bar{c}_n \right|.$$

This is precisely the inequality given by van der Corput in 1931. When we reintroduce the autocorrelation sums and bound the coefficients in the simplest way, we come directly to the inequality (14.17) which was suggested by our challenge problem.

CANCELLATION ON AVERAGE

Many problems pivot on the distinction between phenomena that take place uniformly and phenomena that only take place on average. For example, to make good use of Abel's inequality one needs a uniform bound on the partial sums $|S_k|$, $1 \leq k \leq n$, while van der Corput's inequality can be effective even if we only have a good bound for the average value of $|\rho_N(h)|$ over the fixed range $1 \leq h \leq H$.

It is perhaps most common for problems that have a special role for "cancellation on average" to call on integrals rather than sums. To illustrate this phenomenon, we first recall that a sequence $\{\varphi_k : k \in S\}$ of complex-valued square integrable functions on $[0,1]$ is said to be an

Cancellation and Aggregation

orthonormal sequence provided that for all $j, k \in S$ one has

$$\int_0^1 \varphi_j(x)\overline{\varphi_k(x)}\,dx = \begin{cases} 0 & \text{if } j \neq k \\ 1 & \text{if } j = k. \end{cases} \quad (14.21)$$

The leading example of such a sequence is $\varphi_k(x) = \mathbf{e}(kx) = \exp(2\pi i k x)$, the sequence of complex exponentials which we have already found to be at the heart of many cancellation phenomena.

For any finite set $A \subset S$, the orthonormality conditions (14.21) and direct expansion lead one to the identity

$$\int_0^1 \left| \sum_{k \in A} c_k \varphi_k(x) \right|^2 dx = \sum_{k \in A} |c_k|^2. \quad (14.22)$$

Thus, for $S_k(x) = c_1\varphi_1(x) + c_2\varphi_2(x) + \cdots + c_k\varphi_k(x)$, the application of Schwarz's inequality gives us

$$\int_0^1 |S_n(x)|\,dx \leq \left\{ \int_0^1 |S_n(x)|^2\,dx \right\}^{\frac{1}{2}} = (|c_1|^2 + |c_2|^2 + \cdots + |c_n|^2)^{\frac{1}{2}}$$

and, if we assume that $|c_k| \leq 1$ for all $1 \leq k \leq n$, then "on average" $|S_n(x)|$ is not larger than $\sqrt{n}$. The next challenge problem provides us with a bound for the maximal sequence $M_n(x) = \max_{1 \leq k \leq n} |S_k(x)|$ which is almost as good.

Problem 14.4 (Rademacher–Menchoff Inequality)

Given that the functions $\varphi_k : [0,1] \to \mathbb{C}$, $1 \leq k \leq n$, are orthonormal, show that the partial sums

$$S_k(x) = c_1\varphi_1(x) + c_2\varphi_2(x) + \cdots + c_k\varphi_k(x) \qquad 1 \leq k \leq n$$

satisfy the maximal inequality

$$\int_0^1 \max_{1 \leq k \leq n} |S_k^2(x)|\,dx \leq \log_2^2(4n) \sum_{k=1}^n |c_k|^2. \quad (14.23)$$

This is known as the Rademacher–Menchoff inequality, and it is surely among the most important results in the theory of orthogonal series. For us, much of the charm of the Rademacher–Menchoff inequality rests in its proof and, without giving away too much of the story, one may say in advance that the proof pivots on an artful application of Cauchy's inequality. Moreover, the proof encourages one to explore some fundamental grouping ideas which have applications in combinatorics, the theory of algorithms, and many other fields.

218   *Cancellation and Aggregation*

POSING A COMBINATORIAL QUESTION

Our goal is to bound the integral of $\max_{1\leq k\leq n} |S_k(x)|^2$, and our only tool is the orthogonality identity (14.22). We need to find some way to exploit the full strength of this identity; in particular, we need to exploit the fact that it holds for all possible choices of $A \subset \{1, 2, \ldots, n\}$. This advice is vague, but it still suggests some relevant combinatorial questions.

For example, is there a "reasonably small" collection $\mathcal{B}$ of subsets of $\{1, 2, \ldots, n\}$ such that each of the initial segments

$$I_k = \{1, 2, \ldots, k\} \qquad 1 \leq k \leq n,$$

can be written as a disjoint union of a "reasonably small" number of elements of $\mathcal{B}$? An affirmative answer would suggest that we might get a useful bound on the integral of $\max_{1\leq k\leq n} |S_k(x)|^2$ by using the identity (14.22) on each element of $\mathcal{B}$.

Our experience with binary representations reminds us that integers have succinct representations as sums of powers of two, so perhaps we should seek an analogous representation for the sets $\{I_k : 1 \leq k \leq n\}$. For example, we might try to show that each I_k can be written as a disjoint union of a small number of blocks with length 2^s where s may run between 0 and $\lfloor \log_2 n \rfloor$.

To translate this suggestion into a formal plan, we first let $[a, b]$ denote the interval of integers $\{a, a+1, \ldots, b\}$, and we let $\mathcal{B}$ denote the set of all integer intervals of the form

$$[r2^s + 1, (r+1)2^s] \qquad \text{where } 0 \leq r < \infty, \ 0 \leq s < \infty$$

and where $[r2^s + 1, (r+1)2^s] \subset [1, n]$.

Now, for any integer $k \in [1, n]$, we can easily produce a collection of sets $C(k) \subset \mathcal{B}$ such that

$$[1, k] = \bigcup_{B \in C(k)} B \qquad (14.24)$$

but, if we exercise some care, we can also keep a tight control on $|C(k)|$, the cardinality of the collection $C(k)$.

A GREEDY ALGORITHM

A natural way to construct the desired collection $C(k)$ of binary intervals is to use a *greedy algorithm*. For example, to represent $[1, k]$, we first take the largest element B of $\mathcal{B}$ that begins with 1, and we remove

Cancellation and Aggregation

the elements of B from $[1, k]$. Except when k is a power of 2, the first step leaves us with a nonempty interval of the form $[x, k]$ where x is equal to $2^s + 1$ for some integer s. We then apply the same greedy idea to $[x, k]$.

On the second step, we find the largest element B in $\mathcal{B}$ that begins with x, and we remove the elements of B from $[x, k]$. This time, if the remaining set is nonempty, its first element must be of the form $r2^s + 1$ for some choice of integers r and s. The greedy removal process then continues until one gets down to the empty set.

If we count the number of steps taken by the greedy algorithm we find that it is simply the number of 1s in the binary expansion of k. Since the number of such 1's is at most $\lceil \log_2(k) \rceil$, we have a useful cardinality bound $|C(k)| \leq \lceil \log_2(k) \rceil \leq \lceil \log_2(n) \rceil$.

For a quick confirmation of the construction, one might consider the interval $I_{27} = [1, 27]$. In base 2 one writes 27 as 11011, and we find that the greedy algorithm provides a representation for I_{27} as a 4-term union

$$[1, 27] = \{1, 2, \ldots, 16\} \cup \{17, 18, \ldots, 24\} \cup \{25, 26\} \cup \{27\}.$$

SUMS AND AN OPPORTUNITY FOR CAUCHY'S INEQUALITY

We now need to see how our set representations are related to partial sums such as those in our challenge problem. Still, to keep the combinatorial essentials in clear view, we keep $\varphi_j(x)$ out of the picture for the moment, and we simply focus on partial sums of complex numbers a_j, $1 \leq j \leq n$.

From our representation of $[1, k]$ as the union of the sets in $C(k)$, we have a representation of a generic partial sum,

$$a_1 + a_2 + \cdots + a_k = \sum_{B \in C(k)} \sum_{j \in B} a_j.$$

The benefit of this representation is that the index set for each of the double sums is reasonably small, so one can apply Cauchy's inequality (and the 1-trick) to the outside sum to find

$$|a_1 + a_2 + \cdots + a_k|^2 \leq |C(k)| \sum_{B \in C(k)} \left| \sum_{j \in B} a_j \right|^2. \qquad (14.25)$$

One should now have high hopes of finding a useful estimate for the last sum; after all, it is a sum of squares, and we have already studied such sums at considerable length. If we prepare for the worst, we have

$$|C(k)| \leq \lceil \log_2(k) \rceil \leq \lceil \log_2(n) \rceil \quad \text{and} \quad C(k) \subset \mathcal{B},$$

so the double sum bound (14.25) gives us

$$\max_{1\leq k\leq n} |a_1 + a_2 + \cdots + a_k|^2 \leq \lceil \log_2(n) \rceil \sum_{B\in\mathcal{B}} \left| \sum_{j\in B} a_j \right|^2, \qquad (14.26)$$

and this offers several signs of progress. On the left side one finds a maximal sequence $\max_{1\leq k\leq n} |a_1 + a_2 + \cdots + a_k|^2$ of the kind we hoped to estimate, while on the right side we find a sum of squares which does not depend on the index value $1 \leq k \leq n$. Honest bookkeeping should carry us the rest of the way.

A FINAL ACCOUNTING

If we simply replace a_j by $c_j \varphi_j(x)$ in the bound (14.26) and recall our notation for the partial sums of the $\varphi_j(x)$, $1 \leq j \leq n$, then we find

$$\max_{1\leq k\leq n} |S_k(x)|^2 \leq \lceil \log_2(n) \rceil \sum_{B\in\mathcal{B}} \left| \sum_{j\in B} c_j \varphi_j(x) \right|^2.$$

Now, if we integrate both sides, then we see that the basic orthonormality conditions (14.22) tell us that

$$\int_0^1 \max_{1\leq k\leq n} |S_k(x)|^2 \, dx \leq \lceil \log_2(n) \rceil \sum_{B\in\mathcal{B}} \sum_{j\in B} |c_j|^2, \qquad (14.27)$$

which is almost our target inequality. For each $j \in [1, n]$ there are at most $1 + \lceil \log_2(n) \rceil$ sets $B \in \mathcal{B}$ such that $j \in B$, so we see that inequality (14.27) gives us the bound

$$\int_0^1 \max_{1\leq k\leq n} |S_k(x)|^2 \, dx \leq \lceil \log_2(n) \rceil (1 + \lceil \log_2(n) \rceil) \sum_{j=1}^n |c_j|^2. \qquad (14.28)$$

This bound is actually a bit stronger than the one asserted by the Rademacher–Menchoff inequality (14.23) since for all $n \geq 1$ we have the bound $\lceil \log_2(n) \rceil (1 + \lceil \log_2(n) \rceil) \leq (2 + \log_2 n)^2 = \log_2^2(4n)$.

CANCELLATION AND AGGREGATION

The Rademacher–Menchoff inequality and van der Corput's inequality provide natural illustrations of the twin themes of cancellation and aggregation. They are also two of history's finest examples of pure "Cauchy–Schwarz technique." They contribute to one's effectiveness as a problem solver, and they provide a fitting end to our class — which is not over just yet. Here, as in all the earlier chapters, the exercises are at the heart of the matter.

Cancellation and Aggregation 221

EXERCISES

The first few exercises lean on Abel's inequality and, among other things, they provide an analog for increasing multipliers and an analog for integrals. To help with the latter, Exercise 14.2 develops the slippery "second mean value formula" for integrals. This handy tool is also used to obtain the so-called van der Corput's lemmas — two elementary bounds which turn out to be of fundamental help when facing cancellation in integrals.

The next few exercises address diverse aspects of cancellation, including the exploitation of complete exponential sums, the dyadic trick, and variations on the Rademacher–Menchoff inequality. Lower bounds for complex sums are entertained for the first time in Exercise 14.9, and Exercise 14.10 provides our first example of a *domination* inequality.

The final exercise develops Selberg's inequality. At first, it may seem to be simply a messy variation on Bessel's inequality, but the added complexity and generality serve a genuine purpose. The applications of Selberg's inequality in combinatorics, number theory, and numerical analysis could fill a book, perhaps even a proper sequel to the *Cauchy Schwarz Master Class*.

Exercise 14.1 (Abel's Second Inequality)
Show that for each nondecreasing sequence of nonnegative real numbers $0 \leq b_1 \leq b_2 \cdots \leq b_n$ one has a bound which differs slightly from Abel's first inequality,

$$|b_1 z_1 + b_2 z_2 + \cdots + b_n z_n| \leq 2 b_n \max_{1 \leq k \leq n} |S_k|. \qquad (14.29)$$

Exercise 14.2 (The Integral Mean Value Formulas)
The first integral mean value formula (IMVF) asserts that for each continuous $f : [a,b] \to \mathbb{R}$ and each integrable $g : [a,b] \to [0,\infty)$, there is a $\xi \in [a,b]$ such that

$$\int_a^b f(x)g(x)\,dx = f(\xi) \int_a^b g(x)\,dx, \qquad (14.30)$$

while the second IMVF is the slightly trickier assertion that for each differentiable nonincreasing function $\psi : [a,b] \to (0,\infty)$ and each integrable function $\phi : [a,b] \to \mathbb{R}$, there is a $\xi_0 \in [a,b]$ such that

$$\int_a^b \psi(x)\phi(x)\,dx = \psi(a) \int_a^{\xi_0} \phi(x)\,dx. \qquad (14.31)$$

Prove these formulas. They are both quite handy, and the second one may be tricker than you might guess.

Exercise 14.3 (A Integral Analog to Abel's Inequality)

If $f : [a,b] \to (0, \infty)$ is a nonincreasing function, then for each integrable function $g : [a,b] \to \mathbb{R}$ one has the bound

$$\left| \int_a^b f(x)g(x)\, dx \right| \leq f(a) \sup_{a \leq y \leq b} \left| \int_a^y g(x)\, dx \right|, \qquad (14.32)$$

which is the natural integral analog of Abel's inequality. Prove the bound (14.32) and show that it implies

$$\left| \int_a^b \frac{\sin x}{x}\, dx \right| \leq \frac{2}{a} \qquad \text{for all } 0 < a < b < \infty. \qquad (14.33)$$

Exercise 14.4 (van der Corput on Oscillatory Integrals)

(a) Given a differentiable function $\theta : [a,b] \to \mathbb{R}$ for which the derivative $\theta'(\cdot)$ is monotonic and satisfies $\theta'(x) \geq \nu > 0$ for all $x \in [a,b]$, show that one has the bound

$$\left| \int_a^b e^{i\theta(x)}\, dx \right| \leq \frac{4}{\nu}. \qquad (14.34)$$

(b) Use the bound (14.34) to show that if $\theta : [a,b] \to \mathbb{R}$ is a twice differentiable function with $\theta''(x) \geq \rho > 0$ for all $x \in [a,b]$, then

$$\left| \int_a^b e^{i\theta(x)}\, dx \right| \leq \frac{8}{\sqrt{\rho}}. \qquad (14.35)$$

These workhorses lie behind many basic cancellation arguments for integrals and sums. They also come to us from the same J. G. van der Corput who gave us our third challenge problem. In fact, these may be the best known of van der Corput many inequalities, even though they are notably less subtle than the bound (14.17).

Exercise 14.5 (The "Extend and Conquer" Paradigm)

First show that for integers m and j one has the formula

$$\sum_{k=1}^{m-1} \mathbf{e}(jk/m) = \begin{cases} 0 & \text{if } m \text{ does not divide } j \\ m & \text{if } m \text{ does divide } j. \end{cases} \qquad (14.36)$$

This formula tells us that for such a *complete sum* one either has total cancellation, or no cancellation at all. There are many remarkable consequences of this elementary observation.

For example, use it to show that for each prime $p \geq 3$, and each pair

Cancellation and Aggregation 223

A and B of subsets of $\mathbb{F}_p = \{0, 1, 2, \ldots, p-1\}$, one has

$$\left|\sum_{j \in A} \sum_{k \in B} \exp\left(\frac{2\pi i j k}{p}\right)\right| \leq p^{\frac{1}{2}} |A|^{\frac{1}{2}} |B|^{\frac{1}{2}}. \tag{14.37}$$

Exercise 14.6 (Another Dyadic Passage)

Sometimes we have an estimate for $f(x)$ and we would like an estimate of $g(x)$, but we cannot show $g(x) \leq f(x)$. We may still be able to get a useful bound on $g(x)$ if we only know that $f(x)$ dominates "half" of g in the sense that

$$g(x) - g(x/2) \leq f(x) \qquad \text{for all } x \geq 0.$$

To be specific, assume such a function is continuous and show that if $f(x) \leq Ax + B$ for $x \geq 0$, then g satisfies the (only slightly worse) bound $g(x) \leq A'x + B' \log_2(x) + C'$ for $x \geq 1$ for appropriate constants A', B', and C'.

Exercise 14.7 (Rademacher–Menchoff with Weights)

Let $\psi_1, \psi_2, \ldots, \psi_n$ be real-valued functions for which

$$\int_0^1 \psi_j^2(x) \, dx = 1 \quad \text{and} \quad \int_0^1 \psi_j(x) \psi_k(x) \, dx = a_{jk}. \tag{14.38}$$

Show that if there exists a constant C such that

$$\left|\sum_{j=1}^n \sum_{k=1}^n a_{jk} y_j y_k\right| \leq C \sum_{j=1}^n y_j^2, \tag{14.39}$$

for any n real numbers $y_1, y_2, \ldots, y_n$, then we also have

$$\int_0^1 \max_{1 \leq k \leq n} \left(\sum_{j=1}^k c_j \psi_j(x)\right)^2 dx \leq C \log_2^2(4n) \sum_{k=1}^n c_k^2 \tag{14.40}$$

for all real $c_1, c_2, \ldots, c_n$.

Exercise 14.8 (Functions with Geometric Dependence)

If the constant ρ satisfies $0 < \rho < 1$ and the sequence of functions $\{\psi_j\}$ satisfies

$$\int_0^1 \psi_j(x) \psi_k(x) \, dx \leq \rho^{|j-k|} \left(\int_0^1 \psi_j^2(x) \, dx\right)^{\frac{1}{2}} \left(\int_0^1 \psi_k^2(x) \, dx\right)^{\frac{1}{2}}$$

for all $1 \leq j, k \leq n$, then there is a constant M depending only on ρ

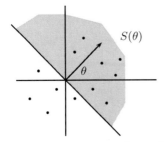
The set $S(\theta)$ consists of all of the elements z_j contained in the half-plane $H(\theta)$ given by $\{z : \operatorname{Re} z e^{-i\theta} \geq 0\}$.

Fig. 14.1. To find a subset of $S = \{z_1, z_2, \ldots, z_n\}$ whose sum has a large absolute value, why not first consider the just the subsets $S(\theta)$ for $\theta \in (0, 2\pi)$?

such that partial sums $S_k(x) = \psi_1(x) + \psi_2(x) + \cdots + \psi_k(x)$, $1 \leq k \leq n$, satisfy the maximal inequality

$$\int_0^1 \max_{1 \leq k \leq n} S_k^2(x)\, dx \leq M \log_2^2(4n) \sum_{k=1}^n \int_0^1 \psi_k^2(x)\, dx.$$

Exercise 14.9 (The Subset Lower Bound)

Show that for complex numbers $z_1, z_2, \ldots, z_n$ one has

$$\frac{1}{\pi} \sum_{j=1}^n |z_j| \leq \max_{I \subset \{1,2,\ldots,n\}} \left| \sum_{j \in I} z_j \right|, \qquad (14.41)$$

and show that the constant factor $1/\pi$ cannot be replace by a larger one. The qualitative message of this cancellation story is that there is always *some subset* with a sum whose modulus is a large fraction of the sum of all the moduli. For a hint one might consider the special subset S_θ defined in Figure 14.1.

Exercise 14.10 (A Domination Principle)

If the complex numbers a_n satisfy the bounds $|a_n| \leq A_n$, $1 \leq n \leq N$, then the complex array $\{y_{nr} : 1 \leq n \leq N, 1 \leq r \leq R\}$ satisfies the bounds

$$\sum_{r=1}^R \sum_{s=1}^R \left| \sum_{n=1}^N a_n y_{nr} \bar{y}_{ns} \right|^2 \leq \sum_{r=1}^R \sum_{s=1}^R \left| \sum_{n=1}^N A_n y_{nr} \bar{y}_{ns} \right|^2. \qquad (14.42)$$

Cancellation and Aggregation

Exercise 14.11 (An Inequality of P. Enflo)

Show that for vectors $\mathbf{u}_m$, $1 \leq m \leq M$, and $\mathbf{v}_n$, $1 \leq n \leq N$, in the inner product space $\mathbb{C}^d$ one has the bound

$$\sum_{m=1}^{M}\sum_{n=1}^{N}|\langle \mathbf{u}_m, \mathbf{v}_n\rangle|^2$$
$$\leq \left\{\sum_{m=1}^{M}\sum_{\mu=1}^{M}|\langle \mathbf{u}_m, \mathbf{u}_\mu\rangle|^2\right\}^{1/2}\left\{\sum_{n=1}^{N}\sum_{\nu=1}^{N}|\langle \mathbf{v}_n, \mathbf{v}_\nu\rangle|^2\right\}^{1/2}.$$

Exercise 14.12 (Selberg's Inequality)

Prove that if $\mathbf{x}$ and $\mathbf{y}_1, \mathbf{y}_2, \ldots, \mathbf{y}_n$ are elements of a real or complex inner product space, then we have

$$\sum_{j=1}^{n}\frac{|\langle \mathbf{x}, \mathbf{y}_j\rangle|^2}{\sum_{k=1}^{n}|\langle \mathbf{y}_j, \mathbf{y}_k\rangle|} \leq \langle \mathbf{x}, \mathbf{x}\rangle. \qquad (14.43)$$

Selberg's inequality can sometimes be used as a replacement for the orthonormality identity (14.22) or Bessel's inequality (4.29) when the elements $\mathbf{y}_1, \mathbf{y}_2, \ldots, \mathbf{y}_n$ are only approximately orthogonal. Techniques for relaxing the requirements of orthonormality have important consequences throughout probability, number theory, and combinatorics.

Solutions to the Exercises

CHAPTER 1: STARTING WITH CAUCHY

SOLUTION FOR EXERCISE 1.1. The first inequality follows by applying Cauchy's inequality to $\{a_k\}$ and $\{b_k\}$ where one takes $b_k = 1$ for all k. In isolation, this "1-trick" is almost trivial, but it is remarkably general: *every sum* can be estimated in this way. The art is rather one of anticipating when the resulting estimate might prove to be helpful.

For the second problem we apply Cauchy's inequality to the product of $\{a_k^{1/3}\}$ and $\{a_k^{2/3}\}$. This is a simple instance of the "splitting trick" where one estimates the sum of the a_k by Cauchy's inequality after writing a_k as a product $a_k = b_k c_k$. Almost every chapter will make some use of the splitting trick, and some of these applications are remarkably subtle.

SOLUTION FOR EXERCISE 1.2. This is another case for the splitting trick; one just applies Cauchy's inequality to the sum

$$1 \leq \sum_{j=1}^{n} \left\{ p_j^{\frac{1}{2}} a_j^{\frac{1}{2}} \right\} \left\{ p_j^{\frac{1}{2}} b_j^{\frac{1}{2}} \right\}.$$

SOLUTION FOR EXERCISE 1.3. The first inequality just requires two applications of Cauchy's inequality according to the grouping $a_k(b_k c_k)$, but one might wander around a bit before hitting on the proof of second inequality.

One key to the proof of the second bound comes from noting that when we substitute $a_k = b_k = c_k = 1$ we get the lackluster bound $n^2 \leq n^3$. This suggests the inequality is not particularly strong, and it encourages us to look for a cheap shot. One might then think to deal with the c_k factors by introducing

$$\hat{c}_k = c_k^2/(c_1^2 + c_2^2 + \cdots + c_n^2),$$

so the target inequality would follow if we could show

$$\sum_{k=1}^{n} |a_k b_k \hat{c}_k| \leq \left(\sum_{k=1}^{n} a_k^2 \right)^{\frac{1}{2}} \left(\sum_{k=1}^{n} b_k^2 \right)^{\frac{1}{2}};$$

but this bound is an immediate consequence of the usual Cauchy inequality and the trivial observation that $|\hat{c}_k| \leq 1$.

SOLUTION FOR EXERCISE 1.4. For part (a) we note by Cauchy's inequality and the 1-trick that we have

$$S^2 \leq (1^2 + 1^2 + 1^2) \left(\frac{x+y}{x+y+z} + \frac{x+z}{x+y+z} + \frac{y+z}{x+y+z} \right) = 6.$$

For part (b) we apply Cauchy's inequality to the splitting

$$x + y + z = \frac{x}{\sqrt{y+z}} \sqrt{y+z} + \frac{y}{\sqrt{x+z}} \sqrt{x+z} + \frac{z}{\sqrt{x+y}} \sqrt{x+y}.$$

SOLUTION FOR EXERCISE 1.5. From Cauchy's inequality, the splitting $p_k = p_k^{1/2} p_k^{1/2}$, and the identity $\cos^2(x) = \{1 + \cos(2x)\}/2$, one finds

$$g^2(x) \leq \sum_{k=1}^{n} p_k \sum_{k=1}^{n} p_k \cos^2(\beta_k x)$$

$$= \sum_{k=1}^{n} p_k \frac{1}{2}(1 + \cos(2\beta_k x)) = \{1 + g(2x)\}/2.$$

SOLUTION FOR EXERCISE 1.6. We first expand the sum

$$\sum_{k=1}^{n} (p_k + 1/p_k)^2 = 2n + \sum_{k=1}^{n} p_k^2 + \sum_{k=1}^{n} 1/p_k^2, \qquad (14.44)$$

and then we estimate the last two terms separately. By the 1-trick and the hypothesis $p_1 + p_2 + \cdots + p_n = 1$, the first of these two sums is at least $1/n$. To estimate the last sum in (14.44), we first apply Cauchy's inequality to the sum of the products $1 = \sqrt{p_k} \cdot (1/\sqrt{p_k})$ to get

$$n^2 \leq \sum_{k=1}^{n} 1/p_k,$$

and to complete the proof we apply Cauchy's inequality to the sum of the products $1/p_k = 1 \cdot 1/p_k$ to get

$$n^3 \leq \sum_{k=1}^n 1/p_k^2.$$

There are several other solutions to this problem, but this one does an especially nice job of illustrating how much can be achieved with just Cauchy's inequality and the 1-trick.

SOLUTION FOR EXERCISE 1.7. The natural candidate for the inner product is given by $\langle \mathbf{x}, \mathbf{y} \rangle = 5x_1y_1 + x_1y_2 + x_2y_1 + 3y_2^2$ where one has set $\mathbf{x} = (x_1, x_2)$ and $\mathbf{y} = (y_1, y_2)$. All of the required inner product properties are immediate, except perhaps for the first two. For these we just need to note that the polynomial $5z^2 + 3z + 3 = 0$ has no real roots.

More generally, if a_{jk}, $1 \leq j, k \leq n$, is a square array of real numbers that is symmetric in the sense that $a_{jk} = a_{kj}$ for all $1 \leq j, k \leq n$, then the sum

$$\langle \mathbf{x}, \mathbf{y} \rangle = \sum_{j=1}^n \sum_{k=1}^n a_{jk} x_j y_k \tag{14.45}$$

provides a candidate for inner products on $\mathbb{R}^n$. The candidate (14.45) yields a legitimate inner product on $\mathbb{R}^n$ if (a) the polynomial defined by $Q(x_1, x_2, \ldots, x_n) = \sum_{j=1}^n \sum_{k=1}^n a_{jk} x_j x_k$ is nonnegative for all vectors $(x_1, x_2, \ldots, x_n) \in \mathbb{R}^n$ and if (b) $Q(x_1, x_2, \ldots, x_n) = 0$ only when $x_j = 0$ for all $1 \leq j \leq n$. A polynomial with these two properties is called a *positive definite quadratic form*, and each such form provides us with potentially useful of Cauchy's inequality.

SOLUTION FOR EXERCISE 1.8. In each case, one applies Cauchy's inequality, and then estimates the resulting sum. In part (a) one uses the sum for a geometric progression: $1 + x^2 + x^4 + x^6 + \cdots = 1/(1-x^2)$, while for part (b), one can use Euler's famous formula

$$\sum_{k=1}^\infty \frac{1}{k^2} = \frac{\pi^2}{6} = 1.6449\ldots < 2,$$

or, alternatively, one can use the nice telescoping argument,

$$\sum_{k=1}^n \frac{1}{k^2} \leq 1 + \sum_{k=2}^n \frac{1}{k(k-1)} = 1 + \sum_{k=2}^n \left(\frac{1}{k-1} - \frac{1}{k} \right) = 2 - \frac{1}{n}.$$

For part (c) one has the integral comparison

$$\frac{1}{n+k} < \int_{n+k-1}^{n+k} \frac{dx}{x} \quad \text{so} \quad \sum_{k=1}^{n} \frac{1}{n+k} < \int_{n}^{2n} \frac{dx}{x} = \log 2.$$

Finally, for part (d) one uses the explicit sum for the squares of the binomial coefficients

$$\sum_{k=0}^{n} \binom{n}{k}^2 = \sum_{k=0}^{n} \binom{n}{k}\binom{n}{n-k} = \binom{2n}{n},$$

which one can prove by a classic counting argument. Specifically, one considers the number of ways to form a committee of n people from a group of n men and n women. The middle sum first counts the number of committees with k men and then sums over $0 \le k \le n$, while the last term directly counts the number of ways to choose n people out of $2n$.

SOLUTION FOR EXERCISE 1.9. If T denotes the left-hand side of the target inequality, then by expansion one gets

$$T = 2\sum_{j=1}^{n} a_j^2 + 4 \sum_{(j,k)\in S} a_j a_k,$$

where S is the set of all (j,k) such that $1 \le j < k \le n$ with $j+k$ even. From the elementary bound $2a_j a_k \le a_j^2 + a_k^2$, one then finds

$$T \le 2\sum_{j=1}^{n} a_j^2 + 2 \sum_{(j,k)\in S} (a_j^2 + a_k^2) \le 2\sum_{j=1}^{n} a_j^2 + 2\sum_{s=1}^{n} n_s a_s^2,$$

where n_s denotes the number of pairs (j,k) in S with $j=s$ or $k=s$. One has $n_s \le \lfloor (n-1)/2 \rfloor$, so

$$T \le \left(2 + 2\lfloor (n-1)/2 \rfloor\right) \sum_{j=1}^{n} a_j^2 \le (n+2) \sum_{j=1}^{n} a_j^2.$$

SOLUTION FOR EXERCISE 1.10. If we apply Cauchy's inequality to the splitting $|c_{jk}|^{\frac{1}{2}}|x_j||c_{jk}|^{\frac{1}{2}}|y_k|$ we find

$$\left|\sum_{j,k} c_{jk} x_j y_k\right| \le \left(\sum_{j,k} |c_{jk}||x_j|^2\right)^{\frac{1}{2}} \cdot \left(\sum_{j,k} |c_{jk}||y_k|^2\right)^{\frac{1}{2}}$$

$$= \left(\sum_{j=1}^{m} \left\{\sum_{k=1}^{n} |c_{jk}|\right\} |x_j|^2\right)^{\frac{1}{2}} \cdot \left(\sum_{k=1}^{n} \left\{\sum_{j=1}^{m} |c_{jk}|\right\} |y_k|^2\right)^{\frac{1}{2}},$$

and the sums in the braces are bounded by C and R respectively.

SOLUTION FOR EXERCISE 1.11. Only a few alterations are needed in Schwarz's original proof (page 11), but the visual impression does shift. First, we apply the hypothesis and the definition of $p(t)$ to find

$$0 \leq p(t) = \langle \mathbf{v}, \mathbf{v} \rangle + 2t \langle \mathbf{v}, \mathbf{w} \rangle + t^2 \langle \mathbf{w}, \mathbf{w} \rangle.$$

The discriminant of $p(t)$ is $D = B^2 - AC = \langle \mathbf{v}, \mathbf{w} \rangle^2 - \langle \mathbf{v}, \mathbf{v} \rangle \langle \mathbf{w}, \mathbf{w} \rangle$, and we deduce that $D \leq 0$, or else $p(t)$ would have two real roots (and therefore $p(t)$ would be strictly negative for some value of t).

SOLUTION FOR EXERCISE 1.12. We define a new inner product space $(V^{[n]}, [\cdot, \cdot])$ by setting $V^{[n]} = \{(\mathbf{v}_1, \mathbf{v}_2, \ldots, \mathbf{v}_n) : \mathbf{v}_j \in V, 1 \leq j \leq n\}$ and by defining $[\mathbf{v}, \mathbf{w}] = \sum_{j=1}^{n} \langle \mathbf{x}_j, \mathbf{y}_j \rangle$ where $\mathbf{v} = (\mathbf{v}_1, \mathbf{v}_2, \ldots, \mathbf{v}_n)$ and where $\mathbf{w} = (\mathbf{w}_1, \mathbf{w}_2, \ldots, \mathbf{w}_n)$. After checking that $[\cdot, \cdot]$ is an honest inner product, one sees that the bound (1.24) is just the Cauchy–Schwarz inequality for the inner product $[\cdot, \cdot]$.

SOLUTION FOR EXERCISE 1.13. If we view $\{x_{jk} : 1 \leq j \leq m, 1 \leq k \leq n\}$ as a vector of length mn then Cauchy's inequality and the one-trick splitting $x_{jk} = x_{jk} \cdot 1$ imply the general bound

$$\left(\sum_{j=1}^{m} \sum_{k=1}^{n} x_{jk} \right)^2 \leq mn \sum_{j=1}^{m} \sum_{k=1}^{n} x_{jk}^2. \qquad (14.46)$$

We apply this bound to $x_{jk} = a_{jk} - r_j/n - c_k/m$ where

$$r_j = \sum_{k=1}^{n} a_{jk}, \quad c_k = \sum_{j=1}^{m} a_{jk}, \quad \text{and if we set} \quad T = \sum_{j=1}^{m} \sum_{k=1}^{n} a_{jk},$$

then the left side of the bound (14.46) works out to be T^2, and the right side works out to be

$$mn \sum_{j=1}^{m} \sum_{k=1}^{n} a_{jk}^2 - m \sum_{j=1}^{m} r_j^2 - n \sum_{k=1}^{n} c_k^2 + 2T^2,$$

so the Cauchy bound (14.46) reduces to our target inequality.

To characterize the case of equality, we note that equality holds in the bound (14.46) if and only if x_{jk} is equal to a constant c in which case one can take $\alpha_j = c + r_j$ and $\beta_k = c_k$ to provide the required representation for a_{jk}. This result is Theorem 1 of van Dam (1998) where one also finds a proof which uses matrix theory as well as some instructive corollaries.

SOLUTION FOR EXERCISE 1.14. More often than one might like to admit, tidiness is important in problem solving, and here the hygienic use of parentheses can make the difference between success and failure. One just carefully computes

$$\sum_{1\leq i,j,k\leq n} a_{ij}^{\frac{1}{2}} b_{jk}^{\frac{1}{2}} c_{ki}^{\frac{1}{2}} = \sum_{1\leq i,k\leq n} c_{ki}^{\frac{1}{2}} \left\{ \sum_{j=1}^{n} a_{ij}^{\frac{1}{2}} b_{jk}^{\frac{1}{2}} \right\}$$

$$\leq \sum_{1\leq i,k\leq n} c_{ki}^{\frac{1}{2}} \left\{ \sum_{j=1}^{n} a_{ij} \right\}^{\frac{1}{2}} \left\{ \sum_{j=1}^{n} b_{jk} \right\}^{\frac{1}{2}}$$

$$= \sum_{k=1}^{n} \left\{ \sum_{j=1}^{n} b_{jk} \right\}^{\frac{1}{2}} \left(\sum_{i=1}^{n} c_{ki}^{\frac{1}{2}} \left\{ \sum_{j=1}^{n} a_{ij} \right\}^{\frac{1}{2}} \right),$$

which is bounded in turn by

$$\sum_{k=1}^{n} \left\{ \sum_{j=1}^{n} b_{jk} \right\}^{\frac{1}{2}} \left\{ \sum_{i=1}^{n} c_{ki} \right\}^{\frac{1}{2}} \left\{ \sum_{1\leq i,j\leq n} a_{ij} \right\}^{\frac{1}{2}}$$

$$= \left\{ \sum_{1\leq i,j\leq n} a_{ij} \right\}^{\frac{1}{2}} \left\{ \sum_{k=1}^{n} \left(\sum_{j=1}^{n} b_{jk} \right)^{\frac{1}{2}} \left(\sum_{i=1}^{n} c_{ki} \right)^{\frac{1}{2}} \right\}$$

$$\leq \left\{ \sum_{1\leq i,j\leq n} a_{ij} \right\}^{\frac{1}{2}} \left\{ \sum_{1\leq j,k\leq n} b_{jk} \right\}^{\frac{1}{2}} \left\{ \sum_{1\leq k,i\leq n} c_{ki} \right\}^{\frac{1}{2}}.$$

This proof of the triple product bound (1.25) follows Tiskin (2002). Incidentally, the corollary (1.26) was posed as a problem on the 33rd International Mathematical Olympiad (Moscow, 1992). More recently, Hammer and Shen (2002) note that the corollary may be obtained as an application of Kolmogorov complexity. George (1984, p. 243) outlines a proof of continuous Loomis–Whitney inequality, a result which can be used to give a third proof of the discrete bound (1.26).

SOLUTION FOR EXERCISE 1.15. If we differentiate the identities (1.27) and (1.28) we find for all $\theta \in \Theta$ that

$$\sum_{k\in D} \dot{p}_\theta(k;\theta) = 0 \quad \text{and} \quad \sum_{k\in D} g(k)\dot{p}_\theta(k;\theta) = 1.$$

Consequently, we have the identity

$$1 = \sum_{k\in D} (g(k) - \theta)\dot{p}_\theta(k;\theta)$$

$$= \sum_{k\in D} \left\{ (g(k) - \theta)p(k;\theta)^{\frac{1}{2}} \right\} \left\{ \left(\dot{p}_\theta(k;\theta)/p(k;\theta)\right)p(k;\theta)^{\frac{1}{2}} \right\},$$

which yields the Cramér–Rao inequality (1.29) when we apply Cauchy's inequality to this sum of bracketed terms.

The derivation of the Cramér–Rao inequality may be the most significant application of the 1-trick in all of applied mathematics. It has been repeated in hundreds of papers and books.

CHAPTER 2: THE AM-GM INEQUALITY

SOLUTION FOR EXERCISE 2.1. For the general step, consider the sum $S_{k+1} = a_1 b_2 + a_2 b_2 + \cdots + a_{2^{k+1}} b_{2^{k+1}} = S'_{k+1} + S''_{k+1}$ where S'_{k+1} is the sum of the first 2^k products and S''_{k+1} is the sum of the second 2^k products. By induction, apply the 2^k-version of Cauchy's inequality to S'_{k+1} and S''_{k+1} to get $S'_{k+1} \leq A'B'$ and $S'_{k+1} \leq A''B''$ where we set $A' = (a_1^2 + \cdots + a_{2^k}^2)^{\frac{1}{2}}$, $A'' = (a_{2^k+1}^2 + \cdots + a_{2^{k+1}}^2)^{\frac{1}{2}}$, and where we define B' and B'' analogously. The 2-version of Cauchy's inequality implies

$$S_{k+1} \leq A'B' + A''B'' \leq (A'^2 + A''^2)^{\frac{1}{2}}(B'^2 + B''^2)^{\frac{1}{2}},$$

and this is the 2^{k+1}-version of Cauchy's inequality. Thus, induction gives us Cauchy's inequality for all 2^k, $k = 1, 2, \ldots$. Finally, to get Cauchy's inequality for $n \leq 2^k$ we just set $a_j = b_j = 0$ for $n < j \leq 2^k$ and apply the 2^k-version.

SOLUTION FOR EXERCISE 2.2. To prove the bound (2.23) by induction, first note that the case $n = 1$ is trivial. Next, take the bound for general n and multiply it by $1 + x$ to get $1 + (n+1)x + x^2 \leq (1+x)^{n+1}$. This is stronger than the bound (2.23) in the case $n + 1$, so the bound (2.23) holds for all $n = 1, 2, \ldots$ by induction. To show $1 + x \leq e^x$, one replaces x by x/n in Bernoulli's inequality and lets n go to infinity. Finally, to prove the relation (2.25), one sets $f(x) = (1+x)^p - (1+px)$ then notes that $f(0) = 0$, $f'(x) \geq 0$ for $x \geq 0$, and $f'(x) \leq 0$ for $-1 < x \leq 0$, so $\min_{x \in [-1, \infty)} f(x) = f(0) = 0$.

SOLUTION FOR EXERCISE 2.3. To prove the bound (2.26) one takes $p_1 = \alpha/(\alpha + \beta)$, $p_2 = \beta/(\alpha + \beta)$, $a_1 = x^{\alpha+\beta}$, and $a_2 = y^{\alpha+\beta}$ and applies the AM-GM bound (2.7). To get the timely bound we specialize (2.26) twice, once with $\alpha = 2004$ and $\beta = 1$ and once with $\alpha = 1$ and $\beta = 2004$. We then sum the two resulting bounds.

SOLUTION FOR EXERCISE 2.4. The target inequality is equivalent to $a^2 bc + ab^2 c + abc^2 \leq a^4 + b^4 + c^4$, a pure power bound. By the AM-GM inequality, we have $a^2 bc = (a^3)^{2/3}(b^3)^{1/3}(c^3)^{1/3} \leq 2a^3/3 + b^3/3 + c^3/3$,

Solutions to the Exercises   233

and analogous bounds hold for ab^2c and abc^2. The sum of these bounds yields the target inequality.

Equality holds in the target inequality if and only equality holds for both of our applications of the AM-GM bound. Thus, equality holds in the target bound if and only if $a = b = c$. Incidentally, three other solutions of this problem are available on website of the Canadian Mathematical Association.

SOLUTION FOR EXERCISE 2.5. For all j and k, the AM-GM inequality gives us $(x^{j+k}y^{j+k})^{\frac{1}{2}} \leq \frac{1}{2}(x^j y^k + x^k y^j)$. Setting $k = n - 1 - j$ and summing over $0 \leq j < n$ yields the bound

$$n(xy)^{(n-1)/2} \leq x^{n-1} + x^{n-1}y + \cdots + xy^{n-2} + y^{n-1} = \frac{x^n - y^n}{x - y}.$$

SOLUTION FOR EXERCISE 2.6. Since $\alpha + \beta = \pi$ we have $\gamma = \alpha$ and $\delta = \beta$ so the triangles $\Delta(ABD)$ and $\Delta(DBC)$ are similar. By proportionality of the corresponding sides we have $h : a = b : h$, and we find $h^2 = ab$, just as required.

SOLUTION FOR EXERCISE 2.7. The product $(1+x)(1+y)(1+z)$ expands as $1 + x + y + z + xy + xz + yz + xyz$ and the AM-GM bound gives us

$$(x + y + z)/3 \geq xyz \geq 1 \quad \text{and}$$

$$(xy + xz + yz)/3 \geq \{(xy)(xz)(yz)\}^{1/3} = (xyz)^{2/3} \geq 1,$$

so the bound (2.28) follows by summing. With persistence, the same idea can be used to show that for all nonnegative a_k, $1 \leq k \leq n$, one has the inference

$$1 \leq \prod_{k=1}^{n} a_k \quad \Longrightarrow \quad 2^n \leq \prod_{k=1}^{n}(1 + a_k). \qquad (14.47)$$

SOLUTION FOR EXERCISE 2.8. The AM-GM inequality tells us

$$\{a_1 x_1 a_2 x_2 \cdots a_n x_n\}^{1/n} \leq \frac{a_1 x_1 + a_2 x_2 + \cdots + a_n x_n}{n},$$

and this yields a relation between the critical quantities of P_1 and P_2,

$$x_1 x_2 \cdots x_n \leq \frac{(a_1 x_1 + a_2 x_2 + \cdots + a_n x_n)^n}{a_1 a_2 \cdots a_n n^n}.$$

We have equality here if and only if $a_1 x_1 = a_2 x_2 = \cdots = a_n x_n$, and nothing more is needed to confirm the stated optimality criterion.

SOLUTION FOR EXERCISE 2.9. By the AM-GM inequality, one has

$$2\{a^2b^2c^2\}^{1/3} = \{(2ab)(2ac)(2bc)\}^{1/3} \leq \frac{2ab + 2ac + 2bc}{3} = A/3,$$

and this gives the bound (2.9). Finally, equality holds here if and only if $ab = ac = bc$. This is possible if and only if $a = b = c$, so the box of maximum volume for a given surface area is indeed the cube.

SOLUTION FOR EXERCISE 2.10. If we set $p = n$ and $y = x - 1$ in Bernoulli's inequality, we find that $y(n-y^{n-1}) \leq n-1$ and equality holds only for $y = 1$. If we now choose y such that $y^{n-1} = a_n/\bar{a}$ where $\bar{a} = (a_1+a_2+\cdots+a_n)/n$, then we have $n-y^{n-1} = (a_1+a_2+\cdots+a_{n-1})/\bar{a}$, and easy arithmetic takes one the rest of the way to the recursion formula.

As a sidebar, one should note that the recursion also follows from the weighted AM-GM inequality $x^{1/n}y^{(n-1)/n} \leq \frac{1}{n}x + \frac{n-1}{n}y$ by taking $x = a_n$ and $y = (a_1 + a_2 + \cdots + a_{n-1})/(n - 1)$.

SOLUTION FOR EXERCISE 2.11. Following the hint, one finds from the AM-GM inequality that

$$\frac{(a_1a_2\cdots a_n)^{1/n} + (b_1b_2\cdots b_n)^{1/n}}{\{(a_1 + b_1)(a_2 + b_2)\cdots(a_n + b_n)\}^{1/n}}$$

$$= \prod_{j=1}^{n}\left\{\frac{a_j}{a_j+b_j}\right\}^{1/n} + \prod_{j=1}^{n}\left\{\frac{b_j}{a_j+b_j}\right\}^{1/n}$$

$$\leq \frac{1}{n}\sum_{j=1}^{n}\frac{a_j}{a_j+b_j} + \frac{1}{n}\sum_{j=1}^{n}\frac{b_j}{a_j+b_j} = 1,$$

and the proof is complete. The division device is decisive here, and as the introduction to the exercise suggests, this is not an isolated instance.

SOLUTION FOR EXERCISE 2.12 As Figure 2.4 suggests, we have the bound $f(x) = x/e^{x-1} \leq 1$ for all $x \geq 0$. In fact, we used this bound long ago (page 24); it was the key to Pólya's proof of the AM-GM inequality. If we now write $c_k = a_k/A$, then we have $c_1 + c_2 + \cdots + c_n = n$, and from this fact we see that for each k we have

$$\prod_{j=1}^{n} c_j = c_k \prod_{j:j\neq k} c_j \leq c_k \prod_{j:j\neq k} e^{c_j-1} = c_k e^{1-c_k} = c_k/e^{c_k-1} = f(c_k).$$

Since $\epsilon = (A - G)/A$ and $c_k = a_k/A$ we have for all $k = 1, 2, \ldots, n$ that

$$(1 - \epsilon)^n = \frac{a_1a_2\cdots a_n}{A^n} \leq \frac{a_k/A}{\exp(a_k/A - 1)} = f(a_k/A).$$

Solutions to the Exercises 235

Now the bounds (2.33) are immediate from the definition of ρ_-, ρ_+, together with the fact that f is strictly increasing on $[0,1)$ and strictly decreasing on $(1,\infty)$.

This solution was given by Gabor Szegő in 1914 in response to a question posed by George Pólya. It is among the earliest of their many joint efforts; at the time, Szegő was just 19.

SOLUTION FOR EXERCISE 2.13. In general one has $|w| \geq |\operatorname{Re} w|$ and $\operatorname{Re}(w+z) = \operatorname{Re}(w) + \operatorname{Re}(z)$, so from $\operatorname{Re} z_j = \rho_j \cos \theta_j$ we find

$$|z_1 + z_2 + \cdots + z_n| \geq |\operatorname{Re}(z_1 + z_2 + \cdots + z_n)|$$
$$= |z_1| \cos \theta_1 + |z_2| \cos \theta_2 + \cdots + |z_n| \cos \theta_n$$
$$\geq \bigl(|z_1| + |z_2| + \cdots + |z_n|\bigr) \cos \psi$$
$$\geq n\bigl(|z_1| |z_2| \cdots |z_n|\bigr)^{1/n} \cos \psi,$$

where we first used the fact that cosine is monotone decreasing on $[0, \pi/2]$ and then we applied the AM-GM inequality to the nonnegative real numbers $|z_j|$, $j = 1, 2, \ldots, n$. This exercise is based on Wilf (1963). Mitrinović (1970) notes that versions of this bound may be traced back at least to Petrovitch (1917). There are also informative generalizations given by Diaz and Metcalf (1966).

SOLUTION FOR EXERCISE 2.14. Take $x \geq 0$ and $y \geq 0$ and consider the hypothesis $H(n)$, $((x+y)/2)^n \leq (x^n + y^n)/2$. To prove $H(n+1)$ we note by $H(n)$ that

$$\left(\frac{x+y}{2}\right)^{n+1} = \left(\frac{x+y}{2}\right)\left(\frac{x+y}{2}\right)^n \leq \left(\frac{x+y}{2}\right)\frac{x^n + y^n}{2}$$
$$= \frac{x^{n+1} + y^{n+1} + xy^n + yx^n}{4}$$
$$= \frac{x^{n+1} + y^{n+1}}{2} - \frac{(x-y)(x^n - y^n)}{4} \leq \frac{x^{n+1} + y^{n+1}}{2}.$$

Induction then confirms the validity of $H(n)$ for all $n \geq 1$.

Now by $H(n)$ applied twice we find

$$\left\{\frac{x_1 + x_2 + x_3 + x_4}{4}\right\}^n \leq \frac{1}{2}\left\{\left(\frac{x_1 + x_2}{2}\right)^n + \left(\frac{x_3 + x_4}{2}\right)^n\right\}$$
$$\leq \frac{1}{2}\left\{\frac{x_1^n + x_2^n}{2} + \frac{x_3^n + x_4^n}{2}\right\}$$
$$= \frac{x_1^n + x_2^n + x_3^n + x_4^n}{4},$$

and this argument can be repeated to show that for each k and each set of 2^k nonnegative real numbers $x_1, x_2, \ldots, x_{2^k}$ we have

$$\left\{ \frac{x_1 + x_2 + \cdots + x_{2^k}}{2^k} \right\}^n \leq \frac{x_1^n + x_2^n + \cdots + x_{2^k}^n}{2^k}. \tag{14.48}$$

Cauchy's trick of padding a sequence of length m with extra terms to get a sequence of length 2^k now runs into difficulty, so a new twist is needed. One idea that works is to use a full backwards induction.

Specifically, we now let $H_{new}(m)$ denote the hypothesis that

$$\left\{ \frac{x_1 + x_2 + \cdots + x_m}{m} \right\}^n \leq \frac{x_1^n + x_2^n + \cdots + x_m^n}{m} \tag{14.49}$$

for any set of m nonnegative real numbers $x_1, x_2, \ldots, x_m$. We already know that $H_{new}(m)$ is valid when m is any power of two, so to prove that $H_{new}(m)$ is valid for all $m = 1, 2, \ldots$ we just need to show that for $m \geq 2$, the hypothesis $H_{new}(m)$ implies $H_{new}(m-1)$.

Given $m-1$ nonnegative reals $S = \{x_1, x_2, \ldots, x_{m-1}\}$, we introduce a new variable y by setting $y = (x_1 + x_2 + \cdots + x_{m-1})/(m-1)$. Since y is equal to $(x_1 + x_2 + \cdots + x_{m-1} + y)/m$, we see that when we apply $H(m)$ to the m-element set $S \cup \{y\}$, we obtain the bound

$$y^n \leq \frac{x_1^n + x_2^n + \cdots + x_{m-1}^n + y^n}{m},$$

and, when we clear y^n to the left side, we find

$$y^n \leq \frac{x_1^n + x_2^n + \cdots + x_{m-1}^n}{m-1}.$$

This inequality is precisely what one needed to establish the validity of $H_{new}(m-1)$, so the solution the problem is complete. This solution is guided by the one given by Shklarsky, Chentzov, and Yaglom (1993, pp. 391–392).

CHAPTER 3: LAGRANGE'S IDENTITY AND MINKOWSKI'S CONJECTURE

SOLUTION FOR EXERCISE 3.1. From the four geometric tautologies,

$$\cos \alpha = \frac{a_1}{\sqrt{a_1^2 + a_2^2}}, \quad \sin \alpha = \frac{a_2}{\sqrt{a_1^2 + a_2^2}},$$

$$\cos \beta = \frac{b_1}{\sqrt{b_1^2 + b_2^2}}, \quad \sin \beta = \frac{-b_2}{\sqrt{b_1^2 + b_2^2}},$$

Solutions to the Exercises

and the two trigonometric identities,

$$\cos(\alpha + \beta) = \cos\alpha \cos\beta - \sin\alpha \sin\beta = \frac{a_1 b_1 + a_2 b_2}{\sqrt{b_1^2 + b_2^2}\sqrt{a_1^2 + a_2^2}},$$

$$\sin(\alpha + \beta) = \sin\alpha \cos\beta + \cos\alpha \sin\beta = \frac{a_2 b_1 - a_1 b_2}{\sqrt{b_1^2 + b_2^2}\sqrt{a_1^2 + a_2^2}},$$

we find the Pythagorean path to the identity of Diophantus:

$$1 = \cos^2(\alpha + \beta) + \sin^2(\alpha + \beta) = \frac{(a_1 b_1 + a_2 b_2)^2 + (a_1 b_2 - a_2 b_1)^2}{(b_1^2 + b_2^2)(a_1^2 + a_2^2)}.$$

SOLUTION FOR EXERCISE 3.2 Here we just prove the identity of Diophantus since Brahmagupta's identity is analogous. As expected, one first factors. What is amusing is how one then *recombines twice*:

$$(x_1^2 + x_2^2)(y_1^2 + y_2^2) = (x_1 - ix_2)(x_1 + ix_2)(y_1 - iy_2)(y_1 + iy_2)$$
$$= \{(x_1 - ix_2)(y_1 + iy_2)\}\{(x_1 + ix_2)(y_1 - iy_2)\}.$$

The first factor is $\{(x_1 y_1 + x_2 y_2) + i(x_1 y_2 - x_2 y_1)\}$ and the second factor is its conjugate $\{(x_1 y_1 + x_2 y_2) - i(x_1 y_2 - x_2 y_1)\}$ so these have product $(x_1 y_1 + x_2 y_2)^2 + (x_1 y_2 - x_2 y_1)^2$, a computation which reveals the power of the factorization $a^2 + b^2 = (a + ib)(a - ib)$ in a most remarkable way.

SOLUTION FOR EXERCISE 3.3. On can pass from the discrete identity to a continuous version by appealing to the definition of the Riemann integral as a limit of sums, but it is both easier and more informative to consider the anti-symmetric form $s(x, y) = f(x)g(y) - g(x)f(y)$ and to integrate $s^2(x, y)$ over the square $[a, b]^2$. In this way one finds

$$\frac{1}{2}\int_a^b \int_a^b \left\{ f(x)g(y) - f(y)g(x) \right\}^2 dxdy$$
$$= \int_a^b f^2(x)\, dx \int_a^b g^2(x)\, dx - \left\{ \int_a^b f(x)g(x)\, dx \right\}^2, \quad (14.50)$$

provided that that all of the indicated integrals are well defined. Incidentally, anti-symmetric forms often merit exploration. Surprisingly often, they lead us to useful algebraic relations.

SOLUTION FOR EXERCISE 3.4. The two sides of the proposed inequality can be written respectively as

$$A = \left\{ x \sum_{j=1}^n a_j \sum_{k=1}^n b_k + (1-x) \sum_{j=1}^n a_j b_j \right\}^2$$

and as

$$B = \left\{ x\left(\sum_{j=1}^{n} a_j\right)^2 + (1-x)\sum_{j=1}^{n} a_j^2 \right\}\left\{ x\left(\sum_{j=1}^{n} b_j\right)^2 + (1-x)\sum_{j=1}^{n} b_j^2 \right\}$$

from which one finds that $B - A$ can be written as the sum of the term

$$x(1-x)\left\{ \sum_{j=1}^{n} \left(b_j \sum_{k=1}^{n} a_k - a_j \sum_{k=1}^{n} b_k \right)^2 \right\} \quad \text{and the term}$$

$$(1-x)^2\left\{ \sum_{j=1}^{n} a_j^2 \sum_{k=1}^{n} b_k^2 - \sum_{j=1}^{n} a_j b_j \right\}^2.$$

The first term is a sum of squares and the second term is nonnegative by Cauchy's inequality. Thus, $B - A$ is the sum of two nonnegative terms, and the solution is complete. The inequality of the problem is from Wagner (1965) and the solution is from Flor (1965).

SOLUTION FOR EXERCISE 3.5. Since f is nonnegative and nondecreasing one has the integral inequality

$$0 \leq \int_0^1 \int_0^1 f(x)f(y)(y-x)\bigl(f(x) - f(y)\bigr)\,dxdy$$

since the integrand is nonnegative. One may now complete the proof by simple expansion. Incidentally, this way of exploiting monotonicity is exceptionally rich, and several variations on this theme are explored at length in Chapter 5.

SOLUTION FOR EXERCISE 3.6. One expands and then factors

$$\begin{aligned} D_{n+1} - D_n &= \sum_{j=1}^{n} a_j b_j + na_{n+1}b_{n+1} - b_{n+1}\sum_{j=1}^{n} a_j - a_{n+1}\sum_{j=1}^{n} b_j \\ &= \sum_{j=1}^{n} a_j(b_j - b_{n+1}) + a_{n+1}\sum_{j=1}^{n}(b_{n+1} - b_j) \\ &= \sum_{j=1}^{n}(a_{n+1} - a_j)(b_{n+1} - b_j) \geq 0. \end{aligned}$$

According to Mitrinović (1970, p. 206) this elegant observation is due to R.R. Janić. The interaction between order relations and quadratic inequalities is developed more extensively in Chapter 5.

SOLUTION FOR EXERCISE 3.7. In the suggested shorthand, Lagrange's identity can be written as

$$\langle \mathbf{a},\mathbf{a}\rangle\langle \mathbf{b},\mathbf{b}\rangle - \langle \mathbf{a},\mathbf{b}\rangle^2 = \sum_{j<k}\begin{vmatrix}a_j & b_j\\ a_k & b_k\end{vmatrix}^2$$

and if we fix $\mathbf{b}$ and polarize $\mathbf{a}$ with $\mathbf{s}$ we find

$$\langle \mathbf{a},\mathbf{s}\rangle\langle \mathbf{b},\mathbf{b}\rangle - \langle \mathbf{a},\mathbf{b}\rangle\langle \mathbf{s},\mathbf{b}\rangle = \sum_{j<k}\begin{vmatrix}a_j & b_j\\ a_k & b_k\end{vmatrix}\begin{vmatrix}s_j & b_j\\ s_k & b_k\end{vmatrix}.$$

Now, if we fix $\mathbf{a}$ and $\mathbf{s}$ and polarize $\mathbf{b}$ with $\mathbf{t}$ we find

$$\langle \mathbf{a},\mathbf{s}\rangle\langle \mathbf{b},\mathbf{t}\rangle - \langle \mathbf{a},\mathbf{t}\rangle\langle \mathbf{s},\mathbf{b}\rangle = \sum_{j<k}\begin{vmatrix}a_j & b_j\\ a_k & b_k\end{vmatrix}\begin{vmatrix}s_j & t_j\\ s_k & t_k\end{vmatrix},$$

which is the shorthand version of the target identity.

SOLUTION FOR EXERCISE 3.8

After expanding the two products, on sees that the difference of the left-hand side and the right-hand side of Milne's inequality (3.17) can be written as a symmetric sum

$$\sum_{1\leq i<j\leq n}\left(a_ib_j + a_jb_i - \frac{(a_jb_j)(a_i+b_i)}{(a_j+b_j)} - \frac{(a_ib_i)(a_j+b_j)}{(a_i+b_i)}\right).$$

When each summand is put over the denominator $(a_i+b_i)(a_j+b_j)$, the numerator may be simplified, and one finds that this difference coincides with the definition (3.16) of R.

CHAPTER 4 ON GEOMETRY AND SUMS OF SQUARES

SOLUTION FOR EXERCISE 4.1. Each case follows by an application of the triangle inequality to an appropriate sum. Those sums are:

(a) $(x+y+z, x+y+z)) = (x,y) + (y,z) + (z,x)$
(b) $(y,z) = (x,x) + (y-x, z-x)$ and
(c) $(2,2,2) \leq (x+1/x, y+1/y, z+1/z) = (x,y,z) + (1/x, 1/y, 1/z)$.

SOLUTION FOR EXERCISE 4.2. The derivative on the left is equal to $\langle \nabla f(\mathbf{x}), \mathbf{u}\rangle$ which is bounded by $\|\nabla f(\mathbf{x})\|\|\mathbf{u}\| = \|\nabla f(\mathbf{x})\|$ by the Cauchy–Schwarz inequality. On the other hand, the derivative on the right is equal to $\langle \nabla f(\mathbf{x}), \mathbf{v}\rangle = \|\nabla f(\mathbf{x})\|$ by direct calculation and the definition of $\mathbf{v}$. These observations yield the inequality (4.21).

We have equality in the application of the Cauchy–Schwarz inequality only if $\mathbf{u}$ and $\nabla f(\mathbf{x})$ are proportional, so the bound (4.21) reduces to an equality if and only if $\mathbf{u} = \lambda \nabla f(\mathbf{x})$. Since $\mathbf{u}$ is a unit vector, this implies $\lambda = \pm 1/\|\nabla f(\mathbf{x})\|$. Only the positive sign can give equality in the bound (4.21), and in that case we have $\mathbf{u} = \mathbf{v}$.

SOLUTION FOR EXERCISE 4.3. Direct expansion proves the representation (4.22). To minimize $P(t)$ we solve $P'(t) = 2t\langle \mathbf{w}, \mathbf{w} \rangle - 2\langle \mathbf{v}, \mathbf{w} \rangle = 0$ and find $P(t) \geq P(t_0)$ where $t_0 = \langle \mathbf{v}, \mathbf{w} \rangle / \langle \mathbf{w}, \mathbf{w} \rangle$. The evaluation of $P(t_0)$ then leads one to the expression (4.22).

SOLUTION FOR EXERCISE 4.4. This exercise provides a reminder that one sometimes needs a more elaborate algebraic identity to deal with the absolute values of complex numbers than to deal with the absolute values of real numbers. Here the key is to use the Cauchy–Binet four letter identity (3.7) on page 49. The proof of that identity was purely algebraic (no absolute values or complex conjugates were used) so the identity is also valid for complex numbers. One then just makes the replacements $a_k \longmapsto \bar{a}_k$, $b_k \longmapsto b_k$, $s_k \longmapsto a_k$, and $t_k \longmapsto \bar{b}_k$.

SOLUTION FOR EXERCISE 4.5. This observation of S.S. Dragomir (2000) shows how the principles behind Lagrange's identity continue to bear fruit. Here one just takes the natural double sum and expands:

$$0 \leq \frac{1}{2} \sum_{j=1}^{n} \sum_{k=1}^{n} p_j p_k \|\alpha_j \mathbf{x}_k - \alpha_k \mathbf{x}_j\|^2$$

$$= \frac{1}{2} \sum_{j=1}^{n} \sum_{k=1}^{n} p_j p_k \left[\alpha_j^2 \|\mathbf{x}_k\|^2 - 2\langle \alpha_j \mathbf{x}_k, \alpha_k \mathbf{x}_j \rangle + \alpha_k^2 \|\mathbf{x}_j\|^2 \right]$$

$$= \sum_{j=1}^{n} \sum_{k=1}^{n} p_j p_k \alpha_j^2 \|\mathbf{x}_k\|^2 - \sum_{j=1}^{n} \sum_{k=1}^{n} p_j p_k \alpha_j \alpha_k \langle \mathbf{x}_k, \mathbf{x}_j \rangle$$

$$= \sum_{j=1}^{n} p_j \alpha_j^2 \sum_{k=1}^{n} p_k \|\mathbf{x}_k\|^2 - \left\| \sum_{j=1}^{n} p_j \alpha_j \mathbf{x}_j \right\|^2.$$

This identity gives us our target bound (4.24) and shows that the inequality is strict unless $\alpha_j \mathbf{x}_k = \alpha_k \mathbf{x}_j$ for all j and k. Finally, one should also note that a corresponding inequality for a complex inner product spaces can obtained by a similar calculation.

SOLUTION FOR EXERCISE 4.6.

There are proofs of this inequality that use only the tools of plane geometry, but there is also an exceptionally interesting proof that uses the transformation $z \mapsto 1/z$ for complex numbers. There is no loss of generality in setting $A = 0, B = z_1, C = z_2$, and $D = z_3$, and the triangle inequality then gives us

$$\left|\frac{1}{z_1} - \frac{1}{z_3}\right| \leq \left|\frac{1}{z_1} - \frac{1}{z_2}\right| + \left|\frac{1}{z_3} - \frac{1}{z_3}\right|,$$

which may be rewritten as $|z_2||z_1 - z_3| \leq |z_3||z_1 - z_2| + |z_1||z_2 - z_3|$. After identifying these terms with help from Figure 4.7, we see that it is precisely Ptolemy's inequality!

To prove the converse, we first note that one has equality in this application of the triangle inequality if and only if the points $z_1^{-1}, z_2^{-1}, z_3^{-1}$ are on line. One then obtains the required characterization by appealing to the fact that $z \mapsto 1/z$ takes a circle through the origin to a line and vice versa.

The transformation $z \mapsto 1/z$ is perhaps the leading example of a Möbius transformation, which more generally are the maps of the form $z \mapsto (az+b)/(cz+d)$. Every book on complex variables examines these transformations, but the treatment of Needham (1997), pages 122–188, is especially attractive. Needham also discusses Ptolemy's result with the help of inversion, but the quick treatment given here is closer to that of Treibergs (2002).

SOLUTION FOR EXERCISE 4.7. To prove the identity (4.26), expand the inner product squares and use $1 + \alpha + \cdots + \alpha^{N-1} = (1 - \alpha^N)/(1 - \alpha) = 0$. For the second identity, just expand and integrate. This exercise is based on D'Angelo (2002, pp. 53–55) where one finds related material.

SOLUTION FOR EXERCISE 4.8. The first part of the recursion (4.28) gives us $\langle \mathbf{z}_k, \mathbf{e}_j \rangle = 0$ for all $1 \leq j < k$, and this gives us $\langle \mathbf{e}_k, \mathbf{e}_j \rangle = 0$ for all $1 \leq j < k$. The normalization $\langle \mathbf{e}_k, \mathbf{e}_k \rangle = 1$ for $1 \leq k \leq n$ is immediate from the second part of the recursion (4.28), and the triangular spanning relations just rewrite the first part of the recursion (4.28).

SOLUTION FOR EXERCISE 4.9. Without loss of generality may we assume that $\|\mathbf{x}\| = 1$. The Gram–Schmidt relations are then given by $\mathbf{x} = \mathbf{e}_1$ and $\mathbf{y} = \mu_1 \mathbf{e}_2 + \mu_2 \mathbf{e}_2$. Orthonormality gives us $\langle \mathbf{x}, \mathbf{y} \rangle = \mu_1$ and $\langle \mathbf{y}, \mathbf{y} \rangle = |\mu_1|^2 + |\mu_2|^2$, and the bound $|\mu_1| \leq (|\mu_1|^2 + |\mu_2|^2)^{\frac{1}{2}}$ is obvious. But this says $|\langle \mathbf{x}, \mathbf{y} \rangle| \leq \langle \mathbf{y}, \mathbf{y} \rangle^{\frac{1}{2}}$ which is the Cauchy–Schwarz inequality when $\|\mathbf{x}\| = 1$.

SOLUTION FOR EXERCISE 4.10. From the Gram–Schmidt process applied to $\{\mathbf{y}_1, \mathbf{y}_2, \ldots, \mathbf{y}_n, \mathbf{x}\}$ one finds $\mathbf{e}_1 = \mathbf{y}_1$, $\mathbf{e}_2 = \mathbf{y}_2, \ldots, \mathbf{e}_n = \mathbf{y}_n$ and $\mathbf{e}_{n+1} = \mathbf{z}/\|\mathbf{z}\|$ where $\mathbf{z} = \mathbf{x} - (\langle \mathbf{x}, \mathbf{e}_1 \rangle \mathbf{e}_1 + \langle \mathbf{x}, \mathbf{e}_2 \rangle \mathbf{e}_2 + \cdots + \langle \mathbf{x}, \mathbf{e}_n \rangle \mathbf{e}_n)$, provided that $\mathbf{z} \neq 0$. Taking inner products and using orthonormality then gives us

$$\langle \mathbf{x}, \mathbf{x} \rangle = \sum_{j=1}^{n+1} |\langle \mathbf{x}, \mathbf{e}_j \rangle|^2 = |\langle \mathbf{x}, \mathbf{e}_{n+1} \rangle|^2 + \sum_{j=1}^{n} |\langle \mathbf{x}, \mathbf{y}_j \rangle|^2,$$

and since $|\langle \mathbf{x}, \mathbf{e}_{n+1} \rangle|^2$ gives us Bessel's inequality when $\mathbf{z} \neq 0$. When $\mathbf{z} = 0$ one finds that Bessel's inequality is in fact an identity.

SOLUTION FOR EXERCISE 4.11. Without loss of generality we can assume that $\mathbf{x}$, $\mathbf{y}$, and $\mathbf{z}$ are linearly independent and $\|\mathbf{x}\| = 1$, so the Gram–Schmidt relations can be written as $\mathbf{x} = \mathbf{e}_1$, $\mathbf{y} = \mu_1 \mathbf{e}_2 + \mu_2 \mathbf{e}_2$, and $\mathbf{z} = \nu_1 \mathbf{e}_1 + \nu_2 \mathbf{e}_2 + \nu_3 \mathbf{e}_3$, from which we find $\langle \mathbf{x}, \mathbf{x} \rangle = 1$, $\langle \mathbf{x}, \mathbf{y} \rangle = \mu_1$, $\langle \mathbf{x}, \mathbf{z} \rangle = \nu_1$ and $\langle \mathbf{y}, \mathbf{z} \rangle = \mu_1 \nu_1 + \mu_2 \nu_2$. The bound (4.30) asserts $\mu_1 \nu_1 \leq \frac{1}{2}(\mu_1 \nu_1 + \mu_2 \nu_2 + (\mu_1^2 + \mu_2^2)^{\frac{1}{2}} (\nu_1^2 + \nu_2^2 + \nu_3^2)^{\frac{1}{2}})$ or $\mu_1 \nu_1 - \mu_2 \nu_2 \leq (\mu_1^2 + \mu_2^2)^{\frac{1}{2}} (\nu_1^2 + \nu_2^2 + \nu_3^2)^{\frac{1}{2}}$, which is immediate from Cauchy's inequality.

SOLUTION FOR EXERCISE 4.12. With the normalization and notation used in the solution of Exercise 4.11, the left side L of the bound (4.31) can be written as

$$|\langle \mathbf{x}, \mathbf{x} \rangle \langle \mathbf{y}, \mathbf{z} \rangle - \langle \mathbf{x}, \mathbf{y} \rangle \langle \mathbf{x}, \mathbf{z} \rangle| = |\{(\mu_1 \bar{\nu}_1 + \mu_2 \bar{\nu}_2) - \mu_1 \bar{\nu}_1\}|^2 = |\mu_2 \bar{\nu}_2|^2,$$

and the right side R can be written as

$$\{\langle \mathbf{x}, \mathbf{x} \rangle^2 - |\langle \mathbf{x}, \mathbf{y} \rangle|^2\} \{\langle \mathbf{x}, \mathbf{x} \rangle^2 - |\langle \mathbf{x}, \mathbf{z} \rangle|^2\}$$
$$= (1 - |\mu_1|^2)(1 - |\nu_1|^2) = |\mu_2|^2 (|\nu_2|^2 + |\nu_3|^2),$$

since we have $1 = \|\mathbf{y}\| = |\mu_1|^2 + |\mu_2|^2$ and $1 = \|\mathbf{z}\| = |\nu_1|^2 + |\nu_2|^2 + |\nu_3|^2$. These formulas for L and R make it evident that $L \leq R$.

Now, to prove the bound (4.32) it similarly reduces to showing

$$|\mu_1 \bar{\nu}_1 + \mu_2 \bar{\nu}_2|^2 + |\mu_1|^2 + |\nu_1|^2$$
$$\leq 1 + (\bar{\mu}_1 \nu_1 + \bar{\mu}_2 \nu_2) \mu_1 \bar{\nu}_1 + (\mu_1 \bar{\nu}_1 + \mu_2 \bar{\nu}_2) \bar{\mu}_1 \nu_1$$

and, by expansion, this is the same as

$$|\mu_1|^2 + |\nu_1|^2 + |\mu_1 \nu_1|^2 + |\mu_2 \nu_2|^2 + 2\text{Re}\{\mu_1 \bar{\nu}_1 \bar{\mu}_2 \nu_2\}$$
$$\leq 1 + 2|\mu_1 \nu_1|^2 + 2\text{Re}\{\mu_1 \bar{\nu}_1 \bar{\mu}_2 \nu_2\}.$$

After cancelling terms, we see it suffices for us to show

$$L \equiv |\mu_1|^2 + |\nu_1|^2 + |\mu_2 \nu_2|^2 \leq 1 + |\mu_1 \nu_1|^2,$$

but the substitution $|\mu_2 \nu_2|^2 = (1 - |\mu_1|^2)(1 - |\nu_1|^2 - |\nu_3|^2)$ gives us $L = 1 + |\mu_1 \nu_1|^2 + |\nu_3|^2 (|\mu_1|^2 - 1) \leq 1 + |\mu_1 \nu_1|^2$ since $|\mu_1|^2 \leq 1$. This exercise is based on Problems 16.50 and 16.51 of Hewitt and Stomberg (1969, p. 254).

SOLUTION FOR EXERCISE 4.13. Following the hint, we first note

$$\|A^T\mathbf{v}\|^2 = \langle A^T\mathbf{v}, A^T\mathbf{v}\rangle = \langle \mathbf{v}, AA^T\mathbf{v}\rangle \leq \|\mathbf{v}\|\|AA^T\mathbf{v}\| = \|\mathbf{v}\|\|A^T\mathbf{v}\|,$$

so by division $\|A^T\mathbf{v}\| \leq \|\mathbf{v}\|$. Next, by the Cauchy–Schwarz inequality and the properties of A and A^T we have the chain

$$\|\mathbf{v}, \mathbf{v}\|^2 = \langle A\mathbf{v}, A\mathbf{v}\rangle = \langle \mathbf{v}, A^T A\mathbf{v}\rangle \leq \|\mathbf{v}\|\|A^T A\mathbf{v}\| \leq \|\mathbf{v}\|\|A\mathbf{v}\| = \|\mathbf{v}, \mathbf{v}\|^2,$$

so we actually have equality where the first inequality is written. This tells us that there is a λ (which possibly depends on $\mathbf{v}$) for which we have $\lambda\mathbf{v} = A^T A\mathbf{v}$. This relation in turn gives us

$$\lambda\langle\mathbf{v}, \mathbf{v}\rangle = \langle\mathbf{v}, A^T A\mathbf{v}\rangle = \langle A\mathbf{v}, A\mathbf{v}\rangle = \langle\mathbf{v}, \mathbf{v}\rangle,$$

so in fact $\lambda = 1$ (and hence it does not actually depend on $\mathbf{v}$)). We therefore find that $\mathbf{v} = A^T A\mathbf{v}$ for all $\mathbf{v}$, so $A^T A = I$ as claimed. This argument follows Sigillito (1968).

CHAPTER 5: CONSEQUENCES OF ORDER

SOLUTION FOR EXERCISE 5.1. The upper bound of (5.17) follows from

$$h_1 + h_2 + \cdots + h_n = \frac{h_1}{b_1}b_1 + \frac{h_2}{b_2}b_2 + \cdots + \frac{h_n}{b_n}b_n$$
$$\leq \{b_1 + b_2 + \cdots + b_n\}\max_k \frac{h_k}{b_k},$$

and the lower bound is analogous. For application, if we set $a_k = c_k x^k$ and $b_k = c_k y^k$, then we have $\min a_k/b_k = (x/y)^n$ and $\max a_k/b_k = 1$.

SOLUTION FOR EXERCISE 5.2. The $n-1$ elements of S have mean A, so by the induction hypothesis $H(n-1)$ we have

$$a_2 a_3 \cdots a_n(a_1 + a_2 - A) \leq A^{n-1}.$$

The betweenness bound already gave us $a_1 a_n/A \leq a_1 + a_2 - A$, and, when we may apply this bound above, we get $H(n)$ which completes the induction.

This proof from Chong (1975) is closely related to a "smoothing" proof of the AM-GM which exploits the algorithm:

(i) if $a_1, a_2, \ldots, a_n$ are not all equal to the mean A, let a_j and a_k denote the smallest and largest, respectively,

(ii) replace a_j by A and replace a_k by $a_j + a_k - A$,

(iii) note that each step of the algorithm increases by one the number of terms equal to the mean, so the algorithm terminates in at most n steps.

The betweenness bound gives us $a_j a_k \leq A(a_j + a_k - A)$ so each step of the algorithm increases the geometric mean of the current sequence. Since we start with the sequence $a_1, a_2, \ldots, a_n$ and terminate with a sequence of n copies of A, we see $a_1 a_2 \cdots a_n \leq A^n$.

SOLUTION FOR EXERCISE 5.3. If one first considers $V = \mathbb{R}$ and sets $a = u$ and $b = v$ then the inequality in question asserts that

$$AB - ab \geq (A^2 - a^2)^{\frac{1}{2}} (B^2 - b^2)^{\frac{1}{2}}. \tag{14.51}$$

By expansion and factorization, this is equivalent to

$$(aB - Ab)^2 \geq 0,$$

so the bound (14.51) is true and equality holds if and only if $aB = Ab$. To address the general problem, we first note by the Cauchy–Schwarz inequality

$$AB - \langle u, v \rangle \geq AB - \langle u, u \rangle^{\frac{1}{2}} \langle v, v \rangle^{\frac{1}{2}},$$

so, by the bound (14.51) with $a = \langle u, u \rangle^{\frac{1}{2}}$ and $b = \langle v, v \rangle^{\frac{1}{2}}$, one has

$$AB - \langle u, v \rangle \geq (A^2 - \langle u, u \rangle)^{\frac{1}{2}} (B^2 - \langle v, v \rangle)^{\frac{1}{2}}, \tag{14.52}$$

which was to be proved. If equality hold in the bound (14.52), this argument shows that we have $\langle u, v \rangle = \langle u, u \rangle^{\frac{1}{2}} \langle v, v \rangle^{\frac{1}{2}}$, so there is a constant λ such that $u = \lambda v$. By substitution one then finds that $\lambda = A/B$.

The bound (14.52) is abstracted from an integral version given in Theorem 9 of Lyusternik (1966) which Lyusternik used in his proof of the Brunn–Minkowski inequality in two dimensions. The idea viewing $V = \mathbb{R}$ as a special inner product space is often useful, but seldom is it as decisive as it proved to be here. One should also notice the easily overlooked fact that the bound (14.52) is actually equivalent to the light cone inequality (4.15).

SOLUTION FOR EXERCISE 5.4. This problem does not come with an order relation, but we can give ourselves one if we note that by the symmetry of the bound we can assume that $0 \leq x \leq y \leq z$. We then get for free the positivity of the first summand $x^\alpha (x - y)(x - z)$, so to

Solutions to the Exercises

complete the proof we just need to show the positivity of the *sum* of the other two. This follows from the factorization

$$y^\alpha(y-x)(y-z) + z^\alpha(z-x)(z-y) = (z-y)\{z^\alpha(z-x) - y^\alpha(y-x)\}$$

and the observation that $z \geq y$ and $z - x \geq y - x$.

This proof illustrates one of the most general methods at our disposal; the positivity of a sum can often be proved by creatively grouping the summands so that the positivity of each group becomes obvious.

SOLUTION FOR EXERCISE 5.5. This is one of the text's few "plug-in" exercises, but the bound is so nice it had to be made explicit. We just note that $m \stackrel{\text{def}}{=} a/A \leq a_k/b_k \leq A/b \stackrel{\text{def}}{=} M$, then we substitute into the formulas (5.6) and (5.7).

SOLUTION FOR EXERCISE 5.6. Without loss of generality, we can assume that $0 < a \leq b \leq c$, and, under this assumption, we also have

$$\frac{1}{b+c} \leq \frac{1}{a+c} \leq \frac{1}{a+b}.$$

The rearrangement inequality then tells us that

$$\frac{b}{b+c} + \frac{c}{a+c} + \frac{a}{a+b} \leq \frac{a}{b+c} + \frac{b}{a+c} + \frac{c}{a+b}$$

and that

$$\frac{c}{b+c} + \frac{a}{a+c} + \frac{b}{a+b} \leq \frac{a}{b+c} + \frac{b}{a+c} + \frac{c}{a+b}.$$

By summing these two bounds we find Nesbitt's inequality.

Engel (1998, pp. 162–168) provides five instructive proofs of Nesbitt's inequality, including the one given here, but, even so, one can add to the list. Tony Cai recently noted that Nesbitt's inequality follows from the bound (1.21), page 13, provided that one sets

$$p_1 = \frac{a}{a+b+c}, \quad p_2 = \frac{b}{a+b+c}, \quad p_3 = \frac{c}{a+b+c},$$

$$a_1 = \frac{a+b+c}{b+c}, \quad a_2 = \frac{a+b+c}{a+c}, \quad a_3 = \frac{a+b+c}{a+b},$$

and sets $b_k = 1/a_k$ for $k = 1, 2, 3$. With these substitutions the bound (1.21) automatically gives us

$$1 \leq \left(\frac{a}{b+c} + \frac{b}{a+c} + \frac{c}{a+b}\right)\left(\frac{(a+b+c)^2 - (a^2+b^2+c^2)}{(a+b+c)^2}\right),$$

which in turn yields Nesbitt's inequality since the second factor is bounded by 2/3 because Cauchy's inequality for (a,b,c) and $(1,1,1)$ tells us that $(a+b+c)^2 \leq 3(a^2+b^2+c^2)$.

SOLUTION FOR EXERCISE 5.7. Since the sequences $\{c_k\}$ and $\{1/c_k\}$ are oppositely ordered, the rearrangement inequality (5.12) tells us that for any permutation σ one has $n \leq c_1/c_{\sigma(1)} + c_2/c_{\sigma(2)} + \cdots + c_n/c_{\sigma(n)}$, and part (a) is a special case of this observation. If we set $c_k = x_1 x_2 \cdots x_k$ in part (a) we get part (b), and if we then replace x_k by ρx_k, we get part (c). Finally, by setting $\rho = (x_1 x_2 \cdots x_n)^{-n}$ and simplifying, we get the AM-GM bound.

SOLUTION FOR EXERCISE 5.8.

The inequality is unaffected if m, M, and x_j, $1 \leq j \leq n$ are multiplied by a positive constant, so we can assume without loss of generality that $\gamma = 1$, in which case, we have $M = m^{-1}$, and it suffices to show that

$$\left\{\sum_{j=1}^{n} p_j x_j\right\}\left\{\sum_{j=1}^{n} p_j \frac{1}{x_j}\right\} \leq \mu^2 \qquad (14.53)$$

where $2\mu = m + M = m + m^{-1}$. Now, one has $x_j \in [m, m^{-1}]$ for all $1 \leq j \leq n$, so we have

$$x_j + x_j^{-1} \leq m + m^{-1} \leq 2\mu \quad \text{and} \quad \left\{\sum_{j=1}^{n} p_j x_j\right\} + \left\{\sum_{j=1}^{n} p_j \frac{1}{x_j}\right\} \leq 2\mu,$$

and these yield the bound (14.53) after one applies the AM-GM inequality to the two bracketed terms. There are many instructive proofs of Kantorovich's inequality; this elegant approach via the AM-GM inequality is due to Pták (1995).

SOLUTION FOR EXERCISE 5.9. One elegant way to make the monotonicity of f_θ evident is to set $c_j = (a_j b_j)^\theta$ and $d_j = \log(a_j/b_j)$ to obtain

$$f_\theta(x) = \sum_{j=1}^{n} c_j e^{d_j x} \sum_{j=1}^{n} c_j e^{-d_j x} = \sum_{j=1}^{n} \left[c_j^2 + 2\sum_{j<k} c_j c_k \cosh\left(d_j - d_k\right)x\right]$$

where $\cosh y = (e^y + e^{-y})/2$. Since $\cosh y$ is symmetric about zero and monotone on $[0, \infty)$, the monotonicity of $f_\theta(\cdot)$ is now immediate. This solution follows Steiger (1969) where a second proof based on Hölder's inequality is also given.

SOLUTION FOR EXERCISE 5.10. We can assume without loss of generality that $a_1 \geq a_2$, $b_1 \geq b_2$, and $a_1 \geq b_1$. Remaining mindful of the

Solutions to the Exercises 247

relation $a_1+a_2 = b_1+b_2$, the proof can be completed by the factorization

$$x^{a_1}y^{a_2} + x^{a_2}y^{a_1} - x^{b_1}y^{b_2} - x^{b_2}y^{b_1}$$
$$= x^{a_2}y^{a_2}(x^{a_1-a_2} + y^{a_1-a_2} - x^{b_1-a_2}y^{b_2-a_2} - x^{b_2-a_2}y^{b_1-a_2})$$
$$= x^{a_2}y^{a_2}(x^{b_1-a_2} - y^{b_1-a_2})(x^{b_2-a_2} - y^{b_2-a_2}) \geq 0,$$

since $b_1 - a_2 \geq b_2 - a_2 = a_1 - b_1 \geq 0$. Lee (2002) notes that the bound (5.22) may be used to prove analogous inequalities with three or more variables. Chapter 13 will developed such inequalities by other methods.

SOLUTION FOR EXERCISE 5.11. Let A denote the event that $|Z - \mu|$ is at least as large as λ. Now, define a random variable χ_A by setting $\chi_A = 1$ if the event A occurs and setting $\chi_A = 0$ otherwise. Note that $E(\chi_A) = P(A) = P(|Z - \mu| \geq \lambda)$. Also note that $\chi_A \leq |Z - \mu|^2/\lambda^2$, since both sides are zero if A does not occur, and the right side is at least as large as 1 if the event A does occur. On taking the expectation of the last bound one gets Chebyshev's tail bound (5.23). Admittedly, the language used in this problem and its solution are special to probability theory, but nevertheless the argument is completely rigorous.

CHAPTER 6: CONVEXITY — THE THIRD PILLAR

SOLUTION FOR EXERCISE 6.1. Cancelling $1/x$ from both sides and adding the fractions, one sees that Mengoli's inequality is equivalent to the trivial bound $x^2 > x^2 - 1$. For a proof using Jensen's inequality, just note that $x \mapsto 1/x$ is convex. Finally, for a modern version of Mengoli's proof that H_n diverges, we assume $H_\infty < \infty$ and write H_∞ as

$$1 + (1/2 + 1/3 + 1/4) + (1/5 + 1/6 + 1/7) + (1/8 + 1/9 + 1/10) + \cdots.$$

Now, by applying Mengoli's inequality within the indicated groups we find the lower bound $1 + 3/3 + 3/6 + 3/9 + \cdots = 1 + H_\infty$, which yields the contradictions $H_\infty > 1 + H_\infty$.

By the way, according to Havil (2003, p. 38) it was Mengoli who in 1650 first posed the corresponding problem of determining the value of the sum $1 + 1/2^2 + 1/3^2 + \cdots$. The problem resisted the efforts of Europe's finest mathematicians until 1731 when L. Euler determined the value to be $\pi^2/6$.

SOLUTION FOR EXERCISE 6.2. The bound follows by applying Jensen's inequality to the function $f(t) = \log(1 + 1/t) = \log(1+t) - \log(t)$, which is convex because

$$f''(t) = -\frac{1}{(1+t)^2} + \frac{1}{t^2} > 0 \quad \text{for } t > 0.$$

SOLUTION FOR EXERCISE 6.3. From the geometry of Figure 6.4, the area A of an inscribed polygon with n sides can be written as

$$A = \frac{1}{2}\sum_{k=1}^{n} \sin\theta_k \quad \text{where} \quad 0 < \theta_k < \pi \quad \text{and} \quad \sum_{k=1}^{n} \theta_k = 2\pi.$$

Since $\sin(\cdot)$ is strictly concave on $[0, \pi]$, we have

$$A = \frac{1}{2}\sum_{k=1}^{n} \sin(\theta_k) \leq \frac{1}{2}n\sin\left(\frac{1}{n}\sum_{k=1}^{n}\theta_k\right) = \frac{1}{2}n\sin(2\pi/n) \stackrel{\text{def}}{=} A',$$

and we have equality if and only if $\theta_k = 2\pi/n$ for all $1 \leq k \leq n$. Since A' is the area of a regular inscribed n-gon, the conjectured optimality is confirmed.

SOLUTION FOR EXERCISE 6.4. The second bound is the AM-GM inequality for $a_k = 1 + r_k$, $k = 1, 2, \ldots, n$. The first bound follows from Jensen's inequality applied to the convex function $x \mapsto \log(1 + e^x)$. Finally, by taking nth roots and subtracting 1, we see that the investment inequality (6.23) refines the AM-GM bound $r_G \leq r_A$ by slipping $V^{1/n} - 1$ between the two means.

SOLUTION FOR EXERCISE 6.5. To build a proof with Jensen's inequality, we first divide by $(a_1 a_2 \cdots a_n)^{1/n}$ and write c_k for b_k/a_k, so the target inequality takes the form

$$1 + (c_1 c_2 \cdots c_n)^{1/n} \leq \left\{(1+c_1)(1+c_2)\cdots(1+c_n)\right\}^{1/n}.$$

Now, if we take logs and write c_j as $\exp(d_j)$, we find it takes the form

$$\log\left(1 + \exp(\bar{d})\right) \leq \frac{1}{n}\sum_{j=1}^{n} \log(1 + \exp(d_j)),$$

where $\bar{d} = (d_1 + d_2 + \cdots + d_n)/n$. Finally, the last inequality is simply Jensen's inequality for the convex function $x \mapsto \log(1 + e^x)$, so the solution is complete. One feature of this solution worth noting is that progress came quickly after division reduced the number of variables from $2n$ to n. This phenomenon is actually rather common, and such reductions are almost always worth a try.

Here it is perhaps worth noting that Minkowski's proof used yet another idea. Specifically, he built his proof on analysis of the polynomial $p(t) = \prod(a_j + tb_j)$. Can you recover his proof?

SOLUTION FOR EXERCISE 6.6. Essentially no change is needed in Cauchy's argument (page 20). First, for the cases $n = 2^k$, $k = 1, 2, \ldots$, one just applies the defining relation (6.25) to successive halves. For the fall-back step, one chooses k such that $n \leq 2^k$ and applies the 2^k result to the padded sequence y_j, $1 \leq j \leq 2^k$ which one defines by taking $y_j = x_j$ for $1 \leq j \leq n$ and by taking $y_j = (x_1 + x_2 + \cdots + x_n)/n$ for $n < j \leq 2^k$.

SOLUTION FOR EXERCISE 6.7. As we noted in the preceding solution, iteration of the defining condition (6.24) gives us for all $k = 1, 2, \ldots$ that

$$f\left(\frac{1}{2^k}\sum_{j=1}^{2^k} x_j\right) \leq \frac{1}{2^k}\sum_{j=1}^{2^k} f(x_j),$$

so setting $x_j = x$ for $1 \leq j \leq m$ and $x_j = y$ for $m < j \leq 2^k$ we also have

$$f\Big((m/2^k)x + (1 - m/2^k)y\Big) \leq (m/2^k)f(x) + (1 - m/2^k)f(y).$$

If we now choose m_t and k_t such that $m_t/2^{k_t} \to p$ as $t \to \infty$, then continuity of f and the preceding bound give us convexity of the kind required by the modern definition (6.1).

SOLUTION FOR EXERCISE 6.8. The function $L(x, y, z)$ is convex in each of its three variables separately and, by the argument detailed below, this implies that L must attain its maximum at one of the vertices of the cube. After eight easy evaluations we find that $L(1, 0, 0) = 2$ and that no other corner has a larger value, so the solution is complete.

It is also easy to show that if a function on the cube is convex in each variable separately then the function must attain its maximum on one of the corner points. In essence one argues by induction but, for the cube in $\mathbb{R}^3$, one may as well give all of the steps.

First, one notes that a convex function on $[0, 1]$ must take its maximum at one of the end points of the interval, so, for any fixed values of y and z, we have the bound $L(x, y, z) \leq \max\{L(0, y, z), L(1, y, z)\}$. Similarly, by convexity of $y \mapsto L(0, y, z)$ and $y \mapsto L(1, y, z)$ so $L(0, y, z)$ is bounded by $\max\{L(0, 0, z), L(0, 1, z)\}$ and $L(1, y, z)$ is bounded by $\max\{L(1, 0, z), L(1, 1, z)\}$. All together, we have for each value of z that $L(x, y, z)$ is bounded by $\max\{L(0, 0, z), L(0, 1, z), L(1, 0, z), L(1, 1, z)\}$.

Convexity of $z \mapsto L(x,y,z)$ applied four times then gives us the final bound $L(x,y,z) \le \max\{L(e_1,e_2,e_3) : e_k = 0 \text{ or } e_k = 1 \text{ for } k = 1,2,3\}$.

One should note that this argument does *not* show that one can find the maximum by the "greedy algorithm" that performs three successive maximums. In fact, the greedy algorithm can fail miserably here, as easy examples show.

SOLUTION FOR EXERCISE 6.9. To prove the first formula, we note
$$a^2 = b^2 + c^2 - 2bc\cos\alpha = (b-c)^2 + 2bc(1-\cos\alpha)$$
$$= (b-c)^2 + 4A(1-\cos\alpha)/\sin\alpha = (b-c)^2 + 4A\tan(\alpha/2),$$
so, by symmetry and summing, we see that $a^2 + b^2 + c^2$ is equal to
$$(a-b)^2 + (b-c)^2 + (c-a)^2 + 4A\big(\tan(\alpha/2) + \tan(\beta/2) + \tan(\gamma/2)\big).$$
Since $x \mapsto \tan x$ is convex on $[0, \pi/2]$, Jensen's inequality gives us
$$\frac{1}{3}\{\tan(\alpha/2) + \tan(\beta/2) + \tan(\gamma/2)\} \ge \tan\left(\frac{\alpha+\beta+\gamma}{6}\right) = \tan(\pi/6)$$
and $\tan(\pi/6) = \sqrt{3}$, so this completes the proof. Engel (1998, p. 173) gives this as the eighth among his eleven amusing proofs of Weitzenböck's inequality and its refinements.

SOLUTION FOR EXERCISE 6.10. The polynomial $Q(x)$ can be written as a sum of three simple quadratics:
$$\frac{(x-x_2)(x-\mu)}{(x_1-x_2)(x_1-\mu)}f(x_1) + \frac{(x-x_1)(x-\mu)}{(x_2-x_1)(x_2-\mu)}f(x_2) + \frac{(x-x_1)(x-x_2)}{(\mu-x_1)(\mu-x_2)}f(\mu).$$
By two applications of Rolle's theorem we see that $Q'(x) - f'(x)$ has a zero in (x_1, μ) and a zero in (μ, x_2), so a third application of Rolle's theorem shows there is an x^* between these zeros for which we have $0 = Q''(x^*) - f''(x^*)$. We therefore have $Q''(x^*) = f''(x^*) \ge 0$, but
$$Q''(x^*) = \frac{2f(x_1)}{(x_1-x_2)(x_1-\mu)} + \frac{2f(x_2)}{(x_2-x_1)(x_2-\mu)} + \frac{2f(\mu)}{(\mu-x_1)(\mu-x_2)}$$
so, by setting $p = (x_2 - \mu)/(x_2 - x_1)$ and $q = (\mu - x_1)/(x_2 - x_1)$ and simplifying, one finds that the last inequality reduces to the definition of the convexity of f.

SOLUTION FOR EXERCISE 6.11. Given the hint, we obviously want to consider the change of variables, $\alpha = \tan^{-1}(a)$, $\beta = \tan^{-1}(b)$, and $\gamma = \tan^{-1}(c)$. The conditions $a > 0$, $b > 0$, $c > 0$, and $a+b+c = abc$ now tell us that $\alpha > 0$, $\beta > 0$, $\gamma > 0$, and $\alpha+\beta+\gamma = \pi$. The target inequality also becomes $\cos\alpha + \cos\beta + \cos\gamma \leq 3/2$, and this follows directly from Jensen's inequality in view of the *concavity* of cosine on $[0,\pi]$ and the evaluation $\cos(\pi/3) = 1/2$. This solution follows Andreescu and Feng (2000, p. 86). Hojoo Lee has given another solution which exploits the homogenization trick which we discuss in Chapter 12 (page 189).

SOLUTION FOR EXERCISE 6.12. If we write

$$P(z) = a_n(z-r_1)^{m_1}(z-r_2)^{m_2}\cdots(z-r_n)^{m_k}$$

where $r_1, r_2, \ldots, r_k$ are the distinct roots of $P(z)$, and $m_1, m_2, \ldots, m_k$ are the corresponding multiplicities, then comparison of $P'(z)$ and $P(z)$ gives us the familiar formula

$$\frac{P'(z)}{P(z)} = \frac{m_1}{z-r_1} + \frac{m_2}{z-r_2} + \cdots + \frac{m_k}{z-r_n}.$$

Now, if z_0 is a root of $P'(z)$ which is also a root of $P(z)$, then z_0 is automatically in H, so without loss of generality, we may assume that z_0 is a root of $P'(z)$ that is not a root of $P(z)$, in which case we find

$$0 = \frac{m_1}{z_0-r_1} + \frac{m_2}{z_0-r_2} + \cdots + \frac{m_k}{z_0-r_k}$$
$$= \frac{m_1(\bar{z}_0-\bar{r}_1)}{|z_0-r_1|^2} + \frac{m_2(\bar{z}_0-\bar{r}_2)}{|z_0-r_2|^2} + \cdots + \frac{m_k(\bar{z}_0-\bar{r}_k)}{|z_0-r_k|^2}.$$

If we set $w_k = m_k/|z_0-r_k|^2$, then we can rewrite this identity as

$$z_0 = \frac{w_1 r_1 + w_2 r_2 + \cdots + w_k r_k}{w_1 + w_2 + \cdots + w_k},$$

which shows z_0 is a convex combination of the roots of $P(z)$.

SOLUTION FOR EXERCISE 6.13. Write $r_1, r_2, \ldots, r_n$ for the roots of P repeated according to their multiplicity, and for a z which is outside of the convex hull H write $z - r_j$ in polar form $z - r_j = \rho_j e^{i\theta_j}$. We then have

$$\frac{1}{z-r_j} = \rho_j^{-1} e^{-\theta_j i} \qquad 1 \leq j \leq n,$$

and the spread in the arguments θ_j, $1 \leq j \leq n$, is not more than 2ψ. Thus, by the complex AM-GM inequality (2.35) one has the bound

$$(\cos\psi)\left|\frac{1}{z-r_1}\frac{1}{z-r_2}\cdots\frac{1}{z-r_n}\right|^{1/n} \leq \frac{1}{n}\left|\sum_{j=1}^n \frac{1}{z-r_j}\right|$$

and, in terms of P and P', this simply says

$$\left|\frac{a_n}{P(z)}\right|^{1/n} \leq \frac{1}{n\cos\psi}\left|\frac{P'(z)}{P(z)}\right| \quad \text{for all } z \notin H, \tag{14.54}$$

just as we hoped to prove.

SOLUTION FOR EXERCISE 6.14. If 2ψ is the viewing angle determined by U when viewed from $z \notin U$, then we have $1 = |z|\sin\psi$, so Pythagoras's theorem tells us that $\cos\psi = (1 - |z|^{-2})^{\frac{1}{2}}$. The target inequality (6.27) then follows directly from Wilf's bound (6.26).

SOLUTION FOR EXERCISE 6.15. This is *American Mathematical Monthly* Problem E10940 posed by Y. Nievergelt. We consider the solution by A. Nakhash. The disk $D_0 = \{z : |1 - z| \leq 1\}$ in polar coordinates is $\{re^{i\theta} : 0 \leq r \leq 2\cos\theta, -\pi/2 < \theta < \pi/2\}$, so for each j we can write $1 + z_j$ as $r_j e^{i\theta_j}$ where $-\pi/2 < \theta < \pi/2$ and where $r_j \leq 2\cos\theta_j$. It is immediate that $z_0 = -1 + (r_1 r_2 \cdots r_n)^{1/n} \exp(i(\theta_1 + \theta_2 + \cdots + \theta_n)/n)$ solves Nievergelt's equation (6.28), and to prove that $z_0 \in D$ it suffices to show $1 + z_0 \in D_0$; equivalently, we need to show

$$(r_1 r_2 \cdots r_n)^{1/n} \leq 2\cos\left(\frac{\theta_1 + \theta_2 + \cdots + \theta_n}{n}\right). \tag{14.55}$$

Since $(r_1 r_2 \cdots r_n)^{1/n}$ is bounded by $((2\cos\theta_1)(2\cos\theta_2)\cdots(2\cos\theta_n))^{1/n}$, it therefore suffices to show that

$$((\cos\theta_1)(\cos\theta_2)\cdots(\cos\theta_n))^{1/n} \leq \cos\left(\frac{\theta_1 + \theta_2 + \cdots + \theta_n}{n}\right),$$

and this follows the concavity of $f(x) = \log(\cos x))$ on $-\pi/2 < \theta < \pi$ together with Jensen's inequality.

SOLUTION FOR EXERCISE 6.16. A nice solution using Jensen's inequality for $f(x) = 1/x$ was given by Robert Israel in the sci.math newsgroup in 1999. If we set $S = a_1 + a_2 + a_3 + a_4$ and let C denotes the sum on the right hand side of the bound (6.29), then Jensen's with $p_j = a_j/S$ and $x_1 = a_2 + a_3$, $x_2 = a_3 + a_4$, $x_3 = a_4 + a_1$, and $x_4 = a_1 + a_2$ gives us $C/S \geq \{D/S\}^{-1}$ or $C \geq S^2/D$, where one has set

$$D = a_1(a_2 + a_3) + a_2(a_3 + a_4) + a_3(a_4 + a_1) + a_4(a_1 + a_2).$$

Now, it is easy to check that $S^2 - 2D = (a_1 - a_3)^2 + (a_2 - a_4)^2 > 0$, and this lucky fact suffices to complete the solution.

Solutions to the Exercises 253

SOLUTION FOR EXERCISE 6.17. By interpolation and convexity one has
$$x = \frac{b-x}{b-a}a + \frac{x-a}{b-a}b \Rightarrow f(x) \leq \frac{b-x}{b-a}f(a) + \frac{x-a}{b-a}f(b)$$
so, after subtracting $f(a)$, we find
$$f(x) - f(a) \leq \frac{x-a}{b-a}\{f(b) - f(a)\}. \tag{14.56}$$
This gives us the second inequality of (6.30), and the second is proved in the same way.

SOLUTION FOR EXERCISE 6.18. Let $g(h) = \{f(x+h) - f(x)\}/h$ and check from the Three Chord Lemma that for $0 < h_1 < h_2$ one has $g(h_1) \leq g(h_2)$. Next choose y with $a < y < x$ and use the Three Chord Lemma to check that $-\infty < \{f(x) - f(y)\}/\{x-y\} \leq g(h)$ for all $h > 0$. The monotonicity and boundedness $g(h)$ guarantee that $g(h)$ has finite limit as $h \to 0$. This gives us the first half of the problem, and the second half almost identical.

SOLUTION FOR EXERCISE 6.19. This is just more handy work of the Three Chord Lemma which gives us for $0 < s$ and $0 < t$ with $y - s \in I$ and $y + t \in I$ that $\{f(y) - f(y-s)\}/s \leq \{f(y+t) - f(y)\}/t$. From Exercise 6.18 we have that finite limits as $s, t \to 0$, and these limits are $f'_-(y)$ and $f'_+(y)$ respectively. This gives us $f'_-(y) \leq f'_+(y)$ and the other bounds are no harder. Incidentally, the bound $f'_-(y) \leq f'_+(y)$ may be regarded as an "infinitesimal" version of the Three Chord Lemma.

For $a < x \leq s \leq t \leq y < b$ and $M = \max\{|f'_+(x)|, |f'_-(y)|\}$ the bound (6.31) gives us $|f(t) - f(s)| \leq M|t - s|$, which is more than we need to say that f is continuous.

CHAPTER 7: INTEGRAL INTERMEZZO

SOLUTION FOR EXERCISE 7.1. The substitution gives us
$$2f(x)f(y)g(x)g(y) \leq f^2(x)g^2(y) + f^2(y)g^2(x),$$
so integration over $[a,b] \times [a,b]$ yields
$$2\int_a^b f(x)g(x)\,dx \int_a^b f(y)g(y)\,dy$$
$$\leq \int_a^b f^2(x)\,dx \int_a^b g^2(y)\,dy + \int_a^b f^2(y)\,dy \int_a^b g^2(x)\,dx,$$
which we recognize to be Schwarz inequality once it is rewritten with only a single dummy variable.

This derivation was suggested by Claude Dellacherie who also notes that the continuous version of Lagrange's identity (14.50) follows by a similar calculation provided that one begins with $(u-v)^2 = u^2+v^2-2uv$.

SOLUTION FOR EXERCISE 7.2. Setting $D(f,g) = A(fg) - A(f)A(g)$ we have the identity

$$D(f,g) = \int_{-\infty}^{\infty} \{f(x) - A(f)\}\, w^{\frac{1}{2}}(x) \{g(x) - A(g)\}\, w^{\frac{1}{2}}(x)\, dx,$$

and Schwarz's inequality gives $D^2(f,g) \leq D(f,f)D(g,g)$ which is our target bound.

SOLUTION FOR EXERCISE 7.3. We first note that without loss of generality we can assume that both of the integrals on the right of Heisenberg's inequality are finite, or else there is nothing to prove. The inequality (7.11) of Problem 7.3 then tells us that $f^2(x) = o(x)$ as $|x| \to \infty$, so starting with the general integration by parts formula

$$\int_{-A}^{B} f^2(x)\, dx = \Big|_{-A}^{B} xf^2(x) - -2\int_{-A}^{B} xf(x)f'(x)\, dx,$$

we can let $A, B \to \infty$ to deduce that

$$\int_{-\infty}^{\infty} |f(x)|^2\, dx = -2\int_{-\infty}^{\infty} xf(x)f'(x)\, dx \leq 2\int_{-\infty}^{\infty} |xf(x)|\, |f'(x)|\, dx.$$

Schwarz's inequality now finishes the job.

SOLUTION FOR EXERCISE 7.4. One applies Jensen's inequality (7.19) to the integrals in turn:

$$\int_b^{b+1} \frac{dx}{x+y} > \frac{1}{b+\frac{1}{2}+y} \quad \text{and} \quad \int_a^{a+1} \frac{dy}{b+\frac{1}{2}+y} > \frac{1}{b+a+1}.$$

SOLUTION FOR EXERCISE 7.5. By differentiation under the integral sign we have

$$\left|\frac{d^4}{dx^4}\frac{\sin t}{t}\right| = \left|\int_0^1 s^4 \cos(st)\, ds\right| \leq \int_0^1 s^4\, ds = \frac{1}{5}.$$

To be complete, one should note that differentiation under the integral sign is legitimate since for $f(t) = \cos(st)$ once can check that the difference quotients $(f(t+h) - f(x))/h$ are uniformly bounded for all $0 \leq s \leq 1$ and $0 < h \leq 1$.

Solutions to the Exercises 255

SOLUTION FOR EXERCISE 7.6. By the pattern of Problem 7.4, we find

$$(B-A)^2 \leq cB^2 \log B \int_A^B \frac{dx}{f(x)},$$

so setting $A = 2^j$ and $B = 2^{j+1}$ one finds

$$\frac{1}{4c(j+1)\log 2} \leq \int_{2^j}^{2^{j+1}} \frac{dx}{f(x)} \quad \text{and} \quad \frac{1}{4c\log 2} \sum_{j=0}^n \frac{1}{j+1} \leq \int_1^{2^{n+1}} \frac{dx}{f(x)}.$$

The conclusion then follows by the divergence of the harmonic series.

SOLUTION FOR EXERCISE 7.7. If we set $\delta = f(t)/|f'(t)|$ then the triangle T determined by the points $(t, f(t))$, $(t, 0)$, and $(t, t+\delta)$ lies below the graph of f, so the integral in the bound (7.26) is at least as large as the area of T which is $\frac{1}{2} f^2(t)/|f'(t)|$.

SOLUTION FOR EXERCISE 7.8. Since $0 \leq \sin t \leq 1$ on $[0, \pi/2]$ we can slip $\sin t$ inside the integral to get a smaller one. Thus, we have

$$I_n \geq \int_0^{\pi/2} (1 + \cos t)^n \sin t \, dt = \int_0^1 (1+u)^n \, du = \frac{2^{n+1} - 1}{n+1}.$$

Similarly, one has $u/x \geq 1$ on $[x, \infty)$, so we have the bound

$$I'_n \leq \frac{1}{x} \int_x^\infty u e^{-u^2/2} \, du = \frac{1}{x} e^{-x^2/2}.$$

In each case one slips in a factor to make life easy. Factors that are bounded between 0 and 1 help us find lower bounds, and factors that are always at least 1 help us find upper bounds.

SOLUTION FOR EXERCISE 7.9. In order to argue by contradiction, we assume without loss of generality that there is a sequence $x_n \to \infty$ such that $f'(x_n) \geq \epsilon > 0$. Now, by Littlewood's Figure 7.2 (or the by triangle lower bound of Exercise 7.7), we note that

$$f(x_n + \delta) - f(x_n) = \int_{x_n}^{x_n+\delta} f'(t) \, dt \geq \frac{1}{2} \epsilon^2/B$$

where $B = \sup |f''(x)| < \infty$ and $\delta = \epsilon/B$. This bound implies that $f(x) \neq o(1)$, so we have our desired contradiction.

SOLUTION FOR EXERCISE 7.10. Differentiation suffices to confirm that on $(0, 1)$ the map $t \mapsto t^{-1} \log t$ is decreasing and $t \mapsto (1 + t^{-1}) \log t$ is increasing so we have the bounds

$$\int_x^1 \log(1+t)\,\frac{dt}{t} < (1-x)x^{-1}\log x$$

$$= \frac{1-x}{1+x}(1+x^{-1})\log(1+x) \le 2\log 2\,\frac{1-x}{1+x}.$$

To show $2\log 2$ cannot be replaced by a smaller constant, note that

$$\lim_{x\to 1}\frac{1}{1-x}\int_x^1 \log(1+t)\,\frac{dt}{t} = \log 2$$

since $|\log 2 - \log(1+t)/t| \le \epsilon$ for all x with $|1-x| \le \delta(\epsilon)$.

Solutions to the Exercises 257

SOLUTION FOR EXERCISE 7.11. If $W(x)$ is the integral of w on $[a,x]$, then $W(a) = 0$, $W(b) = 1$, and $W'(x) = w(x)$, so we have

$$\int_a^b \{\log W(x)\} w(x)\, dx = \int_0^1 \log v\, dv = -1.$$

We then have the relations

$$\exp \int_a^b \{\log f(x)\} w(x)\, dx = e \exp \int_a^b \{\log f(x) W(x)\} w(x)\, dx$$

$$\leq e \exp \int_a^b \{\log f(x)\} w(x)\, dx \leq e \int_a^b f(x) w(x)\, dx$$

where we used first the fact that $0 \leq W(x) \leq 1$ for all $x \in [a,b]$ and then we applied Jensen's inequality.

SOLUTION FOR EXERCISE 7.12. Setting I_f to the integrals of f we have

$$\int_0^1 (f(x) - I_f)^2\, dx = (A - I_f)(I_f - \alpha) - \int_0^1 (A - f(x))(f(x) - \alpha)\, dx$$

$$\leq (A - I_f)(I_f - \alpha),$$

and an analogous inequality holds for g. Schwarz's inequality then gives

$$\left| \int_0^1 f(x) g(x)\, dx - I_f I_g \right|^2 = \left| \int_0^1 (f(x) - I_f)(g(x) - I_g)\, dx \right|^2$$

$$\leq \int_0^1 (f(x) - I_f)^2\, dx \int_0^1 (g(x) - I_g)^2\, dx$$

$$\leq (A - I_f)(I_f - \alpha)(B - I_g)(I_g - \beta)$$

$$\leq \frac{1}{4}(A - \alpha)^2 \frac{1}{4}(B - \beta)^2,$$

where in the last step we used the fact that $(U-x)(x-L) \leq \frac{1}{4}(U-L)^2$ for all $L \leq x \leq U$. Finally, to see that Grüss's inequality is sharp, set $f(x) = 1$ for $0 \leq x \leq 1/2$, set $f(x) = 0$ for $1/2 < x \leq 1$, and set $g(x) = 1 - f(x)$ for all $0 \leq x \leq 1$.

CHAPTER 8: THE CONTINUUM OF MEANS

SOLUTION FOR EXERCISE 8.1. Part (a) follows immediately from the Harmonic Arithmetic inequality for equal weights applied to the 3-vector $(1/(y+z), 1/(x+z), 1/(x+y))$. For part (b), one fist notes by Chebyshev's order inequality that $1/3\{x^p/(y+z) + y^p/(x+z) + x^p(x+y)\}$ is bounded

below by the product
$$\left\{\frac{1}{3}(x^p + y^p + z^p)\right\}\left\{\frac{1}{3}\left(\frac{1}{y+z} + \frac{1}{x+z} + \frac{1}{x+y}\right)\right\}.$$
To complete the proof, one then applies the power mean inequality (with $s = 1$ and $t = p$) to lower bound the first factor, and one uses part (a) to lower bound the second factor.

SOLUTION FOR EXERCISE 8.2. By the upside-down HM-AM inequality (8.16) one has
$$\frac{n^2}{a_1 + a_2 + \cdots + a_n} \leq \frac{1}{a_1} + \frac{1}{a_2} + \cdots + \frac{1}{a_n}.$$
If we set $a_k = 2S - x_k$, then $a_1 + a_2 + \cdots + a_n = 2nS - S = (2n-1)S$, and the HM-AM bound yields
$$\frac{n^2}{(2n-1)S} \leq \frac{1}{2S - x_1} + \frac{1}{2S - x_2} + \cdots + \frac{1}{2S - x_n}.$$

SOLUTION FOR EXERCISE 8.3. Both sides of the bound (8.29) are homogeneous of order one in $(a_1, a_2, \ldots, a_n)$, so we can assume without loss of generality that $a_1^{1/3} + a_2^{1/3} + \cdots + a_n^{1/3} = 1$. Given this, we only need to show $a_1^{1/2} + a_2^{1/2} + \cdots + a_n^{1/2} \leq 1$, and this is remarkably easy. By the normalization, we have $a_k \leq 1$ for all $1 \leq k \leq n$, so we also have $a_k^{1/2} \leq a_k^{1/3}$ for all $1 \leq k \leq n$, and we just take the sum to get our target bound. One might want to reflect on what made this exercise so much easier than the proof of the power mean inequality (8.10). For part (b), if we take $f(x) = x^6$ to minimize arithmetic, then we see that the putative bound (8.30) falsely asserts $1/16 \leq 1/27$.

SOLUTION FOR EXERCISE 8.4. We only need to consider $p \in [a, b]$, and in that case we can write
$$F(p) = \max\left\{\frac{p-a}{a}, \frac{b-p}{b}\right\}.$$
The identity
$$\frac{a}{a+b}\left\{\frac{p-a}{a}\right\} + \frac{b}{a+b}\left\{\frac{b-p}{b}\right\} = \frac{b-a}{a+b}$$
tells us that $(b-a)/(a+b)$ is a weighted mean of $(p-a)/a$ and $(b-p)/b$, so we always have the bound
$$F(p) = \max\left\{\frac{p-a}{a}, \frac{b-p}{b}\right\} \geq \frac{b-a}{a+b}.$$

Moreover, we have strict inequality here unless $(p-a)/a = (b-p)/b$, so, as Pólya (1950) observed, the unique minimum of $F(p)$ is attained at $p^* = 2ab/(a+b)$, which is the harmonic mean of a and b.

SOLUTION FOR EXERCISE 8.5. For all $\mathbf{x} \in D$ we have the bound

$$(a_1 a_2 \cdots a_n)^{1/n} = (a_1 x_1 a_2 x_2 \cdots a_n x_n)^{1/n} \leq \frac{1}{n} \sum_{k=1}^{n} a_k x_k \qquad (14.57)$$

by the AM-GM inequality, and we have equality here if and only if $a_k x_k$ does not depend on k. If we take $x_k = a_k/(a_1 a_2 \cdots a_n)^{1/n}$, then $\mathbf{x} \in D$ and the equality holds in the bound (14.57). This is all one needs to justify the identity (8.33).

Now, to prove the the bound (2.31) on page 34, one now just notes

$$\min_{\mathbf{x} \in D} \frac{1}{n} \sum_{k=1}^{n} a_k x_k + \min_{\mathbf{x} \in D} \frac{1}{n} \sum_{k=1}^{n} b_k x_k \leq \min_{\mathbf{x} \in D} \frac{1}{n} \sum_{k=1}^{n} (a_k + b_k) x_k,$$

since two choices are better than one. Incidentally, this type of argument is exploited systematically in Beckenbach and Bellman (1965) where the formula (8.33) is called the *quasilinear representation* of the geometric mean.

SOLUTION FOR EXERCISE 8.6. The half-angle formula for sine gives

$$\frac{\sin x}{x} = \frac{2\sin(x/2)\cos(x/2)}{x} = \cos(x/2)\left\{\frac{\sin(x/2)}{x/2}\right\}$$

$$= \cos(x/2)\cos(x/4)\left\{\frac{\sin(x/4)}{x/4}\right\}$$

$$= \cos(x/2)\cos(x/4)\cdots\cos(x/2^k)\left\{\frac{\sin(x/2^k)}{x/2^k}\right\},$$

and as $k \to \infty$ the bracketed term goes to 1 since $\sin t = t + O(t^3)$ as $t \to 0$. Upon setting $x = \pi/2$, one gets the second formula after computing the successive values of cosine with help from its half-angle formula. Naor (1998, pp. 139–143) gives a full discussion of Viète's formula, including a fascinating geometric proof.

SOLUTION FOR EXERCISE 8.7. Our assumptions give us the bound $(f(t_0 + h) - f(t_0))/h \geq 0$ for all $h \in (0, \Delta]$, and now we just let $h \to 0$ to prove the first claim. To address the second claim, one first notes by the power mean inequality, or by Jensen's inequality, that one has

$$f(t) = \sum_{k=1}^{n} p_k x_k^t - \left(\sum_{k=1}^{n} p_k x_k\right)^t \geq 0 \qquad \text{for all } t \in [1, \infty).$$

Since $0 = f(1) \leq f(t)$ for $1 \leq t$, we also have $f'(1) \geq 0$, and this is precisely the bound (8.35).

SOLUTION FOR EXERCISE 8.8. We argue by contradiction, and we begin by assuming that $(a_{1k}, a_{2k}, \ldots, a_{nk})$ does not converge to the constant limit $\vec{\mu} = (\mu, \mu, \ldots, \mu)$. For each j, the sequence $\{a_{jk} : k = 1, 2, \ldots\}$ is bounded so we can find a subsequence k_s, $s = 1, 2, \ldots$, such that $(a_{1k_s}, a_{2k_s}, \ldots, a_{nk_s})$ converges to $\vec{\nu} = (\nu_1, \nu_2, \ldots, \nu_n)$ with $\vec{\nu} \neq \vec{\mu}$. Letting $s \to \infty$ and applying hypotheses (i) and (ii), we find

$$\frac{\nu_1 + \nu_2 + \cdots + \nu_n}{n} = \mu \quad \text{and} \quad \frac{\nu_1^p + \nu_2^p + \cdots + \nu_n^p}{n} = \mu^p.$$

Now, by Problem 8.1 we see from these two identities and the case of equality in the power mean inequality that imply $\nu_j = \mu$ for all j, but this contradicts our assumption $\vec{\nu} \neq \vec{\mu}$, so the proof is complete.

Niven and Zuckerman (1951) consider only $p = 2$, and in this case Knuth (1968, p. 135) notes that one can give a very easy proof by considering the sum $\sum (a_{jk} - \mu)^2$. The benefit of the subsequence argument is that it works for all ℓ^p with $p > 1$, and, more generally, it reminds us that there many situations where the characterization of the case of equality can be used to prove a limit theorem.

Subsequence arguments often yield a *qualitative* stability result while assuming little more than the ability to identify the case where equality holds. When more is known, specialized arguments may yield more powerful *quantitative* stability results; here the two leading examples are perhaps the stability result for the AM-GM inequality (page 35) and the stability result for Hölder's inequality (page 144).

SOLUTION FOR EXERCISE 8.9. First notes that the hypothesis yields the telescoping relationship,

$$\sum_{i=1}^{n-k}(x_{i+k} - x_i) = (x_n + x_{n-1} + \cdots + x_{n-k+1}) - (x_1 + x_2 + \cdots + x_k) \leq 2k,$$

so the inverted HM-AM inequality (8.16) gives us the informative bound

$$\sum_{i=1}^{n-k} \frac{1}{x_{i+k} - x_i} \geq \frac{(n-k)^2}{2k}.$$

Now, by summation we have

$$\sum_{1\leq j<k\leq n} \frac{1}{x_j - x_k} = \sum_{k=1}^{n-1}\sum_{i=1}^{n-k} \frac{1}{x_{i+k} - x_i}$$

$$\geq \sum_{k=1}^{n-1} \frac{(n-k)^2}{2k} = \frac{n^2}{2}\left(H_{n-1} - \frac{1}{2} + \frac{1}{2n}\right),$$

so the bound $H_{n-1} = 1 + \frac{1}{2} + \cdots + \frac{1}{n-1} > \int_1^n dx/x = \log n$ completes the first part.

For the second part, we note that for any permutation σ one has

$$\sum_{1\leq j<k\leq n} \frac{1}{x_j - x_k} = \sum_{1<k\leq n}\sum_{j=1}^{k-1} \frac{1}{|x_{\sigma(k)} - x_{\sigma(j)}|}$$

$$\leq (n-1) \max_{1<k\leq n} \sum_{j=1}^{k-1} \frac{1}{|x_{\sigma(k)} - x_{\sigma(j)}|}.$$

This argument of Erdős (1961, p. 237) speaks volumes about the rich possibilities of simple averages.

CHAPTER 9: HÖLDER'S INEQUALITY

SOLUTION FOR EXERCISE 9.1. For the second bound one applies Hölder's inequality with $p = 5/4$ and $q = 5$ and finishes with the telescoping identity $1/(1\cdot 2) + 1/(2\cdot 3) + \cdots + 1/\{n(n+1)\} = 1 - 1/(n+1)$. For the second bound one uses $p = 3/4$ and $q = 4$ and finishes with Euler's classic sum $1 + 1/2^2 + 1/3^2 + \cdots = \pi^2/6$. While for the third bound one uses $p = 3/2$ and $q = 3$ and finishes with the geometric sum $1 + x^3 + x^3 + \cdots = 1/(1-x^3)$.

SOLUTION FOR EXERCISE 9.2. Consider z such that $|z| > 1$ and note by Hölder's inequality that one has the bound

$$\left|\sum_{n=0}^{n-1} a_j z^j\right| \leq A_p \left(\sum_{n=0}^{n-1} |z|^{jq}\right)^{1/q}, \quad \text{so we also have}$$

$$|P(z)| \geq |z|^n \left(1 - A_p \left(\sum_{n=0}^{n-1} \frac{1}{|z|^{(n-j)q}}\right)^{1/q}\right), \quad \text{and by summation}$$

$$\sum_{n=0}^{n-1} \frac{1}{|z|^{(n-j)q}} < \sum_{n=0}^{\infty} \frac{1}{|z|^{jq}} = \frac{1}{|z|^q - 1}.$$

Thus, we have $|P(z)| > 0$ if $A_p/(|z|^q - 1)^{1/q} \leq 1$. That is, we have $|P(z)| > 0$ for all z such that $|z| > (1 + A_p^q)^{1/q}$.

The bound (9.29) for the inclusion radius is due to M. Kuniyeda, and it provides a useful reminder how one can benefit from the flexibility afforded by Hölder's inequality. Here, for a given polynomial, a wise choice of the power p sometimes leads to an inclusion radius that is dramatically smaller than the one given simply by taking $p = 2$. This result and many other bounds for the inclusion radius are developed in Mignotte and Ştefănescu (1999).

SOLUTION FOR EXERCISE 9.3. This method is worth understanding, but the exercise does not leave much to do. First apply Cauchy's inequality to the sum of $\alpha_j \beta_j$ where $\alpha_j = a_j b_j c_j d_j$ and $\beta_j = e_j f_j g_j h_j$, then repeat the natural splitting twice more. It is obvious (but easy to overlook!) that each $p \in [1, \infty)$ can be approximated as closely as we like by a rational number of the form $p = 2^k/j$ where $1 \leq j < 2^k$.

SOLUTION FOR EXERCISE 9.4. Apply the Hölder inequality given by Exercise 9.7 with $D = [0, \infty)$ and $w(x) = \phi(x)$ with the natural choices $f(x) = x^{(1-\alpha)t_0}$, $g(x) = x^{\alpha t_1}$, $p = 1/(1-\alpha)$ and $q = 1/\alpha$. One consequence of this bound is that if the tth moment is infinite, then either t_0th or t_1th moment must be infinite.

SOLUTION FOR EXERCISE 9.5. Equality in the bound (9.30) gives us

$$\left| \sum_{k=1}^n a_k b_k \right| = \sum_{k=1}^n |a_k b_k| = \left(\sum_{k=1}^n |a_k|^p \right)^{1/p} \left(\sum_{k=1}^n |b_k|^q \right)^{1/q}. \quad (14.58)$$

Now, if $|a_1|, |a_2|, \ldots, |a_n|$ is a nonzero sequence, then the real variable characterization on page 136 tells us that the second equality holds if and only if there exists a constant $\lambda \geq 0$ such that $\lambda |a_k|^{1/p} = |b_k|^{1/q}$ for all $1 \leq k \leq n$.

The novel issue here is to discover when the first equality holds. If we set $a_k b_k = \rho_k e^{i\theta}$ where $\rho_k \geq 0$ and $\theta_k \in [0, 2\pi)$ and if we further set $p_k = \rho_k/(\rho_1 + \rho_2 + \cdots + \rho_n)$, then the first equality holds exactly when the average $p_1 e^{i\theta_1} + p_2 e^{i\theta_2} + \cdots + p_n e^{i\theta_n}$ is on the boundary of the unit disk, and this is possible if and only if there exists a θ such that $\theta = \theta_k$ for all k such that $p_k \neq 0$. In other words, the first equality holds if and only if the values $\arg\{a_k b_k\}$ are equal for all k for which $\arg\{a_k b_k\}$ is well defined.

SOLUTION FOR EXERCISE 9.6. One checks by taking derivatives that $\phi''(x) = (1-p)x^{-2+1/p}(1+x^{1/p})^{-2+p}/p$, and this is negative since $p > 1$

Solutions to the Exercises 263

and $x \geq 0$. One then applies Jensen's inequality (for concave functions) to $w_k = |a_k|^p$ and $x_k = |b_k|^p/|a_k|^p$; the rest is arithmetic. This modestly miraculous proof is just one move example of how much one can achieve with Jensen's inequality, given the wisdom to chose the "right" function.

SOLUTION FOR EXERCISE 9.7. Without lost of generality, one can assume that the integrals of the upper bound do not vanish. Call these integrals I_1 and I_2, apply Young's inequality (9.6) to $u = |f(x)|/I_1^{1/p}$ and $|g(x)|/I_2^{1/q}$, multiply by $w(x)$, and integrate. Hölder's inequality then follows by arithmetic. For a thorough job, one may want to retrace this argument to sort out the case of equality.

SOLUTION FOR EXERCISE 9.8. The natural calculus exercise shows the Legendre transform of $f(x) = x^p/p$ is $g(y) = y^q/q$ where $q = p/(p-1)$. Thus, the bound (9.33) simply puts Young's inequality (9.6) into a larger context. Similarly, one finds the Legendre transform pair:

$$f(x) = e^x \mapsto g(y) = y \log y - y \quad \text{and} \quad \phi(x) = x \log x - x \mapsto \gamma(y) = e^y.$$

This example suggests the conjecture that for a convex function, the Legendre transform of its Legendre is the original function. This conjecture is indeed true. Finally, for part (c), we take $0 \leq p \leq 1$ and note that $g(py_1 + (1-p)y_2) = \sup_{x \in D}\{x(py_1 + (1-p)y_2) - f(x)\}$ also equals $\sup_{x \in D} \left(p\{(xy_1 - f(x)\} + (1-p)\{xy_2 - f(x)\} \right)$. Since this is bounded by $\sup_{x \in D} p\{(xy_1 - f(x)\} + \sup_{x \in D}(1-p)\{xy_2 - f(x)\}$ which equals $pg(y_1) + (1-p)g(y_2)$, we see that g is convex.

SOLUTION FOR EXERCISE 9.9. Part (a) follows by applying Hölder's inequality for the conjugate pair $(p/r, q/r)$ to the splitting $a_j^r \cdot b_j^r$. Part (b) can be obtained by two similar applications of Hölder's inequality, but one saves arithmetic and gains insight by following Riesz's pattern. By the AM-GM inequality one has $xyz \leq x^p/p + y^q/q + z^r/r$ and, after applying this to the corresponding normalized values $\hat{a}_j, \hat{b}_j$, and $\hat{c}_j$, one can finish exactly as before.

SOLUTION FOR EXERCISE 9.10. The historical Hölder inequality follows directly from the weighted Jensen inequality (9.31) with $\phi(x) = x^p$, a proof which suggests why Hölder might have viewed the inequality (9.31) as his main result.

To pass from the bound (9.34) to the modern Hölder inequality (9.1), one takes $w_k = b_k^q$ and $y_k = a_k/b_k^{q-1}$. To pass from the bound (9.1) to the historical Hölder inequality, one uses the splitting $a_k b_k = \{w_k^{1/p} y_k\}\{w_k^{1/q}\}$.

264 Solutions to the Exercises

SOLUTION FOR EXERCISE 9.11. The hint leads one to the bound

$$\sum_{k=1}^{n}\left\{\theta a_{k}^{p/s}+(1-\theta)b_{k}^{q/s}\right\}^{s} \leq \left\{\theta\left\{\sum_{k=1}^{n}a_{j}^{p}\right\}^{1/s}+(1-\theta)\left\{\sum_{k=1}^{n}b_{k}^{q}\right\}^{1/s}\right\}^{s}$$

and, if we let $s \to \infty$, then the formula (8.5) gives us

$$\sum_{k=1}^{n}a_{k}^{\theta p}b_{k}^{(1-\theta)q} \leq \left\{\sum_{k=1}^{n}a_{k}^{p}\right\}^{\theta}\left\{\sum_{k=1}^{n}b_{k}^{q}\right\}^{1-\theta}, \qquad (14.59)$$

which is Hölder's inequality after one sets $\theta = 1/p$. This derivation of Kedlaya (1999) and Maligranda (2000) serves as a reminder that the formula (8.5) gives us another general tool for effecting the "additive to multiplicative" transformation which is often needed in the theory of inequalities.

SOLUTION FOR EXERCISE 9.12. Fix m and use induction on n. For $n = 1$ we have $w_1 = 1$, and the inequality is trivial. For the induction step apply Hölder's inequality to $u_1v_1 + u_2v_2 + \cdots + u_mv_m$ where

$$u_j = \prod_{k=1}^{n-1}a_{jk}^{w_k}, \ v_j = a_{jn}^{w_n}, \ p = 1/(w_1 + w_2 + \cdots + w_{n-1}), \text{ and } q = 1/w_n.$$

This gives us the bound

$$\sum_{j=1}^{m}\prod_{k=1}^{n}a_{jk}^{w_k} \leq \left\{\sum_{j=1}^{m}\prod_{k=1}^{n-1}a_{jk}^{w_k/(w_1+\cdots+w_{n-1})}\right\}^{w_1+\cdots+w_{n-1}}\left(\sum_{j=1}^{m}a_{jn}\right)^{w_n},$$

and the proof is then completed by applying the induction hypothesis to the bracketed sum of $(n-1)$-fold products.

To prove the inequality (9.36), we first apply the bound (9.35) with $w_1 = w_2 = w_3 = 1/3$ to the array

$$A = \begin{pmatrix} x & x & x \\ x & \sqrt{xy} & y \\ x & y & z \end{pmatrix}$$

to find $x + (xy)^{\frac{1}{2}} + (xyz)^{\frac{1}{3}} \leq \{(3x)(x + y + \sqrt{xy})(x + y + z)\}^{1/3}$. Now, by applying $\sqrt{xy} \leq (x+y)/2$ one finds

$$3x(x + y + \sqrt{xy})(x + y + z) \leq 27x \cdot \frac{x+y}{2} \cdot \frac{x+y+z}{3},$$

and this completes the proof of the bound (9.36) suggested by Lozansky and Rousseau (1996, p. 127). The bound (9.36) is due to Finbarr Holland

Solutions to the Exercises 265

who also conjectured the natural n-variable analogue, a result which was subsequently proved by Kiran Kedlaya.

SOLUTION FOR EXERCISE 9.13. Taking the bound (9.38) and the inversion of the bound (9.39) one finds

$$\left(S_s/S_r\right)^{S_s/(s-r)} \leq b_1^{a_1 b_1^s} b_2^{a_2 b_2^s} \cdots b_n^{a_n b_n^s} \leq \left(S_t/S_s\right)^{S_s/(t-s)},$$

which we can write more leanly as

$$\left(S_s/S_r\right)^{1/(s-r)} \leq \left(S_t/S_s\right)^{1/(t-s)} \quad \text{or} \quad S_s^{t-r} \leq S_r^{t-s} S_t^{s-r}, \quad \text{as claimed.}$$

SOLUTION FOR EXERCISE 9.14. As in Problem 9.6 we begin by noting that scaling and the Converse Hölder inequality imply that it suffices to show that we have the bound

$$\sum_{j=1}^{m} \sum_{k=1}^{n} c_{jk} x_k y_j \leq M_\theta \quad \text{for all } \|\mathbf{x}\|_s \leq 1 \text{ and } \|\mathbf{y}\|_{t'} \leq 1, \quad (14.60)$$

where $t' = t/(t-1)$ is the conjugate power for t (so $1/t + 1/t' = 1$). Also, just as before, the assumption that $c_{jk} \geq 0$ for all j, k implies that it suffices for us to consider nonnegative values for x_j and y_k. To continue with the earlier pattern, we need to set up the splitting trick. Here there are many possibilities, and an unguided search can be frustrating but there are some observations that can help direct our search.

First, we know that we must end up with the sum of the x_k^s and the sum of the $y_j^{t'}$ separated from the c_{jk} factors; this is the only way we can use the hypotheses that $\|x\|_p \leq 1$ and $\|y\|_{t'} \leq 1$. Also, the definition of the splitting will surely need to exploit the defining relations (9.42) for the three variables s, t, and θ.

When we try to combine these hints, we may note that

$$(1-\theta)\frac{s}{s_0} + \theta\frac{s}{s_1} = s\frac{1}{s} = 1$$

while for the conjugate powers $t' = t/(t-1)$, $t'_0 = t_0/(t_0 - 1)$, and $t'_1 = t_1/(t_1 - 1)$ we have the analogous relation

$$(1-\theta)\frac{t'}{t'_0} + \theta\frac{t'}{t'_1} = t'\left\{(1-\theta)(1 - 1/t'_0) + \theta(1 - 1/t'_1)\right\} = t'\frac{1}{t'} = 1.$$

Now, we just need use these relations to create a splitting of $c_{jk} x_k y_j$ which will bring the sums of x_k^s and $y_j^{t'}$ in to view after an applications of Hölder's inequality. With just a little experimentation, one should then find the bound

$$\sum_{j=1}^{m}\sum_{k=1}^{n} c_{jk}x_k y_j = \sum_{j=1}^{m}\sum_{k=1}^{n}(c_{jk}x_k^{s/s_0} y_j^{t'/t'_0})^{1-\theta}(c_{jk}x_k^{s/s_1} y_j^{t'/t'_1})^{\theta}$$

$$\leq \left(\sum_{j=1}^{m}\left\{\sum_{k=1}^{n} c_{jk}x_k^{s/s_0}\right\} y_j^{t'/t'_0}\right)^{1-\theta}$$

$$\times \left(\sum_{j=1}^{m}\left\{\sum_{k=1}^{n} c_{jk}x_k^{s/s_1}\right\} y_j^{t'/t'_1}\right)^{\theta}.$$

This bound is a grand champion among splitting trick estimates and, after our eyes adjust to the clutter, we see that it is the natural culmination of a line of argument which we have used several times before.

To complete our estimate, we need to bound the last two factors. For the first factor we naturally want to apply Hölder's inequality for the conjugate pair t_0 and $t'_0 = t_0/(t_0 - 1)$. We then find that

$$\sum_{j=1}^{m}\left(\sum_{k=1}^{n} c_{jk}x_k^{s/s_0}\right) y_j^{t'/t'_0} \leq \left(\sum_{j=1}^{m}\left(\sum_{k=1}^{n} c_{jk}x_k^{s/s_0}\right)^{t_0}\right)^{1/t_0}\left(\sum_{j=1}^{m} y_j^{t'}\right)^{1/t'_0}$$

$$\leq M_0\left(\sum_{k=1}^{n} x_k^s\right)^{1/s_0}\left(\sum_{j=1}^{m} y_j^{t'}\right)^{1/t'_0} \leq M_0,$$

where in the second inequality we applied the bound $\|T\mathbf{x}\|_{t_0} \leq M_0\|\mathbf{x}\|_{s_0}$ to the vector $\mathbf{x} = (x_1^s, x_2^s, \ldots, x_n^s)$. We can then bound the second factor in exactly the same way to find

$$\sum_{j=1}^{m}\left\{\sum_{k=1}^{n} c_{jk}x_k^{s/s_1}\right\} y_j^{t'/t'_1} \leq M_1,$$

so when we return to our first bound we have the estimate

$$\sum_{j=1}^{m}\sum_{k=1}^{n} c_{jk}x_k y_j \leq M_1^{\theta} M_0^{1-\theta}.$$

This is exactly what we needed to complete the solution.

SOLUTION FOR EXERCISE 9.15. One can proceed barehanded, but it is also instructive to apply the result of the preceding exercise. From the hypothesis we have $\|T\mathbf{x}\|_2 \leq \|\mathbf{x}\|_2$, and from the definition of M one finds $\|T\mathbf{x}\|_\infty \leq M\|\mathbf{x}\|_1$. Since the linear system

$$\left(\frac{1}{p},\frac{1}{q}\right) = \theta\left(\frac{1}{1},\frac{1}{\infty}\right) + (1-\theta)\left(\frac{1}{2},\frac{1}{2}\right),$$

has $\theta = (2-p)/p \in [0,1]$ as its unique solution, the bound (9.43) is indeed a corollary of Exercise 9.14.

CHAPTER 10: HILBERT'S INEQUALITY

SOLUTION FOR EXERCISE 10.1. The proof fits in a single line: just note

$$\sum_{j,k,=1}^{n} a_j a_k x^j x^k = \left(\sum_{j=1}^{n} a_j x_j\right)^2 \geq 0$$

and integrate over $[0,1]$. For the general case, one naturally uses the representation $1/\lambda_j = \int_0^1 x^{\lambda_j} dx$.

This problem is a reminder that there are many circumstances where dramatic progress is made possible by replacing a number (or a function) with an appropriate integral. Although this example and the one given by Exercise 7.5, page 116, are simple, the basic theme has countless variations. Some of these variation are quite deep.

SOLUTION FOR EXERCISE 10.2. We substitute, switch order, apply the bound (10.18), switch again, and finish with Cauchy's inequality to find

$$\left|\sum_{j,k} a_{jk} h_{jk} x_j y_k\right|$$

$$= \left|\int_D \left(\sum_{j,k} a_{jk} x_j f_j(x) y_k g_k(x)\right) dx\right|$$

$$\leq \int_D M \left(\sum_j |x_j f_j(x)|^2\right)^{1/2} \left(\sum_k |y_k g_k(x)|^2\right)^{1/2} dx$$

$$\leq M \left(\sum_j x_j^2 \int_D |f_j(x)|^2 dx\right)^{1/2} \left(\sum_k y_k^2 \int_D |g_k(x)|^2 dx\right)^{1/2}.$$

The bound $\alpha\beta M \|\mathbf{x}\|_2 \|\mathbf{y}\|_2$ now follows from the assumption (10.21).

SOLUTION FOR EXERCISE 10.3. To mimic our proof of Hilbert's inequality we take $\lambda > 0$ and use the analogous splitting to find

$$\sum_{m=1}^{\infty} \sum_{n=1}^{\infty} \frac{a_m b_n}{\max(m,n)} = \sum_{m=1}^{\infty} \sum_{n=1}^{\infty} \frac{a_m b_n}{\max(m,n)} \left(\frac{m}{n}\right)^\lambda \left(\frac{n}{m}\right)^\lambda$$

$$= \sum_{m=1}^{\infty} \sum_{n=1}^{\infty} \frac{a_m}{\max^{\frac{1}{2}}(m,n)} \left(\frac{m}{n}\right)^\lambda \frac{b_n}{\max^{\frac{1}{2}}(m,n)} \left(\frac{n}{m}\right)^\lambda.$$

By Cauchy's inequality, the square of the double sum is bounded by the product of the sum given by

$$\sum_{m=1}^{\infty}\sum_{n=1}^{\infty} \frac{a_m^2}{\max(m,n)} \left(\frac{m}{n}\right)^{2\lambda} = \sum_{m=1}^{\infty} a_m^2 \sum_{n=1}^{\infty} \frac{1}{\max(m,n)} \left(\frac{m}{n}\right)^{2\lambda}$$

and the corresponding sum containing $\{b_n^2\}$. If we take $\lambda = 1/4$, we have

$$\sum_{n=1}^{\infty} \frac{1}{\max(m,n)} \left(\frac{m}{n}\right)^{\frac{1}{2}} = \sum_{n=1}^{m} \frac{1}{m} \left(\frac{m}{n}\right)^{\frac{1}{2}} + \sum_{n=m+1}^{\infty} \frac{1}{n} \left(\frac{m}{n}\right)^{\frac{1}{2}}$$

$$\le \frac{1}{\sqrt{m}} \sum_{n=1}^{m} \frac{1}{\sqrt{n}} + \sqrt{m} \sum_{n=m+1}^{\infty} \frac{1}{n^{3/2}}$$

$$\le \frac{1}{\sqrt{m}} 2\sqrt{m} + \sqrt{m}\, 2\frac{1}{\sqrt{m}} \le 4$$

so, to complete the proof we only need to note that the $\{b_n^2\}$ sum satisfies an exactly analogous bound.

Finally, the usual "stress testing" method shows that 4 cannot be replaced with a smaller value. After setting $a_n = b_n = n^{-\frac{1}{2}-\epsilon}$, one checks that

$$\sum_{n=1}^{\infty} a_n = \frac{1}{2\epsilon} + O(1) \quad \text{and} \quad \sum_{m=1}^{\infty}\sum_{n=1}^{\infty} \frac{a_m a_n}{\max(m,n)} = \frac{2}{\epsilon} + O(1).$$

Peeking ahead and taking $K(x,y) = 1/\max(x,y)$ in Exercise 10.5, one finds that the constant 4 in the bound (10.3) is perhaps best understood when interpreted as the integral

$$4 = \int_0^{\infty} \frac{1}{\sqrt{u}} \frac{1}{\max(1,u)}\, du.$$

SOLUTION FOR EXERCISE 10.4. One can repeat the proof of the discrete case line-by-line, and to do so is worth one's time. The parallel between the discrete and continuous problems is really quite striking.

SOLUTION FOR EXERCISE 10.5. The first step exploits the homogeneity

Solutions to the Exercises

condition of $K(x,y)$ by a homogeneous change of variables $y = ux$:

$$I = \int_0^\infty \int_0^\infty f(x)K(x,y)g(y)\,dx dy$$

$$= \int_0^\infty f(x) \left\{ \int_0^\infty K(x,y)g(y)\,dy \right\} dx$$

$$= \int_0^\infty f(x) \left\{ \int_0^\infty K(x,ux)g(ux)x\,du \right\} dx$$

$$= \int_0^\infty f(x) \left\{ \int_0^\infty K(1,u)g(ux)\,du \right\} dx$$

$$= \int_0^\infty K(1,u) \left\{ \int_0^\infty f(x)g(ux)\,dx \right\} du.$$

Now, once K has been pulled outside, we can apply Schwarz's inequality to the inside integral to find

$$\int_0^\infty f(x)g(ux)\,dx \leq \left(\int_0^\infty |f(x)|^2\,dx \right)^{\frac{1}{2}} \left(\int_0^\infty |g(ux)|^2\,dx \right)^{\frac{1}{2}}$$

$$= \left(\int_0^\infty |f(x)|^2\,dx \right)^{\frac{1}{2}} \frac{1}{\sqrt{u}} \left(\int_0^\infty |g(v)|^2\,dv \right)^{\frac{1}{2}},$$

so we see at last that

$$I \leq \int_0^\infty K(1,u) \frac{1}{\sqrt{u}}\,du \cdot \left(\int_0^\infty |f(x)|^2\,dx \right)^{\frac{1}{2}} \left(\int_0^\infty |g(v)|^2\,dv \right)^{\frac{1}{2}}.$$

This completes the solution of the exercise with c given by the first of the three indicated integrals, and we can make a simple change of variables to check that all three of the integrals are equal.

This argument is yet another of the gems from Schur's remarkable 1911 paper. Actually, Schur proves the trickier finite range result,

$$\int_a^b f(x)K(x,y)g(y)\,dx dy \leq c \left(\int_a^b |f(x)|^2\,dx \right)^{\frac{1}{2}} \left(\int_a^b |g(y)|^2\,dy \right)^{\frac{1}{2}},$$

where $0 \leq a < b < \infty$. In this case, the domain of integration changes with the change of variables, but the original plan still works.

SOLUTION FOR EXERCISE 10.6. Integral comparison gives part (a) by

$$\frac{t}{t^2+1^2} + \frac{t}{t^2+2^2} + \cdots + \frac{t}{t^2+n^2} < \int_0^n \frac{t}{t^2+x^2}\,dx \leq \int_0^\infty \frac{dy}{1+y^2} = \pi/2,$$

270   Solutions to the Exercises

and for part (b) we note that

$$\sum_{k=1}^n a_k^2 w_k(t) = t \sum_{k=1}^n a_k^2 + \frac{1}{t}\sum_{k=1}^n k^2 a_k^2 = tA + \frac{1}{t}B$$

is minimized by taking $t = (B/A)^{\frac{1}{2}}$. Part (c) just assembles the pieces.

SOLUTION FOR EXERCISE 10.7. Since $|t - \pi| \leq \pi$ for $t \in [0, 2\pi]$ we have

$$|I| \leq \pi\left\{\frac{1}{2\pi}\int_0^{2\pi}\left|\sum_{k=1}^N a_k e^{ikt}\right|\left|\sum_{k=1}^N b_k e^{ikt}\right|dt\right\}$$

$$\leq \pi\left\{\frac{1}{2\pi}\int_0^{2\pi}\left|\sum_{k=1}^N a_k e^{ikt}\right|^2 dt\right\}^{1/2}\left\{\frac{1}{2\pi}\int_0^{2\pi}\left|\sum_{k=1}^N b_k e^{ikt}\right|^2 dt\right\}^{1/2}$$

$$= \pi\left\{\sum_{k=1}^n a_k{}^2\right\}^{1/2}\left\{\sum_{k=1}^n b_k{}^2\right\}^{1/2}.$$

This remarkably quick way of obtaining Hilbert's inequality is known as Toeplitz's method. Hilbert's original proof also used trigonometric integrals, but those used by Hilbert were not quite as efficient. Toeplitz's argument tells us more generally that if φ is any bounded function on $[0, 2\pi]$ with Fourier coefficients c_n, $-\infty < n < \infty$, then one has the bound

$$\left|\sum_{m=1}^N\sum_{n=1}^N c_{m+n}\, a_m\, b_n\right| \leq \|\varphi\|_\infty \|a\|_2 \|b\|_2.$$

Integral representation can also be used to prove more distinctive generalizations of Hilbert's inequality. For example, if $\alpha \notin \mathbb{Z}$ one finds

$$\frac{1}{2\pi}\int_0^{2\pi} e^{i(n+\alpha)t}dt = \frac{1}{\pi(n+\alpha)}e^{i\alpha\pi}\sin\alpha\pi,$$

and this representation can be used to show

$$\left|\sum_{m=1}^N\sum_{n=1}^N \frac{a_m b_n}{m+n+\alpha}\right| \leq \frac{\pi}{|\sin\alpha\pi|}\|a\|_2 \|b\|_2.$$

Solutions to the Exercises 271

SOLUTION FOR EXERCISE 10.8. We substitute and change orders:

$$\int_0^\infty \frac{1}{1+y} \frac{1}{y^{2\lambda}} \, dy = \int_0^\infty \left\{ \int_0^\infty e^{-t(1+y)} \, dt \right\} \frac{1}{y^{2\lambda}} \, dy$$

$$= \int_0^\infty e^{-t} \left\{ \int_0^\infty e^{-ty} \frac{1}{y^{2\lambda}} \, dy \right\} dt$$

$$= \int_0^\infty e^{-t} \left\{ \frac{\Gamma(1-2\lambda)}{t^{1-2\lambda}} \right\} dt = \Gamma(2\lambda)\Gamma(1-2\lambda).$$

CHAPTER 11 HARDY'S INEQUALITY AND THE FLOP

SOLUTION FOR EXERCISE 11.1. By applying Hölder's inequality with $p = \beta/\alpha$ and $q = \beta/(\beta - \alpha)$ to the right side of the bound (11.20) we obtain the inequality

$$\int_0^T \varphi^\beta(x) \, dx \leq C \left(\int_0^T \varphi^\beta(x) \, dx \right)^{\alpha/\beta} \left(\int_0^T \psi^{\beta/(\beta-\alpha)}(x) \, dx \right)^{(\beta-\alpha)/\beta}.$$

There is no loss of generality if we assume that the first integral factor on the right is nonzero, so we may divide both sides by that factor. If we then raise both sides of the resulting bound to the power $\beta/(\beta - \alpha)$ to get our target bound (11.20).

It is only in the division step where we use the condition that φ is bounded. The inequality for bounded functions can then be used to prove a corresponding inequality for functions that need not be bounded. It is quite common in arguments that call on the flop for one to first consider bounded functions so that one can sidestep any inappropriate arithmetic with integrals that might be infinite.

SOLUTION FOR EXERCISE 11.2. By the AM-GM inequality we have $2x^3 \leq y^3 + y^2 x + yx^2 \leq y^3 + 2y^3/3 + x^3/3 + y^3/3 + 2x^3/3 = 2y^3 + x^3$. In this example, the higher power on the left made the transformation possible, but for the transformation to be nontrivial one also needed cooperation from the constant factors. If we replace 2 by 1/2 in the original problem, we only obtain the trivial bound $-x^3 \leq 4y^3$.

SOLUTION FOR EXERCISE 11.3. The hypothesis (11.21) and Schwarz's inequality give us

$$\int_{-\pi}^\pi v^4(\theta) \, d\theta \leq \int_{-\pi}^\pi u^4(\theta) \, d\theta + 6 \left\{ \int_{-\pi}^\pi u^4(\theta) \, d\theta \right\}^{\frac{1}{2}} \left\{ \int_{-\pi}^\pi v^4(\theta) \, d\theta \right\}^{\frac{1}{2}}$$

which is $x^2 \leq c^2 + 6cx$ with the natural identifications. If we solve this for the case of equality, we find $x = c(6 \pm \sqrt{40})/2$, so the hypothesis (11.21) implies the bound (11.22) if we take $A = \{(6 + \sqrt{40})/2\}^2$.

SOLUTION FOR EXERCISE 11.4. Only a few obvious changes are needed to convert the proof of the L^2 bound (11.1) to a proof for the corresponding L^p bound (11.23). By that analogy, we first note

$$I = \int_0^T \left\{ \frac{1}{x} \int_0^x f(u)\,du \right\}^p dx = -\frac{1}{p-1} \int_0^T \left\{ \int_0^x f(u)\,du \right\}^p (x^{1-p})' dx,$$

so integration by parts gives us the identity

$$I = \frac{p}{p-1} \int_0^T f(x) \left\{ \frac{1}{x} \int_0^x f(u)\,du \right\}^{p-1} dx - \left. \frac{x^{1-p}}{p-1} \left\{ \int_0^x f(u)\,du \right\}^p \right|_0^T.$$

As before, the boundary contribution at zero is zero, and the contribution at T is nonpositive; therefore, we have the bound

$$I \leq \frac{p}{p-1} \int_0^T f(x) \left\{ \frac{1}{x} \int_0^x f(u)\,du \right\}^{p-1} dx,$$

which is the L^p analog of the preflop L^2 bound (11.4). One now finishes with the L^p flop precisely as in Exercise 11.1 provided that one sets $\alpha = p - 1$ and $\beta = p$.

SOLUTION FOR EXERCISE 11.5. Without loss of generality we assume that $a_n \geq 0$ for all $n = 1, 2, \ldots$. We then set $A_n = a_1 + a_2 + \cdots + a_n$ and apply Cauchy's inequality followed by Hardy's inequality (11.9) to get

$$T = \sum_{n=1}^\infty a_n (A_n/n) \leq 2 \sum_{n=1}^\infty a_n^2.$$

We then finish with the even simpler bound

$$T \geq \sum_{n=1}^\infty a_n \sum_{m=1}^n \frac{a_m}{m+n} = \sum_{1 \leq m \leq n < \infty} \frac{a_m a_n}{m+n} \geq \frac{1}{2} S.$$

SOLUTION FOR EXERCISE 11.6. The solution of this problem is not easy, and here we follow the one provided by Richberg (1993) that begins with the observation that

$$\int_x^1 \int_x^1 \frac{1-(st)^N}{1-st}\,ds\,dt = \sum_{j=1}^N \int_x^1 \int_x^1 (st)^{j-1}\,ds\,dt = \sum_{j=1}^N \left(\frac{1-x^j}{j} \right)^2,$$

so our target inequality is equivalent to

$$\int_x^1 \int_x^1 \frac{1-(st)^N}{1-x^{2N}} \frac{ds\,dt}{1-st} < (4\log 2) \frac{1-x}{1+x}.$$

Solutions to the Exercises 273

This bound would follow from
$$\int_x^1 \int_x^1 \frac{ds\,dt}{1-st} < (4\log 2)\frac{1-x}{1+x},$$
and by a direct calculation one finds
$$\int_x^1 \int_x^1 \frac{ds\,dt}{1-st} = 2\int_x^1 \log(1+t)\frac{dt}{t},$$
so the proof of the our target inequality is reduced to showing
$$\int_x^1 \log(1+t)\frac{dt}{t} < (2\log 2)\frac{1-x}{1+x}.$$

This bound and the fact that it is sharp was already addressed in Exercise 7.10, so the solution is complete.

SOLUTION FOR EXERCISE 11.7. This observation is painfully obvious, but it seems necessary for completeness. The hypothesis gives us the bounds $b_1 \leq a_1, b_2 \leq a_2, \ldots, b_N \leq a_N$; thus, for all $1 \leq n \leq N$ we have $(b_1 b_2 \cdots b_n)^{1/n} \leq (a_1 a_2 \cdots a_n)^{1/n}$, which is more than we need. There are questions on infinite rearrangements which are subtle, but this is not one of them.

SOLUTION FOR EXERCISE 11.8.

From the convergence of the sum, we know that the sequence of remainders $r_n = a_{n+1}/(n+1) + a_{n+2}/(n+2) + a_{n+3}/(n+3) + \cdots$ must converge to zero as $n \to \infty$. When we write these terms in longhand,

$$\begin{aligned}
r_0 &= a_1 &+ a_2/2 &+ a_3/3 &\cdots + \cdots + \cdots &+ a_n/n &+ r_n \\
r_1 &= &a_2/2 &+ a_3/3 &\cdots + \cdots + \cdots &+ a_n/n &+ r_n \\
r_2 &= & &a_3/3 &\cdots + \cdots + \cdots &+ a_n/n &+ r_n \\
&\vdots & & & & & \\
r_{n-2} &= & & &a_{n-1}/(n-1) &+ a_n/n &+ r_n \\
r_{n-1} &= & & & &a_n/n &+ r_n,
\end{aligned}$$

we see they may be summed to yield the nice identity

$$(a_1 + a_2 + \cdots + a_n)/n = -r_n + (r_0 + r_1 + \cdots + r_{n-1})/n, \quad (14.61)$$

which makes the limit (11.25) routine.

CHAPTER 12: SYMMETRIC SUMS

SOLUTION FOR EXERCISE 12.1. If the roots of $P(x)$ are $x_1, x_2, \ldots, x_n$, then $a_{n-1}/a_n = (1/x_1 + 1/x_2 + \cdots + 1/x_n)$ and $a_1 = x_1 + x_2 + \cdots + x_n$

so we have $(a_{n-1}/a_n)^{-1} \le a_1/n$ by the HM-AM inequality (8.14). This exercise offers a basic reminder: facts for polynomial coefficients and facts for symmetric sums are almost in a one-to-one correspondence.

SOLUTION FOR EXERCISE 12.2. (a) By expansion and simplification, we see that we need to prove
$$6abc \le ac^2 + ab^2 + ba^2 + bc^2 + ca^2 + cb^2 \stackrel{\text{def}}{=} R,$$
and after setting $a = x_1$, $b = x_2$, and $c = x_3$ we also have
$$\sum_{\sigma \in S(3)} x_{\sigma(1)} x_{\sigma(2)} x_{\sigma(3)} = 6abc \quad \text{and} \quad \sum_{\sigma \in S(3)} x_{\sigma(1)} x_{\sigma(2)}^2 = R.$$

Since $(1,1,1) = \frac{1}{6}(2,1,0) + \frac{1}{6}(2,0,1) + \cdots + \frac{1}{6}(0,1,2)$ we have $(1,1,1)$ is in $H[(2,1,0)]$, so we may apply Muirhead's inequality.

(b) We have $(1,1,0,\ldots,0) = \frac{1}{2}(2,0,0,\ldots,0) + \frac{1}{2}(0,2,0,\ldots,0)$ so we have $(1,1,0,\ldots,0) \in H[(2,0,0,\ldots,0)]$, and by Muirhead's inequality it suffices to note that
$$\sum_{\sigma \in S(n)} a_{\sigma(1)} a_{\sigma(2)} = 2(n-2)! \sum_{1 \le j < k \le n} a_j a_k \quad \text{and} \quad \sum_{\sigma \in S(n)} a_{\sigma(1)}^2 = (n-1)! \sum_{j=1}^n a_j^2.$$

(c) Since the average $\{(1/2,1/2,0,\ldots,0) + \cdots + (0,\ldots,0,1/2,1/2)\}/\binom{n}{2}$ equals $(1/n, 1/n, \ldots, 1/n)$, it suffices by Muirhead's inequality to note
$$\sum_{\sigma \in S(n)} a_{\sigma(1)}^{1/n} \cdots a_{\sigma(n)}^{1/n} = n!(a_1 a_2 \cdots a_n)^{1/n} \quad \text{and}$$
$$\sum_{\sigma \in S(n)} a_{\sigma(1)}^{1/2} a_{\sigma(2)}^{1/2} = 2(n-2)! \sum_{1 \le j < k \le n} \sqrt{a_j a_k}.$$

SOLUTION FOR EXERCISE 12.3. Multiply the left side of the bound (12.23) by $(xyz)^{1/3}$ and consider the candidate inequality
$$x^{7/3} y^{1/3} z^{1/3} + x^{1/3} y^{7/3} z^{1/3} + x^{1/3} y^{1/3} z^{7/3} \le x^3 + y^3 + z^3. \quad (14.62)$$

This generalizes our original problem in the sense that if we can prove that the bound (14.62) holds for all nonnegative x, y, z then the bound (12.23) must hold when $xyz = 1$. Fortunately, the new bound (14.62) is a corollary of Muirhead's inequality and the relationship
$$(7/3, 1/3, 1/3) = \frac{7}{9}(3,0,0) + \frac{1}{9}(0,3,0) + \frac{1}{9}(0,0,3).$$

Kedlaya (1999) presents several more sophisticated examples of the homogenization device.

Solutions to the Exercises

SOLUTION FOR EXERCISE 12.4. By expanding the bound (12.24) we see after simplification that it is equivalent to the assertion that

$$\sum_{(j,k):j\neq k} x_j^m x_k^m \leq \sum_{(j,k):j\neq k} x_j^{m-1} x_k^{m+1}$$

but $(m, m, 0, \ldots, 0) = \frac{1}{2}(m+1, m-1, 0, \ldots, 0) + \frac{1}{2}(m-1, m+1, 0, \ldots, 0)$ so the bound (12.24) follows from Muirhead's inequality.

SOLUTION FOR EXERCISE 12.5. With surprising frequency, solvers of this exercise find the same example discovered by Bunyakovsky (1854):

$$p(x, y) = \{x^2 + (1-y)^2\}\{y^2 + (1-x)^2\}.$$

Here one has $p(1, 0) = 0$ and $p(0, 1) = 0$ but otherwise $p(x, y)$ is strictly positive. Thus, despite the symmetry of p, the minimum of p is not on the diagonal $D = \{(x, y) : x = y\}$. Incidentally, this problem reminds us that whenever we are in pursuit of some conjecture, it is important to allocate time to the search for counterexamples. One often discovers quite quickly that the conjecture must be refined — or even rejected.

SOLUTION FOR EXERCISE 12.6. First, by (cyclical) symmetry, we can assume that $x \geq y$ and $x \geq z$. This makes x "special," so it is then natural to consider the symmetry properties of y and z. If we consider the difference

$$f(x, y, z) - f(x, z, y) = (y - z)(x - y)(x - z),$$

we see it is negative when y is less than z, so we can assume without loss of generality that $y \geq z$. Finally, assuming $x \geq y \geq z$ we note $f(x + z, y, 0) - f(x, y, z) = z^2 y + yz(x - y) + xy(y - z) \geq 0$, so we may also assume without loss of generality that $z = 0$. We can now finish with calculus as suggested by the hint or, alternatively, we can use the AM-GM inequality check that for $x + y = 1$ we have

$$f(x, y, 0) = \frac{x^2 y}{2} \leq \frac{1}{2}\left(\frac{x + x + 2y}{3}\right)^3 = 4/27.$$

One lesson to take away from this exercise is that it is often possible to make step-by-step progress by considering how a function changes when subjected to simple transformations such as the interchange of two variables.

SOLUTION FOR EXERCISE 12.7. Pitman solves his Problem 3.1.24 by first expanding $1 = (x+y+z)^3$ and then noting that it suffices to show

$$Q = x^2y + x^2z + y^2x + y^2z + z^2x + z^2y \leq 1/4 \quad \text{when } x+y+z = 1.$$

If we write $Q = x\{x(y+z)\} + y\{y(x+z)\} + z\{z(x+y)\}$, then it now suffices to notice that each of the three braced expressions is bounded below by 1/4 by the AM-GM inequality. Other solutions can be based on the homogenization trick of Exercise 12.3, or Schur's inequality (page 83), or the reduction devices of Exercise 12.6.

SOLUTION FOR EXERCISE 12.8. This elementary (but very useful!) inequality serves as a reminder that symmetry is often the key to successful telescoping. Here the telescoping identity

$$a_1 a_2 \cdots a_n - b_1 b_2 \cdots b_n = \sum_{j=1}^{n} a_1 \cdots a_{j-1}(a_j - b_j) b_{j+1} \cdots b_n$$

makes the Weierstrass inequality immediate. Naturally, generalizations of this identity lead one to more elaborate versions of Weierstrass inequality.

CHAPTER 13: MAJORIZATION AND SCHUR CONVEXITY

SOLUTION FOR EXERCISE 13.1. From each of the representations

$$\begin{pmatrix} a \\ b \\ c \end{pmatrix} = \begin{pmatrix} 1/2 & 1/3 & 1/6 \\ 1/3 & 2/3 & 0 \\ 1/6 & 0 & 5/6 \end{pmatrix} \begin{pmatrix} x \\ y \\ z \end{pmatrix} \quad \begin{pmatrix} a \\ b \\ c \end{pmatrix} = \begin{pmatrix} 0 & 1/2 & 1/2 \\ 1/2 & 1/6 & 1/3 \\ 1/2 & 1/3 & 1/6 \end{pmatrix} \begin{pmatrix} x \\ y \\ z \end{pmatrix}$$

one gets $(a, b, c) \prec (x, y, z)$. The inequalities of the exercise then follow from the Schur concavity of the map $(x, y, z) \mapsto xyz$ and the Schur convexity of the map $(x, y, z) \mapsto 1/x^5 + 1/y^5 + 1/z^5$.

SOLUTION FOR EXERCISE 13.2. If we set $s = (x_1 + x_2 + \cdots + x_n)/k$ we have $(x_1, x_2, \ldots, x_n) \prec (s, s, \ldots, s, 0, 0, \ldots, 0)$ when we take k copies of s. Thus, for convex $\phi : [0, \infty) \to \mathbb{R}$, Schur's majorization inequality (13.18) gives us $\phi(x_1) + \phi(x_2) + \cdots + \phi(x_n) \leq (n-k)\phi(0) + k\phi(s)$, and we can set $\phi(x) = 1/(1+x)$ to obtain the bound (13.21).

SOLUTION FOR EXERCISE 13.3. If one sets

$$y_k = \begin{cases} \mu + \delta/m & \text{for } 1 \leq k \leq m \\ \mu - \delta/(n-m) & \text{for } m < k \leq n \end{cases}$$

where $\mu = (x_1 + x_2 + \cdots + x_n)/n$, then from the condition (13.22) it follows easily that $\mathbf{y} \prec \mathbf{x}$. The map $f(\mathbf{x}) = x_1^2 + x_2^2 + \cdots + x_n^2$ is Schur convex, so we have $f(\mathbf{y}) \prec f(\mathbf{x})$, and, after expansion, this is precisely the target inequality (13.23). For the connection to Szemerédi's Regularity Lemma, see Komlós and Simonovits (1996).

SOLUTION FOR EXERCISE 13.4. Two applications of cancellation identity (13.24) permit one to reduce Schur's differential (13.4) to

$$-(x_s - x_t)^2 e_{k-1}(x_1, x_2, \ldots, x_{s-1}, x_{s+1}, \ldots, x_{t-1}, x_{t+1}, \ldots x_n),$$

and this polynomial is obviously nonpositive for $\mathbf{x} \in [0, \infty)^n$.

SOLUTION FOR EXERCISE 13.5. Use Schur's criterion (13.4) and note

$$(x_j - x_k)(s_{x_j}(\mathbf{x}) - s_{x_k}(\mathbf{x})) = 2(x_j - x_k)^2/(n-1) \geq 0 \quad \text{and}$$

$$(p_j - p_k)(h_{p_j}(\mathbf{p}) - h_{p_k}(\mathbf{p})) = (p_j - p_k)(\log p_j - \log p_k) \geq 0,$$

where the subscripts connote partial derivatives. Incidentally, the second formula verifies that $h(\mathbf{p})$ is Schur convex on all of $(0, \infty)^n$, not just the subset of $(0, \infty)^n$ where $\mathbf{p}$ has sum equal to one.

SOLUTION FOR EXERCISE 13.6. Since $(1/n, 1/n, \ldots, 1/n) \prec \mathbf{p}$, this is a special case of the bound (13.18) for $\phi(x) = (x + 1/x)^\alpha$ since

$$\phi''(x) = \alpha(x + 1/x)^\alpha (x + x^3)^{-2} \{(1 + x^2 - x^4) + \alpha(1 - x^2)^2\}$$

must be positive for $0 \leq x \leq 1$ and $\alpha > 0$. The relevance of Schur convexity to this problem was noted by Marshall and Olkin (1979, p. 72); a proof using Lagrange multipliers is given by Mitrinović (1970, p. 282).

SOLUTION FOR EXERCISE 13.7. In the uniform case the probability is $1 - (1 - 1/365) \cdot (1 - 2/365) \cdots (1 - 22/365) \sim 0.5079 \ldots$. In the general case the probability is $1 - e_n(p_1, p_2, \ldots, p_{365})$ where $e_n(\mathbf{p})$ is the nth symmetric polynomial and p_k is the probability that a randomly chosen person is born on day k. By Exercise 13.4 the polynomial $e_n(\mathbf{p})$ is Schur concave, and this is even more than one needs. The connection between majorization and the birthday problem has been made in Clevenson and Watkins (1991) and Proschan and Joag–Dev (1992); McConnell (2001) gives a treatment for nonuniform probabilities without explicit recourse to majorization.

SOLUTION FOR EXERCISE 13.8. The necessity of the condition is immediate, so we just need to prove sufficiency. In Weyl's terms, girl j knows precisely the boys in the set S_j, so for a given set A of girls, every boy in the set $\cup_{j \in A} S_j$ will be known by some girl in A. We now consider two cases.

In Case I, we assume that the inequality (13.26) is strict for all A with $|A| < n$. Girl n then marries any boy b she knows. Since the condition (13.26) continues to hold for all $A \subset \{1, 2, \ldots, n-1\}$ when each S_j, $1 \leq j \leq n-1$, is replaced by $S_j \setminus \{b\}$, the remaining girls can be married by induction to the remaining boys.

In Case II, we assume that equality holds in the bound (13.26) for some A_0 with $|A_0| < n$. We then let

$$B = \bigcup_{j \in A_0} S_j \quad \text{and set} \quad S'_j = S_j \setminus B \quad \text{for all } j \in A_0^c.$$

The girls in A_0 can be married to the boys in B by induction, and it remains to show that the girls in A_0^c can be married to the boys in B^c. We now take any $A \subset A_0^c$ and note that

$$\left| \bigcup_{j \in A_0 \cup A} S_j \right| \geq |A_0 \cup A| = |A| + |A_0|.$$

We also have the identity

$$\left| \bigcup_{j \in A_0 \cup A} S_j \right| = \left| \left\{ \bigcup_{j \in A_0} S_j \right\} \cup \left\{ \bigcup_{j \in A} S'_j \right\} \right| = |A_0| + \left| \bigcup_{j \in A} S'_j \right|.$$

Thus, we find for all $A \subset A_0^c$ that we have

$$\left| \bigcup_{j \in A} S'_j \right| \geq |A|;$$

that is, every set of k girls in A_0^c knows at least k boys in B^c. By induction the girls in A_0^c can be married to the boys in B^c. This proof is essentially the one given by Halmos and Vaughan (1950). The marriage lemma is a cornerstone of the large and active field of matching theory which is beautifully surveyed by Lovász and Plummer (1986).

SOLUTION FOR EXERCISE 13.9. One can argue by induction on the number of nonzero entries of D, but it is perhaps more concrete to look for an algorithm to compute the required convex combination. Either way, the basic idea is to use the marriage lemma to make step-by-step progress.

For each $1 \leq j \leq n$, we let S_j denote the set of all k such that $d_{jk} > 0$,

Solutions to the Exercises 279

and we note that for each $A \subset \{1, 2, \ldots, n\}$ one has

$$|A| = \sum_{j \in A} \sum_{k \in S_j} d_{jk} \leq \sum_{k \in \cup_{j \in A} S_j} \sum_{1 \leq j \leq n} d_{jk} = \left| \bigcup_{j \in A} S_j \right|.$$

By the marriage lemma, there is a system of SDRs of $\{S_1, S_2, \ldots, S_n\}$, so we can define a permutation σ by taking $\sigma(j)$ to be the representative from S_j for each $j = 1, 2, \ldots, n$. Now, we let P_σ be the permutation matrix associated with σ and set $\alpha = \min d_{j\sigma(j)} > 0$. If $\alpha = 1$ then D is a permutation matrix, and there is nothing left to prove. On the other hand, if $\alpha < 1$ consider the new matrix D' defined by setting $D' = (1 - \alpha)^{-1}(D - \alpha P_\sigma)$. We then have $D = \alpha P_\sigma + (1 - \alpha)D'$ and D' is a doubly stochastic matrix with more zero entries than D. The proof may now be completed by applying the induction hypothesis to D'. Alternatively, one can compute the required summands by repeating the analogous steps until the representation is complete; at most n^2 steps will be needed.

CHAPTER 14: CANCELLATION AND AGGREGATION

SOLUTION FOR EXERCISE 14.1. To prove the second bound (14.29), we again sum $b_1 z_1 + b_2 z_2 + \cdots + b_n z_n$ by parts to get

$$S_1(b_1 - b_2) + S_2(b_2 - b_3) + \cdots + S_{n-1}(b_{n-1} - b_n) + S_n b_n,$$

but this time we bound the sum $|b_1 z_1 + b_2 z_2 + \cdots + b_n z_n|$ by noting

$$|S_1||b_1 - b_2| + |S_2||b_2 - b_3| + \cdots + |S_{n-1}||b_{n-1} - b_n| + |S_n|b_n$$
$$\leq \max_{1 \leq k \leq n} |S_k| \{(b_2 - b_1) + (b_3 - b_2) + \cdots + (b_n - b_{n-1}) + b_n\}$$
$$= \{(b_n - b_1) + b_n\} \max_{1 \leq k \leq n} |S_k| \leq 2 b_n \max_{1 \leq k \leq n} |S_k|.$$

SOLUTION FOR EXERCISE 14.2. From the nonnegativity of g one has the bounds

$$\min_{a \leq y \leq b} f(y) \int_a^b g(x)\,dx \leq \int_a^b f(x) g(x)\,dx \leq \max_{a \leq y \leq b} f(y) \int_a^b g(x)\,dx,$$

and by the continuity of f it takes on all values between its minimum and its maximum. These observations give us the first IMVF (14.30).

To prove the second, choose Φ with $\Phi(a) = 0$ such that $\Phi'(x) = \phi(x)$, then integrate by parts and apply the first IMVF with $f(x) = \Phi(x)$ and

$g(x) = -\psi'(x) \geq 0$ to find

$$\int_a^b \psi(x)\phi(x)\,dx = \int_a^b \psi(x)\Phi'(x)\,dx = \psi(b)\Phi(b) - \int_a^b \psi'(x)\Phi(x)\,dx$$

$$= \psi(b)\Phi(b) - \Phi(\xi)\int_a^b \psi'(x)\,dx$$

$$= \psi(a)\left\{\frac{\psi(a) - \psi(b)}{\psi(a)}\Phi(\xi) + \frac{\psi(b)}{\psi(a)}\Phi(b)\right\}.$$

Since $0 < \psi(b) \leq \psi(a)$ the bracketed quantity is an average of $\Phi(b)$ and $\Phi(\xi)$ so it must be equal to $\Phi(\xi_0)$ for some $\xi_0 \in [\xi, b] \subset [a, b]$ by the continuity of Φ.

SOLUTION FOR EXERCISE 14.32. The bound (14.32) is immediate from the second IMVF (14.31). The sine bound (14.33) then follows by taking $f(x) = 1/x$, $g(x) = \sin x$ and by noting that the integral of g over $[a, b]$ is $\cos b - \cos a$ which is bounded by 2 in absolute value.

SOLUTION FOR EXERCISE 14.4. We are given that $\theta'(\cdot)$ is monotonic, and, without loss of generality, we assume it is nondecreasing. From the second IMVP of Exercise 14.2, we find that

$$\int_a^b \cos\theta(x)\,dx = \int_a^b \frac{\theta'(x)\cos\theta(x)}{\theta'(x)}\,dx$$

$$= \frac{1}{\theta'(a)}\int_a^\xi \{\cos\theta(x)\}\theta'(x)\,dx = \frac{\sin\theta(\xi) - \sin\theta(a)}{\theta'(a)}.$$

The last ratio has modulus bounded by $2/\nu$, so to complete the proof, one only needs to check that an exactly analogous argument applies to the imaginary part of the integral in our target inequality (14.34).

Since $\theta'(x)$ is strictly monotone, it vanishes at most once in the interval $[a, b]$, and, for the moment, suppose it vanishes at c. To prove the second bound (14.35), we write the integral I over $[a, b]$ as the sum $I_1 + I_2 + I_3$ of integrals over $[a, c-\delta]$, $[c-\delta, c+\delta]$ and $[c+\delta, b]$. In the interval $[c+\delta, b]$, one has $\theta'(x) \geq \rho\delta$, so by the bound (14.34), we have $|I_3| \leq 4/\rho\delta$. An analogous bound applies to I_1, while for the integral I_2 we have the trivial bound $|I_2| \leq 2\delta$. In sum we have

$$|I| \leq |I_1| + |I_2| + |I_3| \leq \frac{8}{\rho\delta} + 2\delta,$$

which we can minimize by setting $\delta = 2/\sqrt{\rho}$ to obtain the target bound (14.35). To be 100% complete, one finally needs to note that the target bound continues to hold if $c \pm 2/\sqrt{\rho} \notin [a,b]$, or, indeed, if $\theta'(x)$ does not vanish in $[a,b]$.

SOLUTION FOR EXERCISE 14.5. To begin let W denote the target sum, and note that

$$|W| = \left|\sum_{j \in A}\sum_{k \in B} \exp\left(\frac{2\pi i j k}{p}\right)\right| \leq \sum_{j \in A}\left|\sum_{k \in B} \exp\left(\frac{2\pi i j k}{p}\right)\right|$$

so Cauchy's inequality gives us

$$|W|^2 \leq |A| \sum_{j \in A}\left|\sum_{k \in B} \exp\left(\frac{2\pi i j k}{p}\right)\right|^2.$$

Now we come to a devilish trick: we extend the outside sum to all of $\mathbb{F}_p = \{0, 1, \ldots, p-1\}$. This is *feasible* because we are just adding positive terms, and it is *sensible* because it sets up the application of the cancellation identity (14.36). To put the algebra neatly, we first define the function $\delta(x)$ by setting $\delta(0) = 1$ and $\delta(x) = 0$ for $x \neq 0$, then we note

$$|W|^2 \leq |A| \sum_{j \in \mathbb{F}_p}\left|\sum_{k \in B} \exp\left(\frac{2\pi i j k}{p}\right)\right|^2$$

$$= |A| \sum_{j \in \mathbb{F}_p} \sum_{k_1, k_2 \in B} \exp\left(\frac{2\pi i j (k_1 - k_2)}{p}\right)$$

$$= |A| \sum_{k_1, k_2 \in B} \sum_{j \in \mathbb{F}_p} \exp\left(\frac{2\pi i j (k_1 - k_2)}{p}\right)$$

$$= |A| p \sum_{k_1, k_2 \in B} \delta(k_1 - k_2) = p|A||B|.$$

This problem and the description "extend and conquer" are from the informative exposition of Shparlinski (2002) where one finds several further examples of the ways to exploit complete sums. Shparlinski links bounds of the type (14.37) back to the work of I.M. Vinogradov; in particular, Exercise 14 of Vinogradov (1954, p. 128) is of this kind.

SOLUTION FOR EXERCISE 14.6. For each $1 \leq k \leq \lceil \log_2(x) \rceil = K$ we have the bound $g(x/2^{k-1}) - g(/2^k) \leq Ax/2^k + B$. Summing these gives us $g(x) - g(x/2^K) \leq Ax(1 + 1/2 + 1/2^2 + \cdots + 1/2^K) + KB$, or $g(x) \leq 2Ax + B\lceil \log_2(x)\rceil + \max_{0 \leq t \leq 1} g(t)$, so we can take $A' = 2A$, $B' = B$ and $C' = B + \max_{0 \leq t \leq 1} g(t)$.

SOLUTION FOR EXERCISE 14.7. For any $A \subset \{1, 2, \ldots, n\}$ we have

$$\int_0^1 \left(\sum_{j \in A} c_j \psi_j(x) \, dx \right)^2 dx = \sum_{j \in A} \sum_{k \in A} c_j c_k a_{jk} \leq C \sum_{j \in A} c_j^2, \qquad (14.63)$$

where the last inequality comes from applying the hypothesis (14.39) where $y_j = c_j$ if $j \in A$ and $y_j = 0$ if $j \notin A$. Next, if we replace a_i by $c_i \psi_i(x)$ in the real-variable inequality (14.26) and integrate, we find

$$\int_0^1 \max_{1 \leq k \leq n} \left(\sum_{i=1}^k c_i \psi_i(x) \right)^2 dx \leq \lceil \log_2(n) \rceil \sum_{B \in \mathcal{B}} \int_0^1 \left(\sum_{i \in B} \psi_i(x) \right)^2 dx$$

$$\leq \lceil \log_2(n) \rceil \sum_{B \in \mathcal{B}} c \sum_{i \in B} c_i^2$$

$$\leq \lceil \log_2(n) \rceil \lceil 1 + \log_2(n) \rceil c \sum_{i=1}^n c_i^2,$$

which is slightly stronger than the target inequality (14.40).

SOLUTION FOR EXERCISE 14.8. From the splitting

$$\rho^{-|j-k|} y_j y_k = \rho^{-|j-k|/2} y_j \cdot \rho^{-|j-k|/2} y_k,$$

we see that Cauchy's inequality gives us

$$\left(\sum_{j=1}^n \sum_{k=1}^n \rho^{-|j-k|} y_j y_k \right)^2$$

$$\leq \sum_{j=1}^n \sum_{k=1}^n \rho^{-|j-k|} y_j^2 \cdot \sum_{j=1}^n \sum_{k=1}^n \rho^{-|j-k|} y_k^2$$

$$\leq \sum_{j=1}^n y_j^2 \left(\max_{1 \leq j \leq n} \sum_{k=1}^n \rho^{-|j-k|} \right) \cdot \sum_{k=1}^n y_k^2 \left(\max_{1 \leq k \leq n} \sum_{j=1}^n \rho^{-|j-k|} \right).$$

Next, geometric summation shows that we have

$$\max_{1 \leq k \leq n} \sum_{j=1}^n \rho^{-|j-k|} \leq \sum_{j \in \mathbb{Z}} \rho^{-|j|} = \frac{1+\rho}{1-\rho},$$

so our Cauchy estimate may be reduced to the simple bound

$$\left| \sum_{j=1}^n \sum_{k=1}^n \rho^{-|j-k|} y_j y_k \right| \leq \frac{1+\rho}{1-\rho} \sum_{k=1}^n y_k^2. \qquad (14.64)$$

Given the inequality (14.64), the conclusion of Exercise 14.8 with the value $M = (1+\rho)/(1-\rho)$ follows from Exercise 14.7.

Solutions to the Exercises 283

SOLUTION FOR EXERCISE 14.9. From the definition of S_θ one finds

$$f(\theta) \stackrel{\text{def}}{=} \left| \sum_{z_k \in S_\theta} z_k \right| = \left| \sum_{z_k \in S_\theta} z_k e^{-i\theta} \right| \geq \left| \sum_{z_k \in S_\theta} \text{Re}\left(z_k e^{-i\theta}\right) \right|$$

$$= \left| \sum_{z_k \in S_\theta} |z_k| \cos(\theta - \arg z_k) \right| = \sum_{z_k \in S_\theta} |z_k| \cos(\theta - \arg z_k).$$

It suffices to show that max $f(\theta)$ is as large as the left side of the bound (14.41). To do this we compute the average,

$$\frac{1}{2\pi} \int_0^{2\pi} f(\theta)\, d\theta \geq \frac{1}{2\pi} \int_0^{2\pi} \sum_{z_k \in S_\theta} |z_k| \cos(\theta - \arg z_k)\, d\theta$$

$$= \sum_{k=1}^n \frac{|z_k|}{2\pi} \int_{\arg(z_k) - \pi/2}^{\arg(z_k) + \pi/2} \cos(\theta - \arg z_k)\, d\theta = \frac{1}{\pi} \sum_{k=1}^n |z_k|$$

so, indeed, there must exist some value θ^* for which $f(\theta^*)$ is at least as large as the last sum. By taking $\{z_k = \exp(ik2\pi/N) : 0 \leq k < N\}$ for large N one can show that the constant $1/\pi$ cannot be improved. This argument follows W.W. Bledsoe (1970); Mitrinović (1970, p. 331) notes that similar results were obtained earlier by D.Ž. Djoković.

SOLUTION FOR EXERCISE 14.10. If L and R denote the left and right sides of the target bound (14.42), then by squaring and changing order, one finds the representation

$$L = \sum_{r=1}^R \sum_{s=1}^R \sum_{n=1}^N \sum_{m=1}^N \bar{a}_n\, \bar{y}_{nr}\, y_{ns}\, a_m\, y_{mr}\, \bar{y}_{ms}$$

$$= \sum_{n=1}^N \sum_{m=1}^N a_m \bar{a}_n \left\{ \sum_{r=1}^R \sum_{s=1}^R y_{mr}\, \bar{y}_{ms}\, \bar{y}_{nr}\, y_{ms} \right\}$$

$$= \sum_{n=1}^N \sum_{m=1}^N a_m\, \bar{a}_n \sum_{r=1}^R y_{mr}\, \bar{y}_{nr} \sum_{s=1}^R \bar{y}_{ms}\, y_{ms}$$

$$= \sum_{n=1}^N \sum_{m=1}^N a_m\, \bar{a}_n \left| \sum_{r=1}^R y_{mr}\, \bar{y}_{nr} \right|^2,$$

and the identical calculation from the right side R shows

$$R = \sum_{n=1}^N \sum_{m=1}^N A_m\, A_n \left| \sum_{r=1}^R y_{mr}\, \bar{y}_{nr} \right|^2,$$

so our hypothesis gives us $L \leq R$. The bound (14.42) provides a generic example of a class of inequalities called *majorant principles*, and the treatment given here follows Theorem 4 of Montgomery (1994, p. 132).

SOLUTION FOR EXERCISE 14.11. The most direct proof just requires a big piece of paper and a timely application of Cauchy's inequality. First expand the squares $|\langle \mathbf{u}_m, \mathbf{v}_n \rangle|^2$ in terms of the vector components u_{mj}, $1 \leq j \leq d$ and v_{nk}, $1 \leq k \leq d$. Next, change the order of summation so that the double sum over j and k is outermost, and only now apply Cauchy's inequality. Finally, within each of the two resulting rooted expressions, you change the order of summation within each of the braces and reinterpret the sums innermost sums as inner products.

This solution amplifies the remark of Montgomery (1994, p. 144) that manipulations like those used in the solution of Exercise 14.10 can be used to prove Enflo's inequality. An alternative solution may be based on the observation that the functions $\phi_{n,m}(x, y) = \mathbf{e}(mx)\mathbf{e}(ny)$ are orthonormal on the square $[0, 1]^2$. One then introduces the function

$$f(x, y) \stackrel{\text{def}}{=} \sum_{m=1}^{M} \sum_{n=1}^{N} \langle \mathbf{u}_m, \mathbf{v}_n \rangle \mathbf{e}(mx)\mathbf{e}(ny)$$

and exploits the fact that the integral of $|f(x, y)|^2$ over $[0, 1]^2$ gives the left side of Enflo's inequality.

SOLUTION FOR EXERCISE 14.12. One always has $\langle \mathbf{z}, \mathbf{z} \rangle \geq 0$ so if we set $\mathbf{z} = \mathbf{x} - (c_1 \mathbf{y}_1 + c_2 \mathbf{y}_2 + \cdots + c_n \mathbf{y}_n)$, we find for all c_j, $1 \leq j \leq n$ that

$$0 \leq \langle \mathbf{x}, \mathbf{x} \rangle - \sum_{j=1}^{n} c_j \overline{\langle \mathbf{x}, \mathbf{y}_j \rangle} - \sum_{j=1}^{n} \bar{c}_j \langle \mathbf{x}, \mathbf{y}_j \rangle + \sum_{j=1}^{n} \sum_{k=1}^{n} c_j \bar{c}_k \langle \mathbf{y}_j, \mathbf{y}_k \rangle.$$

The so-called humble bound $|c_j \bar{c}_k| \leq \frac{1}{2}|c_j|^2 + \frac{1}{2}|c_k|^2$ gives us

$$0 \leq \langle \mathbf{x}, \mathbf{x} \rangle - \sum_{j=1}^{n} c_j \overline{\langle \mathbf{x}, \mathbf{y}_j \rangle} - \sum_{j=1}^{n} \bar{c}_j \langle \mathbf{x}, \mathbf{y}_j \rangle$$
$$+ \frac{1}{2} \sum_{j=1}^{n} \sum_{k=1}^{n} |c_j|^2 |\langle \mathbf{y}_j, \mathbf{y}_k \rangle| + \frac{1}{2} \sum_{j=1}^{n} \sum_{k=1}^{n} |c_k|^2 |\langle \mathbf{y}_j, \mathbf{y}_k \rangle|,$$

and if we set $c_j = \langle \mathbf{x}, \mathbf{y}_j \rangle / \sum_{k=1}^{n} |\langle \mathbf{y}_j, \mathbf{y}_k \rangle|$ simple algebra bring us to the inequality (14.43). This argument is based on the classic exposition of E. Bombieri (1974).

Chapter Notes

CHAPTER 1: STARTING WITH CAUCHY

Bunyakovsky's 1859 *Mémoire* was eighteen pages long, and it sold as a self-standing piece for 25 kopecks, a sum which was then represented by a silver coin roughly the size of a modern US quarter. Yale University library has one of the few extant copies of the *Mémoire*. On the title page the author used the French transliteration of his name, Bouniakowsky; here this spelling is used in the references, but elsewhere in the text the more common spelling Bunyakovsky is used.

The volume containing Schwarz's 1885 article was issued in honor of the 60th birthday of Karl Weierstrass. In due course, Schwarz came to occupy the chair of mathematics in Berlin which had been long held by Weierstrass.

Dubeau (1990) is one of the few articles to advocate the inductive approach to Cauchy's inequality that is favored in this chapter.

The Cramér–Rao inequality of Exercise 1.15 illustrates one way that the Cauchy–Schwarz inequality can be used to prove lower bounds. Chapter 6 of Matoušek (1999) gives an insightful development of several deeper examples from the theory of geometric discrepancy. The recent monograph of Dragomir (2003) provides an extensive survey of discrete inequalities which refine and extend Cauchy's inequality.

CHAPTER 2: THE AM-GM INEQUALITY

The AM-GM inequality is arguably the world's oldest nontrivial inequality. As Exercise 2.6 observes, for two variables it was known even to the ancients. By the dawn of the era of calculus it was known for n variables, and there were even subtle refinements such as Maclaurin's inequality of 1729. Bullen, Mitrinović, and Vasić (1987, pp. 56–89) give fifty-two proofs of the AM-GM inequality in (essentially) their chronological order.

Duncan and McGregor (2004) survey several proofs of Carleman's inequality including Carleman's original, and Pečarić and Stolarsky (2001) provide a comprehensive historical review.

Pólya's 1926 article proves in one page what his 1949 article proves in eight, but Pólya's 1949 explanation of how he found his proof is one of the great classics of mathematical exposition. It is hard to imagine a better way to demonstrate how the possibilities for exploiting an inequality are enhanced by understanding the cases where equality holds. The quote from Pólya on page 23 is from Alexanderson (2000, p. 75).

CHAPTER 3: LAGRANGE'S IDENTITY AND MINKOWSKI'S CONJECTURE

Stillwell (1998, p. 116) gives the critical quote from *Arithmetica*, Book III, Problem 19, which suggests that Diophantus knew the case $n = 2$ of Lagrange's identity. Stillwell also gives related facts and references that are relevant here — including connections to Fibonacci, Brahmagupta, and Abu Ja'far al-Khazin. Exercise 3.2 is motivated by a similar exercise of Stillwell (1998, p. 218). Bashmakova (1997) provides an enjoyable introduction to Diophantus and his namesake equations.

Lagrange (1771, pp. 662–663) contains Lagrange's identity for the case $n = 3$, but it is only barely visible behind the camouflage of a repetitive system of analogous identities. For the contemporary reader, the most striking feature of Lagrange's article may be the wild proliferation of expressions such as $ab - cd$ which nowadays one would contain within determinants or wedge products.

The treatment of Motzkin's trick in Rudin (2000) helped frame the discussion given here, and the theory of representation by a sum of squares now has an extensive literature which is surveyed by Rajwade (1993) and by Prestel and Delzell (2001). Problem 3.5 was on the 1957 Putnam Exam which is reprised in Bush (1957).

CHAPTER 4: ON GEOMETRY AND SUMS OF SQUARES

The von Neumann quote (page 51) is from G. Zukav (1979, p. 226 footnote). A long oral tradition precedes the example of Figure 4.1, but this may be the first time it has found its way into print. The bound (4.8) is developed for complex inner products in Buzano (1971/1973) which cites an earlier result for real inner product spaces by R.U. Richards. Magiropoulos and Karayannakis (2002) give another proof which depends more explicitly on the Gram–Schmidt process, but the argument given here is closest to that of Fuji and Kubo (1993) where one also finds an interesting application of the linear product bound to the exclusion region for polynomial zeros.

The proof of the light cone inequality (page 63) is based on the discussion of Aczél (1961, p. 243). A generalization of the light cone inequality is given in van Lint and Wilson (1992, pp. 96–98), where it is used to give a stunning proof of the van der Waerden permanent conjecture. *Hilbert's pause* (page 55) is an oft-repeated folktale. It must have multiple print sources, but none has been found.

CHAPTER 5: CONSEQUENCES OF ORDER

The bound (5.5) is known as the Diaz–Metcalf inequality, and the discussion here is based on Diaz–Metcalf (1963) and the comments by Mitrinović (1970, p. 61). The original method used by Pólya and Szegö is more complicated, but, as the paper of Henrici (1961) suggests, it may be applied somewhat more broadly.

The Thread by Philip Davis escorts one through a scholar's inquiry into the origins and transliterations of the name "Pafnuty Chebyshev."

The order-to-quadratic conversion (page 77) also yields the traditional proof of the Neyman–Pearson Lemma, a result which many consider to be one of the cornerstones of statistical decision theory.

CHAPTER 6: CONVEXITY — THE THIRD PILLAR

Hölder clearly viewed his version of Jensen's inequality as the main contribution of his 1888 paper. Hölder also cites Rogers's 1887 paper quite generously, but, even then, Hölder seems to view Rogers's main contribution to be the weighted version of the AM-GM inequality. Everyone who works in relative obscurity may take heart from the fact that neither Hölder nor Rogers seems to have had any inkling that their inequality would someday become a mathematical mainstay. Pečarić, Proschan, and Tong (1992, p. 44) provide further details on the early history of convexity.

This chapter on inequalities for convex *functions* provides little information on inequalities for convex *sets*, and the omission of the Prékopa-Leindler and the Brunn-Minkowski inequalities is particularly regrettable. In a longer and slightly more advanced book, each of these would deserve its own chapter. Fortunately, Ball (1997) provides a well motivated introductory treatment of these inequalities, and there are definitive treatments in the volumes of Burago and Zalgaller (1988) and Scheidner (1993).

CHAPTER 7: INTEGRAL INTERMEZZO

Hardy, Littlewood, and Pólya (1952, p. 228) note that the case $\alpha = 0$, $\beta = 2$ of inequality (7.4) is due to C.F. Gauss (1777-1855), though presumably Gauss used an argument that did not call on the inequality of Schwarz (1885) or Bunyakovsky (1859). Problem 7.1 is based on Exercise 18 of Bennett and Sharpley (1988, p. 91). Problem 7.3 (page 110) and Exercise 7.3 (page 116) slice up and expand Exercise 7.132 of George (1984, p. 297). The bound of Exercise 7.3 is sometimes called Heisenberg's Uncertainty Principle, but one might note that there are several other inequalities (and identities!) with that very same name. The discrete analog of Problem 7.4 was used by Weyl (1909, p. 239) to illustrate a more general lemma.

CHAPTER 8: THE LADDER OF POWER MEANS

Narkiewicz (2000, p. xi) notes that Landau (1909) did indeed introduce the notation $o(\cdot)$, but Narkiewicz also makes the point that Landau only popularized the related notation $O(\cdot)$ which had been introduced earlier by P. Bachmann. Bullen, Mitrinović, and Vasić (1987) provide extensive coverage of the theory of power means, including extensive references to original sources.

CHAPTER 9: HÖLDER'S INEQUALITY

Maligranda and Persson (1992, p. 193) prove for complex $a_1, a_2, \ldots, a_n$ and $p \geq 2$ that one has the inequality

$$\left|\sum_{j=1}^{n} a_j\right|^p + \sum_{1 \leq j < k \leq n} |a_j - a_k|^p \leq n^{p-1} \sum_{j=1}^{n} |a_j|^p. \qquad (14.65)$$

This refines the 1-trick bound $\delta(\mathbf{a}) \geq 0$ which is given on page 144, and it leads automatically to stability results for Hölder's inequality which complement Problem 9.5 (page 145).

Problem 9.6 and the follow-up Exercises 9.14 and 9.15 open the door to the theory of interpolation of linear operators, which is one of the most extensive and most important branches of the theory of inequalities. In these problems we considered the interpolation bounds for any reciprocal pairs $(1/s_1, 1/t_1)$ and $(1/s_0, 1/t_0)$ anywhere in $S = [0, 1] \times [0, 1]$, but we also made the strong assumption that $c_{jk} \geq 0$ for all j, k.

In 1927, Marcel Riesz, the brother of Frigyes Riesz (whose work we have seen in several chapters), proved that the assumption that the c_{jk} are nonnegative can be dropped provided that one assumes that the reciprocal pairs $(1/s_1, 1/t_1)$ and $(1/s_0, 1/t_0)$ are from the "clear" upper triangle of Figure 9.3. M. Riesz's proof used only elementary methods, but it was undeniably subtle. It was also unsettling that Riesz's argument did not apply to the whole rectangle, but this was inevitable. Easy examples show that the interpolation bound (9.41) can fail for reciprocal pairs from the "gray" lower half of the unit square S.

Some years after M. Riesz proved his interpolation theorem, Riesz's student G.O. Thorin made a remarkable breakthrough by proving that the interpolation bound is valid for the whole square S under one important proviso: it is essential to consider the *complex* normed linear spaces ℓ^p in lieu of the real ℓ^p spaces.

Thorin's key insight was to draw a link between the interpolation problem and the maximum modulus theorem from the theory of analytic functions. Over the years, this link has become one of the most robust tools in the theory of inequalities, and it has been exploited in hundreds of papers. Bennett and Sharpley (1988, pp. 185–216) provide an instructive discussion of the arguments of Riesz and Thorin in a contemporary setting.

CHAPTER 10: HILBERT'S INEQUALITY

Hilbert's inequality has a direct connection to the eigenvalues of a special integral equations which de Bruijn and Wilf (1961) used to show that for an n by n array one can replace the π in Hilbert's inequality with the smaller value $\lambda_n = \pi - \pi^5/\{2(\log n)^2\} + O(\log \log n / \log n)^2)$. The finite sections of many inequalities are addressed systematically by Wilf (1970).

Mingzhe and Bichen (1998) show that the Euler–Maclaurin expansions can be used to obtain instructive refinements of the estimates on page 158. Such refinements are almost always a possibility when integrals are used to estimate sums, but there can be many devils in the details.

The notion of "stressing" an inequality is motivated by the discussion of Hardy, Littlewood, and Pólya (1952, pp. 232–233). The method works so often that its failures are more surprising than its successes.

Chung, Hajela, and Seymour (1988) exploit the inequality (10.22) in the analysis of self-organizing lists, a topic of importance in theoretical computer science. Exercise 10.6 elaborates on an argument which is given quite succinctly in Hardy (1936). Maligranda and Person (1993) note that Carlson suggested in his original paper that the bound (10.24) could not be derived from Hölder's inequality (or Cauchy's), yet Hardy was quick to find a path.

CHAPTER 11: HARDY'S INEQUALITY AND THE FLOP

In 1920 Hardy gave only an imperfect version of the discrete inequality (11.2), and his primary point at the time was to record the quantitative Hilbert's inequality described in Exercise 11.5. Hardy raised but did not resolve the issue of the best constant, although Hardy gives a footnote citing a letter of Issai Schur which comes very close.

Hardy (1920, p. 316) has another intriguing footnote which cites the inequality of Rogers (1888) and Hölder (1889) in its pre-Riesz form (9.34). In this note, Hardy says "the well-known inequality...seems to be due to Hölder." In support of his statement, Hardy refers to Landau (1907), and this may be the critical point at which Rogers's contribution lapsed into obscurity. By the time Hardy, Littlewood, and Pólya wrote *Inequalities*, they had read Hölder's paper, and they knew that Hölder did not claim the inequality as his own. Unfortunately, by the time *Inequalities* was to appear, it was Rogers who became a footnote.

The argument given here for the inequality (11.1) is a modest simplification of the L^p argument of Elliot (1926). The proof of the discrete Hardy inequality can be greatly shortened, especially (as Claude Dellacherie notes) if one appeals to ideas of Stieltjes integration. The volumes of B. Opic and A. Kufner (1990) and Grosse–Erdmann (1998) show how the problems discussed in this chapter have grown into a field.

CHAPTER 12: SYMMETRIC SUMS

The treatment of Newton's inequalities follows the argument of Rosset (1989) which is elegantly developed in Niculescu (2000). Waterhouse (1983) discusses the symmetry questions which evolve from questions such as the one posed in Exercise 12.5. Symmetric polynomials are at the heart of many important results in algebra and analysis, so the literature is understandably enormous. Even the first few chapters of Macdonald (1995) reveal hundreds of identities.

CHAPTER 13: SCHUR CONVEXITY AND MAJORIZATION

The Schur criterion developed in Problem 13.1 relies mainly on the treatment of Olkin and Marshall (1979, pp. 54–58).

The development of the HLP representation is a colloquial rendering of the proof given by Hardy, Littlewood, and Pólya in *Inequalities*.

CHAPTER 14: CANCELLATION AND AGGREGATION

Exponential sums have a long rich history, but few would dispute that

Chapter Notes 291

the 1916 paper of Hermann Weyl created the estimation of exponential sums as a mathematical specialty. Weyl's paper contained several seminal results, and, in particular, it pioneered what is now called *Weyl's method*, where one applies the bound (14.10) recursively to estimate the exponential sum associated with a general polynomial.

The discussion of the quadratic bound (14.7) introduces some of the most basic ideas of Weyl's method, but it can only hint at the delicacy of the general case. The inequality of van der Corput's inequality (14.17) is more special, but van der Corput's 1931 argument must be one of history's finest examples of pure Cauchy–Schwarz artistry.

Nowadays, the form (14.23) of the Rademacher–Menchoff inequality is quite standard, but it is not given so explicitly in the fundamental works of Rademacher (1922) and Menchoff (1923). Instead, this form seems to come to us from Kazmarz and Steinhaus. One finds the inequality in (essentially) its modern form as Lemma 534 in the 1951 second edition of their famous monograph of 1935, and searches have not yielded an earlier source.

References

Aczél, J. (1961/1962). Ungleichungen und ihre Verwendung zur elementaren Lösung von Maximum- und Minimumaufgaben, *L'Enseignement Math.* **2**, 214–219.

Alexanderson, G. (2000). *The Random Walks of George Pólya*, Math. Assoc. America, Washington, D.C.

Andreescu, T. and Feng, Z. (2000). *Mathematical Olympiads: Problems and Solutions from Around the World*, Mathematical Association of America, Washington, DC.

Andrica, D. and Badea, C. (1988). Grüss' inequality for positive linear functional, *Period. Math. Hungar.*, **19**, 155–167.

Artin, E. (1927). Über die Zerlegung definiter Funktionen in Quadrate, *Abh. Math. Sem. Univ. Hamburg*, **5**, 100-115.

Bak, J. and Newman, D.J. (1997). *Complex Analysis*, 2nd Edition, Springer-Verlag, Berlin.

Ball, K. (1997). An elementary introduction to modern convex geometry, in *Flavors of Geometry* (S. Levy, ed.), MSRI Publications **31** 1–58, Cambridge University Press, Cambridge, UK.

Bashmakova, I.G. (1997). *Diophantus and Diophantine Equations* (translated by Abe Shenitzer, originally published by Nauka, Moscow, 1972), Mathematical Association of America, Washington D.C.

Beckenbach, E.F. and Bellman, R. (1965). *Inequalities* (2nd revised printing), Springer-Verlag, Berlin.

Bennett, C. and Sharpley, R. (1988). *Interpolation of Operators*, Academic Press, Orlando, FL.

Bradley, D. (2000). Using integral transforms to estimate higher order derivatives, *Amer. Math. Monthly*, **107**, 923–931.

Bombieri, E. (1974). *Le grand crible dans la théorie analytique des nombres*, 2nd Edition, *Astérisque*, **18**, Soc. Math. de France, Paris.

Bouniakowsky, V. (1859). Sur quelques inégalités concernant les intégrales ordinaires et les intégrales aux différences finies. *Mémoires de l'Acad. de St.-Pétersbourg* (ser. 7) 1, No. 9.

Bullen P.S., Mitrinović, D.S., and Vasić, P.M. (1988). *Means and Their Inequalities*, Reidel Publishers, Boston.

Burago, Yu. D. and Zalgaller, V.A. (1988). *Geometric Inequalities*, Springer-Verlag, Berlin.

Bush, L.E. (1957). The William Lowell Putnam Mathematical Competition, *Amer. Math. Monthly*, **64**, 649–654.

Bushell, P. J. (1994). Shapiro's "Cyclic Sums," *Bulletin of the London Math. Soc.*, **26**, 564–574.

Buzano, M.L. (1971/1973). Generalizzazione della disegualianza di Cauchy–Schwarz, *Rend. Sem. Mat. Univ. e Politech. Trimo*, **31**, 405–409.

Cartan, H. (1995). *Elementary Theory of Analytic Functions of One or Several Complex Variables* (translation of *Théorie élémentaire des fonctions analytiques d'une ou plusieurs vairable complexes*, Hermann, Paris, 1961) reprinted by Dover Publications, Mineola, New York.

Carleman, T. (1923). Sur les fonctions quasi-analytiques, in the *Proc. of the 5th Scand. Math. Congress*, 181–196, Helsinki, Finland.

Carleson, L. (1954). A proof of an inequality of Carleman, *Proc. Amer. Math. Soc.*, **5**, 932–933.

Cauchy, A. (1821). *Cours d'analyse de l'École Royale Polytechnique, Première Partie. Analyse algébrique*, Debure frères, Paris. (Also in *Oeuvres complètes d'Augustin Cauchy, Série 2, Tome 3*, Gauthier-Villars et Fils, Paris, 1897.)

Cauchy, A. (1829). *Leçon sur le calcul différentiel, Note sur la détermination approximative des racine d'une équation algébrique ou transcendante*. (Also in *Oeuvres complètes d'Augustin Cauchy* (*Série 2, Tome 4*), 573–609, Gauthier-Villars et Fils, Paris, 1897.)

Chong, K.-M. (1969). An inductive proof of the A.M.–G.M. inequality, *Amer. Math. Monthly*, **83**, 369.

Chung, F.R.K., Hajela, D., and Seymour, P.D. (1988). Self-organizing sequential search and Hilbert's inequalities, *J. Computing and Systems Science*, **36**, 148–157.

Clevenson, M.L., and Watkins, W. (1991). Majorization and the Birthday Problem, *Math. Magazine*, **64**, 183–188.

D'Angelo, J.P. (2002). *Inequalities from Complex Analysis*, Mathematical Association of America, Washington, D.C.

de Bruijn, N.G. and Wilf, H.S. (1962). On Hilbert's inequality in n dimensions, *Bull. Amer. Math. Soc.*, **69**, 70–73.

Davis, P.J. (1989). *The Thread: A Mathematical Yarn*, 2nd Edition, Harcourt, Brace, Javonovich, New York.

Diaz, J.B. and Metcalf, R.T. (1963). Stronger forms of a class of inequalities of G. Pólya–G. Szegő, and L.V. Kantorovich, *Bulletin of the Amer. Math. Soc.*, **69**, 415–418.

Diaz, J.B. and Metcalf, R.T. (1966). A complementary triangle inequality in Hilbert and Banach spaces, *Proc. Amer. Math. Soc.*, **17**, 88–97

Dragomir, S.S. (2000). On the Cauchy–Buniakowsky–Schwarz inequality for sequences in inner product spaces, *Math. Inequalities Appl.*, **3**, 385–398.

Dragomir, S.S. (2003). *A Survey on Cauchy–Buniakowsky–Schwarz Type Discrete Inequalities*, School of Computer Science and Mathematics, Victoria University, Melbourne, Australia (monograph preprint).

Duncan, J. and McGregor, M.A. (2003). Carleman's inequality, *Amer. Math. Monthly*, **101** 424–431.

Dubeau, F. (1990). Cauchy–Bunyakowski–Schwarz Revisited, *Amer. Math. Monthly*, **97**, 419–421.

Dunham, W. (1990). *Journey Through Genius: The Great Theorems of Mathematics*, John Wiley and Sons, New York.

Elliot, E.B. (1926). A simple extension of some recently proved facts as to convergency, *J. London Math. Soc.*, **1**, 93–96.

Engle, A. (1998). *Problem-Solving Strategies*, Springer-Verlag, Berlin.

Erdős, P. (1961). Problems and results on the theory of interpolation II, *Acta Math. Acad. Sci. Hungar.*, **12**, 235–244.

Flor, P. (1965). Über eine Ungleichung von S.S. Wagner, *Elem. Math.*, **20**, 165.

Fujii, M. and Kubo, F. (1993). Buzano's inequality and bounds for roots of algebraic equations, *Proc. Amer. Math. Soc.*, **117**, 359–361.

George, C. (1984). *Exercises in Integration*, Springer-Verlag, New York.

Grosse-Erdmann, K.-G. (1998). *The Blocking Technique, Weighted Mean*

Operators, and Hardy's Inequality, Lecture Notes in Mathematics No. 1679, Springer-Verlag, Berlin.

Halmos, P.R. and Vaughan, H.E. (1950). The Marriage Problem, *Amer. J. Math.*, **72**, 214–215.

Hammer, D. and Shen, A. (2002). A Strange Application of Kolmogorov Complexity, Technical Report, Institute of Problems of Information Technology, Moscow, Russia.

Hardy, G.H. (1920). Note on a theorem of Hilbert, *Math. Zeitschr.*, **6**, 314–317.

Hardy, G.H. (1925). Notes on some points in the integral calculus (LX), *Messenger of Math.*, **54**, 150–156.

Hardy, G.H. (1936). A Note on Two Inequalities, *J. London Math. Soc.*, **11**, 167–170.

Hardy, G.H. and Littlewood, J.E. (1928). Remarks on three recent notes in the Journal, *J. London Math. Soc.*, **3**, 166–169.

Hardy, G.H., Littlewood, J.E., and Pólya, G. (1928/1929). Some simple inequalities satisfied by convex functions, *Messenger of Math.*, **58**, 145–152.

Hardy G.H., Littlewood, J.E., and Pólya, G. (1952). *Inequalities*, 2nd Edition, Cambridge University Press, Cambridge, UK.

Havil, J. (2003). *Gamma: Exploring Euler's Constant*, Princeton University Press, Princeton, NJ.

Henrici, P. (1961). Two remarks on the Kantorovich inequality, *Amer. Math. Monthly*, **68**, 904–906.

Hewitt, E. and Stromberg, K. (1969). *Real and Abstract Analysis*, Springer-Verlag, Berlin.

Hilbert, D. (1888). Über die Darstellung definiter Formen als Summe von Formenquadraten, *Math. Ann.*, **32**, 342–350. (Also in *Gesammelte Abhandlungen*, Volume 2, 154–161, Springer-Verlag, Berlin, 1933; reprinted by Chelsea Publishing, New York, 1981.)

Hilbert, D. (1901). Mathematische Probleme, *Arch. Math. Phys.*, **3**, 44–63, 213–237. (Also in *Gesammelte Abhandlungen*, **3**, 290–329, Springer-Verlag, Berlin, 1935; reprinted by Chelsea, New York, 1981; translated by M.W. Newson in *Bull. Amer. Math. Soc.*, 1902, **8**, 427–479.)

Hölder, O. (1889). Über einen Mittelwerthssatz, *Nachr. Akad. Wiss. Göttingen Math.-Phys. Kl.* 38–47.

Jensen, J.L.W.V. (1906). Sur les fonctions convexes et les inégalités entre les valeurs moyennes, *Acta Math.*, **30**, 175–193.

Joag-Dev, K. and Proschan, F. (1992). Birthday problem with unlike probabilities, *Amer. Math. Monthly*, **99**, 10–12.

Kaijser, J., Persson, L-E., and Örberg, A. (2003). On Carleman's and Knopp's inequalities, Technical Report, Department of Mathematics, Uppsala University.

Kaczmarz, S. and Steinhaus, H. (1951). *Theorie der Orthogonalreihen*, second Edition, Chelsea Publishing, New York (1st Edition, Warsaw, 1935).

Kedlaya, K. (1999). $A < B$ (A is less than B), manuscript based on notes for the Math Olympiad Program, MIT, Cambridge MA.

Komlós, J. and Simomovits, M. (1996). Szemeredi's regularity lemma and its applications in graph theory, in *Combinatorics, Paul Erdős is Eighty*, Vol. II (D. Miklós, V.T. Sos, and T. Szőnyi, eds.), 295–352, János Bolyai Mathematical Society, Budapest.

Knuth, D. (1969). *The Art of Computer Programming: Seminumerical Algorithms, Vol. 2*, Addison Wesley, Menlo Park, CA.

Knopp, K. (1928). Über Reihen mit positive Gliedern, *J. London Math. Soc.*, **3**, 205–211.

Lagrange, J. (1773). Solutions analytiques de quelques problèmes sur les pyramides triangulaires, *Acad. Sci. Berlin.*

Landau, E. (1907). Über einen Konvergenzsatz, *Göttinger Nachrichten*, 25–27.

Landau, E. (1909). *Handbuch der Lehre von der Verteilung der Primzahlen*, Leipzig, Germany. (Reprinted 1953, Chelsea, New York).

Lee, H. (2002). Note on Muirhead's Theorem, Technical Report, Department of Mathematics, Kwangwoon University, Seoul, Korea.

Littlewood, J.E. (1988). *Littlewood's Miscellany* (B. Bollobás, ed.), Cambridge University Press, Cambridge, UK.

Lovász, L. and Plummer M.D. (1986). *Matching Theory*, North-Holland Mathematics Studies, Annals of Discrete Mathematics, vol. 29, Elsevier Science Publishers, Amsterdam, and Akadémiai Kiadó, Budapest.

Lozansky, E. and Rousseau, C. (1996). *Winning Solutions*, Springer-Verlag, Berlin.

Love, E.R. (1991). Inequalities related to Carleman's inequality, in *In-*

equalities: *Fifty Years on from Hardy, Littlewood, and Pólya* (W.N. Everitt, ed.), Chapter 8, Marcel Decker, New York.

Lyusternik, L.A. (1966). *Convex Figures and Polyhedra* (translated from the 1st Russian edition, 1956, by D.L. Barnett), D.C. Heath and Company, Boston.

Macdonald, I.G. (1995). *Symmetric Functions and Hall Polynomials*, 2nd Edition, Clarendon Press, Oxford.

Maclaurin, C. (1729). A second letter to Martin Folkes, Esq.; concerning the roots of equations, with the demonstration of other rules in algebra, *Phil. Transactions*, **36**, 59–96.

Magiropoulos, M. and Karayannakis, D. (2002). A "Double" Cauchy–Schwarz type inequality, Technical Report, Technical and Educational Institute of Crete, Heraklion, Greece.

Maligranda, L. (1998). Why Hölder's inequality should be called Rogers' inequality, *Mathematical Inequalities and Applications*, **1**, 69–83.

Maligranda, L. (2000). Equivalence of the Hölder–Rogers and Minkowski inequalities, *Mathematical Inequalities and Applications*, **4**, 203–207.

Maligranda, L. and Persson, L.E (1992). On Clarkson's inequalities and interpolation, *Math. Nachr.*, **155**, 187–197.

Maligranda, L. and Persson, L.E (1993). Inequalities and Interpolation, *Collect. Math.*, **44**, 181–199.

Maor, E. (1998). *Trigonometric Delights*, Princeton University Press, Princeton, NJ.

Matoušek, J. (1999). *Geometric Discrepancy — An Illustrated Guide*, Springer-Verlag, Berlin.

Mazur, M. (2002). Problem Number 10944, *Amer. Math. Monthly*, **109**, 475.

McConnell, T.R. (2001). An inequality related to the birthday problem. Technical Report, Department of Mathematics, University of Syracuse.

Menchoff, D. (1923). Sur les séries de fonctions orthogonales (première partie), *Fundamenta Mathematicae*, **4**, 82–105.

Mignotte, M. and Ştefănescu, S. (1999). *Polynomials: An Algorithmic Approach*, Springer-Verlag, Berlin.

Mingzhe, G. and Bichen, Y. (1998). On the extended Hilbert's inequality, *Proc. Amer. Math. Soc.*, **126**, 751–759.

Mitrinović, D.S. and Vasić, P.M. (1974). History, variations and generalizations of the Čebyšev inequality and the question of some priorities, *Univ. Beograd Publ. Electrotehn. Fak. Ser. Mat. Fiz.*, **461**, 1–30.

Mitrinović, D.S. (with Vasić, P.M.) (1970). *Analytic Inequalities*, Springer-Verlag, Berlin.

Montgomery, H.L. (1994). *Ten Lectures on the Interface Between Analytic Number Theory and Harmonic Analysis*, CBMS Regional Conference Number 84, Conference Board of the Mathematical Sciences, American Mathematical Society, Providence, RI.

Motzkin, T.S. (1967). The arithmetic-geometric inequality, in *Inequalities* (O. Shisha, ed.), 483–489, Academic Press, Boston.

Nakhash, A. (2003). Solution of a Problem 10940 proposed by Y. Nievergelt, *Amer. Math. Monthly*, **110**, 546–547.

Needham, T. (1997). *Visual Complex Analysis*, Oxford University Press, Oxford, UK.

Nesbitt, A.M. (1903). Problem 15114, *Educational Times*, **2**, 37–38.

Niculescu, C.P. (2000). A new look at Newton's inequalities, *J. Inequalities in Pure and Appl. Math.*, **1**, 1–14.

Niven, I. and Zuckerman, H.S. (1951). On the definition of normal numbers, *Pacific J. Math.*, **1**, 103–109.

Nivergelt, Y. (2002). Problem 10940 The complex geometric mean, *Amer. Math. Monthly*, **109**, 393.

Norris, N. (1935). Inequalities among averages, *Ann. Math. Statistics*, **6**, 27–29.

Oleszkiewicz, K. (1993). An elementary proof of Hilbert's inequality, *Amer. Math. Monthly*, **100**, 276–280.

Olkin, I. and Marshall, A.W. (1979). *Inequalities:Theory of Majorization and Its Applications*, Academic Press, New York.

Opic, B. and Kufner, A. (1990). *Hardy-Type Inequalities*, Longman-Harlow, New York.

Petrovitch, M. (1917). Module d'une somme, *L'Enseignement Math.*, **19**, 53–56.

Pečarić, J.E., Proschan, F., and Tong, Y.L. (1992). *Convex Functions, Partial Orderings, and Statistical Applications*, Academic Press, New York.

Pečarić, J.E. and Stolarsky, K.B. (2001). Carleman's inequality: history and new generalizations, *Aequationes Math.*, **61**, 49–62.

Pitman, J. (1997). *Probability*, Springer-Verlag, Berlin.

Pólya, G. (1926). Proof of an inequality, *Proc. London Math. Soc.*, **24**, 57.

Pólya, G. (1949). With, or without, motivation, *Amer. Math. Monthly*, **56**, 684–691, 1949.

Pólya, G. (1950). On the harmonic mean of two numbers, *Amer. Math. Monthly*, **57**, 26–28.

Pólya, G. and Szegő, G. (1954). *Aufgaben und Lehrsätze aus der Analysis, Vol. I*, 2nd edition, Springer-Verlag, Berlin.

Polya, G. (1957). *How to Solve It*, 2nd edition, Princeton University Press, Princeton, NJ.

Prestel, A. and Delzell, C.N. (2001). *Positive Polynomials: From Hilbert's 17th Problem to Real Algebra*, Springer-Verlag, Berlin.

Rademacher, H. (1922). Einige Sätze über Reihen von allgemeinen Orthogonalfunktionen, *Math. Annalen*, **87**, 112–138.

Rajwade, A.R. (1993). *Squares*, London Mathematical Lecture Series, **171**, Cambridge University Press, Cambridge, UK.

Richberg, R. (1993). Hardy's inequality for geometric series (solution of a problem by Walther Janous), *Amer. Math. Monthly*, **100**, 592–593.

Riesz, F. (1909). Sur les suites de fonctions mesurables, *C. R. Acad. Sci. Paris*, **148**, 1303–1305.

Riesz, F. (1910). Untersuchungen über Systeme integrierbare Funktionen, *Math. Annalen*, **69**, 449–497.

Riesz, M. (1927). Sur les maxima des formes bilinéaires et sur les fonctionnelles linéaires, *Acta Math.*, **49**, 465–487.

Roberts, A.W. and Varberg, D.E. (1973). *Convex Functions*, Academic Press, New York.

Rogers, L.J. (1888). An extension of a certain theorem in inequalities, *Messenger of Math.*, **17**, 145–150.

Rosset, S. (1989). Normalized symmetric functions, Newton's inequalities, and a new set of stronger inequalities, *Amer. Math. Monthly*, **96**, 815–819.

Rudin, W. (1987). *Real and Complex Analysis*, 3rd edition, McGraw-Hill, Boston.

Rudin, W. (2000). Sums of Squares of Polynomials, *Amer. Math. Monthly*, **107**, 813–821.

Schur, I. (1911). Bemerkungen zur Theorie der beschränkten Bilinearformen mit unendlich vielen Veränderlichen, *J. Mathematik*, **140**, 1–28.

Schneider, R. (1993). *Convex Bodies: The Brunn-Minkowski Theory*, Cambridge University Press, Cambridge, UK.

Schwarz, H.A. (1885). Über ein die Flächen kleinsten Flächeninhalts betreffendes Problem der Variationsrechnung. *Acta Soc. Scient. Fenn.*, **15**, 315–362.

Shklarsky, D.O., Chentzov, N.N., and Yaglom, I.M. (1993). *The USSR Olympiad Problem Book: Selected Problems and Theorems of Elementary Mathematics*, Dover Publications, Mineola, N.Y.

Sigillito, V.G. (1968). An application of Schwarz inequality, *Amer. Math. Monthly*, **75**, 656–658.

Shparlinski, I.E. (2002). Exponential Sums in Coding Theory, Cryptology, and Algorithms, in *Coding Theory and Cryptology* (H. Niederreiter, ed.), Lecture Notes Series, Institute for Mathematical Sciences, National University of Singapore, Vol. I., World Scientific Publishing, Singapore.

Siegel, C.L. (1989). *Lectures on the Geometry of Numbers* (Notes by B. Friedman rewritten by K. Chandrasekharan with assistance of R. Suter), Springer-Verlag, Berlin.

Steiger, W. L. (1969). On a generalization of the Cauchy–Schwarz inequality, *Amer. Math. Monthly*, **76**, 815–816.

Szegő, G. (1914). Lösung eine Aufgabe von G. Pólya, *Archiv für Mathematik u. Physik*, Ser. 3, **22**, 361–362.

Székeley, G.J., editor (1996). *Contests in Higher Mathematics: Miklós Schweitzer Competition 1962–1991*, Springer-Verlag, Berlin.

Tiskin, A. (2002). A generalization of the Cauchy and Loomis–Whitney inequalities, Technical Report, Oxford University Computing Laboratory, Oxford, UK.

Toeplitz, O. (1910). Zur Theorie der quadratischen Formen von unendlische vielen Veränderlichen, *Göttinger Nach.*, 489–506.

Treibergs, A. (2002). Inequalities that Imply the Isoperimetric Inequality, Technical Report, Department of Mathematics, University of Utah.

van Dam, E.R. (1998). A Cauchy–Khinchin matrix inequality, *Linear Algebra and its Applications*, **280**, 163–172.

van Lint, J.H. and Wilson, R.M. (1992). *A Course in Combinatorics*, Cambridge University Press, Cambridge, UK.

Vince, A. (1990). A rearrangement inequality and the permutahedron, *Amer. Math. Monthly*, **97**, 319–323.

Wagner, S.S. (1965). Untitled, *Notices Amer. Math. Soc.*, **12**, 20.

Waterhouse, W. (1983). Do symmetric problems have symmetric solutions? *Amer. Math. Monthly*, **90**, 378–387.

Weyl, H. (1909). Über die Konvergenz von Reihen, die nach Orthogonalfunktionen fortschreiten, *Math. Ann.*, **67**, 225–245.

Weyl, H. (1916). Über die Gleichverteilung von Zahlen mod Eins, *Math. Ann.*, **77**, 312–352.

Weyl, H. (1949). Almost periodic invariant vector sets in a metric vector space, *Amer. J. Math.*, **71**, 178–205.

Wilf, H.S. (1963). Some applications of the inequality of arithmetic and geometric means to polynomial equations, *Proc. Amer. Math. Soc.*, **14**, 263–265.

Wilf, H.S. (1970). *Finite Sections of Some Classical Inequalities*, Ergebnisse der Mathematik und ihrer Grenzgebiete, **52**, Springer-Verlag, Berlin.

Zukav, G. (1979). *The Dancing Wu Li Masters: An Overview of the New Physics*, William Morrow and Company, New York.

van der Corput, J.G. (1931). Diophantische Ungleichungen I. Zur Gleicheverteilung Modulo Eins, *Acta Math.*, **56**, 373–456.

Vinogradov, I.M. (1954). *Elements of Number Theory* (translated by Saul Kravetz from the 5th revised Russian edition, 1949), Dover Publications, New York.

Index

1-trick, 110, 144, 146, 215, 219, 227, 231
   defined, 226
   first used, 12
   refinement, 205, 288

Abel inequality, 208, 221
Abel, Niels Henrik, 208
Aczél, J., 287
additive bound, 4, 9, 66, 106
   and Hölder, 137, 264
Åkerberg's refinement, 34
al-Khazin, Abu Ja'far, 286
Alexanderson, G., 286
AM-GM inequality, 20, 21
   and Kantorovich, 247
   Åkerberg's refinement, 34
   algorithmic proof, 245
   and investments, 100
   and the exponential, 91
   betweeness induction, 83
   general weights, 23
   geometric interpretation, 33
   integral analog, 113
   Pólya's proof, 23
   rational weights, 22
   smoothing proof, 244
   stability, 35
   via cyclic shifts, 84
   via integral analog, 114
   via rearrangement, 84
   via Schur concavity, 194
American Mathematical Monthly, 96, 192, 253
Andreescu, T., 251
anti-symmetric forms, 237
arithmetic mean-geometric mean inequality, *see* AM-GM inequality
Arithmetica of Diophantus, 40
arrangement of spheres, 52
Artin, Emil, 47

backtracking, 137
backwards induction, 236
baseball inequality, 82
Belentepe, C., vi
Bennett, C., 288, 289
Bernoulli inequality, 31
Bessel inequality, 71, 225
betweeness, exploitation of, 74
Birkhoff's Theorem, 207
birthday problem, 206
Bledsoe, W.W., 283
Bombieri, E., 284
box, thinking outside, 52
Brahmagupta, 286
   identity, 47
Brunn–Minkowski inequality, 245, 287
Bunyakovsky, Victor Yacovlevich, 10, 190, 285
   and AM-GM inequality, 115
   Chebyshev contact, 76
   vis-a-vis Schwarz, 11
Buzano, M.L., 286

Cai, T., vi, 246
cancellation, origins of, 210
Carleman inequality, 118
   Carleson proof, 173
   Pólya's proof, 27
   refined via Knopp, 128
Carleson inequality, 173
Carlson inequality, 165
Cauchy inequality, 1
   and quadratic forms, 14
   beating, 15
   by induction, 2
   case of equality, 37
   in an array, 16
   interpolation, 48
   Pólya–Szegö converse, 83
   three term, 13

302

vector-scalar melange, 69
  via inner product identity, 68
  via monotone interpolation, 85
Cauchy's induction argument
  AM-GM inequality, 20
  Cauchy inequality, 31
  Jensen inequality, 101
Cauchy, Augustin-Louis, 10
Cauchy–Binet identity, 49
Cauchy–Schwarz inequality, 8
  as accidental corollary, 57
  cross term defects, 83
  geometric proof, 58
  self-generalization, 16
  via Gram–Schmidt, 71
centered inequality, 115
Chebyshev
  order inequality, 76
  tail inequality, 86
Chebyshev, Pafnuty Lvovich, 76, 287
Chong, K.M., 244
Chung, F.R.K., 289
Clevenson, M.L., 277
closest point problem, 56
completing the square, 57
conjugate powers, 137
consistency principles, 134
convex functions, continuity of, 254
convex minorant, 98
convexity
  defined, 87
  differential criterion, 90
  geometric characterizations, 87
  strict, 89
  versus J-convexity, 101
  via transformation, 102
Cours d'Analyse Algébrique, 10
Cramér–Rao inequality, 18
Cronin–Kardon, C., vi
crystallographic inequality, 13
cyclic shifts, 84
cyclic sum inequality, 104

D'Angelo, J.P, 242
Davis, P.J., 287
Debeau, F., 285
Dellacherie, C., vi, 254, 290
determinant, 49
Diaconis, P., vi
Diophantus, 40, 286
Diophantus identity
  and Brahmagupta, 47, 237
  and complex factorization, 237
  and Pythagoras, 47, 237
dissection of integrals, 106
doubly stochastic, 196
Dragomir, S.S., 241, 285

Dudley, R.M., vi

elementary symmetric function, 178
Elliot, E.B., 290
Enflo inequality, 225
Engel, A., 94, 246, 251
equality
  in AM-GM inequality, 22
  in Cauchy inequality, 5, 37
  in Cauchy–Schwarz, 8
  in Hölder, 137
  in Jensen inequality, 89
Erdős, P., 261
Euclidean distance
  in $\mathbb{R}^d$, 51
  triangle inequality, 53
exponential sums, 210
extend and conquer, 222

Feng, Z., 251
Fermat, Pierre de, 40
Fibonacci, 286
Fisher information, 18
Flor, P., 238
four letter identity, 49
Fuji, M., 287

gamma function, 165
Gauss, C.F., 288
general means, 120
generic improvement, 94
geometric mean
  as a limit, 120
  as minimum, 133
  superadditivity, 100, 133
George, C., 231, 288
Gram–Schmidt process, 70
Gross–Erdmann, K.-G., 290
Grüss inequality, 119

Hadwiger–Finsler inequality, 102
Hajela, D., 289
Halmos, P., 278
Hammer, D., 231
Hardy's inequality, 166
  and Hilbert, 176
  discrete, 169
  geometric refinement, 177
  in L^p, 176
  special instance, 177
Hardy, G.H., 197, 290
Harker–Kasper inequality, 14
harmonic mean, 126
  minimax characterization, 132
harmonic series
  divergence, 99, 255
Heisenberg principle, 288

304   Index

Hewitt, E., 243
Hilbert inequality, 155
  homogeneous kernel, 164
  integral version, 163
  max version, 163
  via Toeplitz method, 165
Hilbert's 17th problem, 46
Hilbert, David, 46, 55
HM-GM and HM-AM inequalities, 126
Hölder inequality, 135, 136
  case of equality, 137
  converse, 139
  defect estimate, 94
  historical form, 151
  stability, 144, 145
Hölder, Otto Ludwig, 94, 135, 263, 290
homogeneity in Σ, 132
homogenization trick, 189
How to Solve It, 30
humble bound, 284

inclusion radius, 148
inner product space
  Cauchy–Schwarz inequality, 8
  definition, 7
  integral representations, 116
interpolation
  in Cauchy inequality, 48
intuition
  how much, 55
  refining, 53
investment inequalities, 100
isometry, 60
isoperimetric property, 19
  for the cube, 34
Israel, R., 253
iteration
  and discovery, 149

J-convexity, 101
Janous, Walther, 177
Jensen inequality, 87, 263
  and Schur convexity, 201
  case of equality, 89
  for integrals, 113
  geometric applications, 93
  Hölder's defect estimate, 94
  implies Minkowski, 150
  via Cauchy's induction, 101
Jensen, J.L.W.V., 101
Joag–Dev, K., 277

Kahane, J.-P., vi
Karayannakis, D., 286
Kedlaya, K., vi, 152, 264, 274
Knuth, D., 260
Komlós, J., 277

Körner, T., vi
Kronecker's lemma, 177
Kubo, F., 287
Kufner, A., 290
Kuniyeda, M., 262

Lagrange identity, 39
  continuous analogue, 48
  simplest case, 42
Lagrange, Joseph Louis de, 40
Landau's notation, 120
leap-forward fall-back induction, 20, 31, 36, 101
Lee, H., vi, 247, 251
light cone defined, 62
light cone inequality, 63, 245
Littlewood, J.E., 197
looking back, 25, 26
Loomis–Whitney inequality, 16
Lorentz product, 62
Lovász, L., 278
Lozansky, E., 264
Lyusternik, L.A., 245

Maclaurin inequality, 285
Magiropoulos, M., 286
majorant principles, 284
majorization, 191
Maligranda, L., vi, 264, 288, 289
marriage lemma, 206
Marshall, A., 277
Matoušek, J., vi, 285
McConnell, T.R., 277
Meng, X., vi
Mengoli, P., 99, 248
method
  of halves, 122
  of parameterized parameters, 164
  metric space, 54
Mignotte, M., 262
Milne inequality, 50
minimal surfaces, 10
Minkowski inequality, 141
  Riesz proof, 141
  via Jensen, 150
Minkowski's
  conjecture, 44, 46
  light cone, 62
Minkowski, Hermann, 44
Mitrinović, D.S., 277, 283
Möbius transformation, 242
moment sequences, 149
monotonicity
  and integral estimates, 118
Montgomery, H.L., 284
Motzkin, T.S., 46, 286
Muirhead condition

and majorization, 195

Nakhash, A., 253
names of inequalities, 11
Naor, E., 259
Needham, T., 242
Nesbitt inequality, 84, 131, 246
Neyman–Pearson lemma, 287
Niculescu, C.P., 290
Nievergelt, Y., 253
Niven, I., 260
nonnegative polynomials, 43
norm
  p-norm, or ℓ^p-norm, 140
  defined, 55
normalization method, 5, 25, 26, 66
normed linear space, 55

obvious
  and not, 56
  Hilbert story, 56
  triangle inequality, 56
Olkin, I., 277
one-trick, see 1-trick
Opic, B., 290
optimality principles, 33
order inequality, 76
order relationship
  systematic exploitation, 73
  to quadratic, 76
order-to-quadratic conversion, 78, 287
orthogonality, definition, 58
orthonormal, 217
orthonormal sequence, 70
other inequality of Chebyshev, 77

parameterized parameters, 164
Persson, L.E., 288, 289
pillars, three great, 87
Pitman, J., 190, 276
Plummer, M.D., 278
polarization identity, 49, 70
Pólya's
  dream, 23
  questions, 30
Pólya, George, 30, 197, 286
Pólya–Szegö converse, 83
positive definite, 228
power mean continuity relations, 127
power mean curve, 124
power mean inequality, 123
  simplest, 36
Prékopa–Leindler inequality, 287
principle of maximal effectiveness, 27
principle of similar sides, 139
probability model, 17
product of linear forms, 59

projection formula, 56
  guessing, 58
proportionality, 50
  gages of, 39
Proschan, F., 277
Pták, V., 247
Ptolemy inequality, 69
Pythagorean theorem, 47, 51

Qian, Z., vi
quadratic form, 228
quadratic inequality, 76
qualitative inference principle, 3, 27
quasilinear representation
  geometric mean, 259

Rademacher–Menchoff inequality, 217, 223
ratio monotone, 189
reflection, 60
Reznick, B., vi
Richberg, R., 272
Riesz proof of Minkowski inequality, 141
Riesz, F., 288
Riesz, M., 288
Rogers inequality, 152, 153
Rogers, L.C., 135, 152, 290
Rolle's theorem, 102, 251
Rosset, S., 290
Rousseau, C., 264
rule of first finder, 12

Schur convexity, 191
  defined, 192
Schur criterion, 193
Schur inequality, 83
Schur's lemma, 15
Schur, Issai, 192
Schwarz inequality, 10, 11
  centered, 115
  pointwise proof, 115
Schwarz's argument, 11, 63
  failure, 136
  in inner product space, 15
  in light cone, 63
Schwarz, Hermann Amandus, 10
Selberg inequality, 225
self-generalization, 21
  Hölder's inequality, 151
  Cauchy inequality, 16
  Cauchy–Schwarz inequality, 66
Seymour, P.D., 289
Shaman, P., vi
Sharpley, R., 288, 289
Shen, A., 231
Shepp, L., vi
Shparlinski, I.E., vi, 281

Siegel's method of halves, 122
Siegel, Carl Ludwig, 122
Sigillito, V.G., 244
Simonovits, M., 277
Skillen, S., vi
slip-in trick, 117
sphere arrangement, 52
splitting trick, 106, 123, 147, 154, 227, 263, 267
   defined, 226
   first used, 12
   grand champion, 266
stability
   in Hölder inequality, 145
   of AM-GM inequality, 35
steepest ascent, 67
Ştefănescu, D., 262
stress testing, 268
Stromberg, K., 243
sum of squares, 42–44, 46
superadditivity
   geometric mean, 34, 100, 133
symmetry and reflection, 62
Szegö, Gabor, 234
Szeméredi Regularity Lemma, 205

Tang, H., vi
telescoping, 29
thinking outside the box, 52
Three Chord Lemma, 104
tilde transformation, 193
Tiskin, A., 231
Toeplitz method, 165
Treibergs, A., 242
triangle inequality, 54

unbiased estimator, 17

van Dam, E.R., 230
van der Corput inequality, 214
van der Corput, J.G., 214
variance, 18, 116
Vaughn, H.E., 278
Viète identity, 133
Vince, A., 81
Vinogradov, I.M., 281
Vitale, R., vi
von Neumann, John, 51, 286

Wagoner, S.S., 238
Walker, A.W., 192
Ward, N., vi
Watkins, W., 277
Weierstrass inequality, 190
Weitzenböck inequality, 93, 102
Weyl, H., 206, 288
Wiles, Andrew, 40

Wilf, H., vi, 235

Young inequality, 136

Zuckerman, H.S., 260
Zukav, G., 286

Made in United States
North Haven, CT
11 October 2024